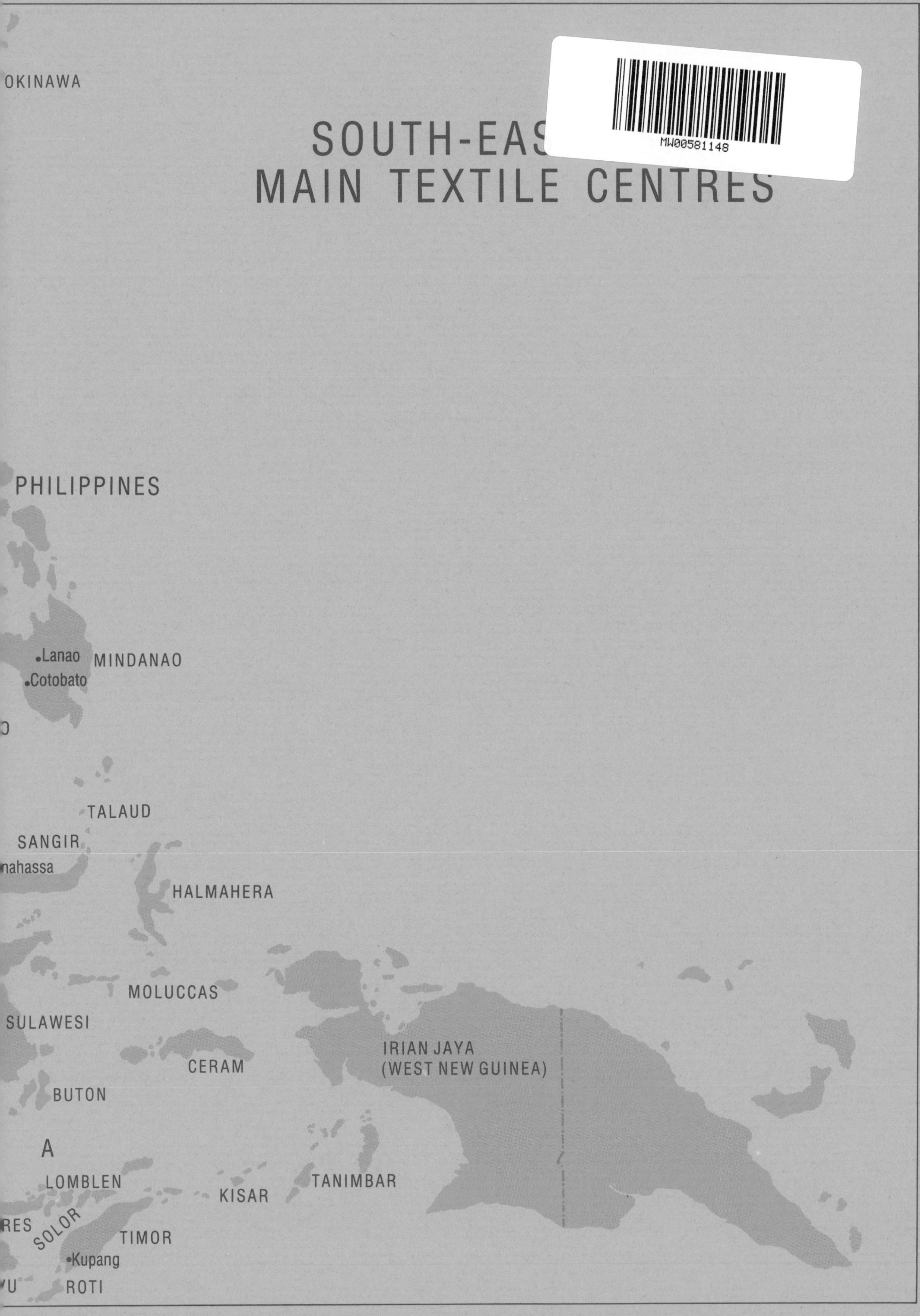

OKINAWA

SOUTH-EAS
MAIN TEXTILE CENTRES

PHILIPPINES

•Lanao MINDANAO
•Cotobato

TALAUD

SANGIR

nahassa

HALMAHERA

MOLUCCAS

SULAWESI

IRIAN JAYA
(WEST NEW GUINEA)

CERAM

BUTON

A

LOMBLEN

TANIMBAR

KISAR

RES

SOLOR

TIMOR

•Kupang

U

ROTI

HANDWOVEN TEXTILES OF SOUTH-EAST ASIA

HANDWOVEN TEXTILES OF SOUTH-EAST ASIA

SYLVIA FRASER-LU

SINGAPORE
OXFORD UNIVERSITY PRESS
OXFORD NEW YORK

Oxford University Press

Oxford New York Toronto
Delhi Bombay Calcutta Madras Karachi
Petaling Jaya Singapore Hong Kong Tokyo
Nairobi Dar es Salaam Cape Town
Melbourne Auckland
and associated companies in
Berlin Ibadan

Oxford is a trade mark of Oxford University Press

First published 1988
First issued as an Oxford University Press paperback 1989
Third impression 1992

ISBN 0 19 588954 1

Printed in Singapore by Kyodo Printing Co. (S) Pte. Ltd.
Published by Oxford University Press Pte. Ltd.,
Unit 221, Ubi Avenue 4, Singapore 1440

To the women weavers of South-East Asia
Long may they continue to delight us with beautiful textiles

Preface

COMING from an ancestry that includes knitters of fine Shetland fair-isle jerseys and weavers of Donegal tweed, I have always been fascinated by handwoven textiles. My interest in South-East Asian textiles was first aroused twenty years ago when, as a volunteer teacher in Sarawak, I was introduced to the wonderful world of Iban warp *ikat* textiles, Malaysian *kain songket*, and Indonesian batik. Sojourns in Burma, Thailand, and Indonesia, and visits to Kampuchea, Brunei, and the Philippines have served to substantially deepen that interest.

This book was originally conceived by the editors of Oxford University Press as a title in the *Images of Asia* series. It started out as a modest essay of 20,000 words to 'acquaint' the general reader with the folk-weaving of South-East Asia. After some preliminary research and visits to many of the weaving areas, it was felt that it would not be possible to do justice to such an important, fascinating, and hitherto little-known subject in such a small publication. Thus, the present volume evolved.

Despite its larger size and greater detail, the focus of this book has not changed. Its primary purpose is to make known to the general public the beauty and scope of South-East Asian handwoven textiles, both past and present. The book begins with a general historical overview and moves on to describe materials, equipment, weaving techniques, and the use of textiles, before taking a look at the current textile scene in each country. To keep the book within manageable proportions, it was decided to concentrate on loom-woven fabrics. Textile techniques concerned with decorating already woven cloth, such as batik, *pelangi*, *tritik*, embroidery, appliqué, beadwork, and the application of gold substances, are mentioned but are not described at great length. (They could well be the topic of another book!)

The book does not aim to be a definitive work, but one merely to open up the topic. Due to the sheer size of South-East Asia, it was not possible to travel to every corner. I was unable to visit Indo-China, the southern Philippines, and parts of Burma due to the political situation. Transportation difficulties made it impossible to adequately cover a number of weaving areas in Borneo, Timor, Flores, Sulawesi, and Sumatra. With the exception of Indonesia, written information pertaining to South-East Asian textiles is very fragmentary and difficult to obtain. Much is in the form of ephemeral articles in the local vernacular, or in obscure learned journals. Apart from the general background information, much of the material contained in this book has been gleaned from observations of, and interviews with, weavers and collectors of South-East Asian textiles, and visits to craft centres, weaving institutes, and museums. There is still much work that needs to be done to increase our knowledge of South-East Asian textiles. The origins of weaving and its historical development are obscure. The evolution of various motifs and the development of regional styles require a lot more research, as does the influence of South-East Asia on other textile traditions.

Over the last few years there has been a gradual awakening of interest in South-East Asian textiles, both in the countries of origin and abroad. Thanks to the efforts of Dr Monni Adams, Dr Mattiebelle Gittinger, the Solyoms, the Maxwells, and Mary Hunt Kahlenburg, Indonesian textiles are now known to a wider public. The recent work of Patricia Cheesman is creating a greater interest in Lao weaving. Khun Uab Sanasen, Khun Nisa Sheanakul, HM Queen Sirikit's SUPPORT Foundation, the Hill Tribe Foundation, and Jim Thompson's Thai Silk Company have all been active in promoting the handwoven textiles of Thailand. The writings of Marian Pastor Roces and a 1981 Stateside exhibition on 'People and Art of the Philippines' sponsored by the Museum of Cultural History, University of California, have been opening windows on Filipino textiles. According to reports, a number of scholars are currently working on monographs covering various aspects of South-East Asian textiles. Hopefully, this publication will help bridge the gap until these new, more detailed works become available.

Acknowledgements

MANY people have given generously of their time and knowledge.

In Thailand, particular thanks should go to Khun Nisa Sheanakul, Khun Uab Sanasen, Khun Vanida, Mrs Van Buren, Khun Naiyanee Srikanthimarak, Mrs Cherie Aung Khin, Mrs Nancy Charles, and Mrs Kaeo Everingham of Bangkok, and Khun Duangjitt Thavisri of Chiangmai who made their textile collections available to me; Mr and Mrs Cort Boylan for their hospitality; Khun Suvan Boonthae of Nongkhai, Khun Chantira Pisarnsin of Khon Kaen, and Khun Sathorn Sorajpobsob Sunti of Hat Sieo, for arranging visits to local weavers. Ms Susan McCauley, Mrs Rose Wanchupela of Esarn Weaving, Archarn Vitti Parnchapak of Chiangmai University, Khun Euayporn Kerdchouay and Mr Michael Smithies of the Siam Society, and Mrs Ardis Wilwerth and Mrs Paulette de Schaller of the National Museum Volunteers, also provided much useful information.

In Burma: U Than Aung, Chairman of the Mandalay Division Cooperative Syndicate, for arranging visits to weaving concerns in the Amarapura area; U Win Maung, artist of Mandalay, who very kindly made available information on *acheik* weaving; U Sein Myint of Mandalay, Khun Pang, Curator of the Taunggyi Museum, Khin Myo Chit and Mr and Mrs Vum Ko Hao, all of whom brought a number of interesting textiles to my attention. Daw Amy Nyunt Myint of the UN Library, Rangoon, and U Thaw Kaung of Rangoon University Library also provided me with photographs and references.

In Malaysia: Raja Fuziah bte Tun Raja Uda, Director-General of Kraftangan, very kindly put the resources of her department at my disposal. Norwani M. D. Norwawi, Mohamed Kassim bin Haji Ali, and Puan Mainsurawati of Kraftangan, were particularly helpful. Dato and Datin Leo Moggie, in addition to providing hospitality, shared their wonderful collection of Iban textiles with me. In Sarawak, Mr Lucas Chin, Curator of the Sarawak Museum, and his assistant, Angkin Kuding, allowed me to study their excellent collection of Iban and local Malay textiles. Puan Fatimah of the Sarawak Arts Council took me to visit local weavers and craftswomen.

In Brunei, Haji Mohamed Yassim very kindly showed me over the Brunei Arts and Handicrafts Centre, while Matussin bin Omar, Acting Director of the Museum, provided me with photographs of the Brunei Museum's collection.

In Indonesia: Mrs Susan Market, in addition to providing me with hospitality, introduced me to the wonderful textile collections of Mrs Marta Ostwald and Mrs James Limanjuntak. H. A. Muddarijah of Sengkang, the Fung Family of Waingapu and Kuta, Verra Darwiko of Bali, Father Piet Petu of Flores, and Mrs Eiko Kusuma of Jakarta, also made their textile collections available to me. Father Bollen of Maumere arranged a special demonstration of traditional weaving techniques.

In the Philippines: Roland Goh of Baguio, and Ricardo Baylosis and Ana M. Juco of Manila, showed me a number of rare antique local textiles from their collections and answered numerous questions. Mrs Lelita Klainatorn provided much of the information on Yakan textiles of Basilan Island.

Members of various institutions abroad were most helpful. The National Museum of Natural History of the Smithsonian Institution allowed me to study and photograph a number of textiles in their collection. The library staff of the Textile Museum, Washington, DC, very kindly provided me with research materials, while Dr Gittinger, Research Associate of that institution, has been most supportive of this project and has provided very useful information on the structures of some textiles. Ms Patricia Herbert of the British Library was also instrumental in providing many reference materials.

In conclusion, I would like to thank Mr Robert Retka of Bangkok for carefully reading the manuscript and making a number of excellent suggestions for improvement, and Mrs Carolyn Reagan of Tokyo who helped with selecting slides. The assistance of my husband George was considerable. Without his support and patience this book would not have been complete.

It goes without saying that any errors and shortcomings are entirely my own.

Tokyo
May 1987

SYLVIA FRASER-LU

Contents

Plates

Figures

Maps

PART I

A General Introduction to Weaving

CHAPTER ONE

Setting the Scene

The Physical Setting

SOUTH-EAST ASIA is the generally accepted name for a series of islands and peninsulas which lie east of India and west of China. South-East Asia extends approximately 4 900 km from east to west, and 3 250 km north to south. The total land mass is about 2 500 000 sq km. It is fairly evenly divided between the mainland, which comprises the present-day nations of Burma, Thailand, Laos, Kampuchea, Vietnam, Malaysia, Singapore, and the archipelagos of Indonesia and the Philippines.

Mainland South-East Asia is bordered in the north by the lofty foothills of the Himalayas, which form the eastern rim of Tibet, and the south-west edge of China. Moving eastwards from Burma, the mountains shift to a north–south alignment and give a ridge-and-furrow effect to the topography. Past volcanic activity, wind and rain erosion, and inundation by the sea have combined to produce a very fragmented landscape. The topography is further dissected in the interior by well-fed rivers which have deposited vast amounts of sediment to form coastal plains and deltas in the lower reaches.

The mainland mountain chains continue to run in an arc-like fashion through Indonesia and the Philippines, bequeathing almost all islands a mountainous backbone crowned in some places by imposing volcanic cones.[1] The islands of Sumatra, Java, and Borneo lie on the Sunda Shelf, which at one time was joined to the mainland of South-East Asia. The Philippines was also formerly connected by land bridges to Indonesia, Taiwan, and possibly mainland Asia.[2] The melting of the polar ice cap 10,000–20,000 years ago caused the sea to rise and inundate the areas now known as the Gulf of Siam, the Straits of Malacca, and the Java Sea.

Bisected in the south by the equator, South-East Asia is blessed with abundant sunshine and plentiful rain. With the exception of North Vietnam and a few mountain locations, virtually all of South-East Asia falls within the 27–36 °C annual temperature range. There are, however, seasonal variations in rainfall and temperature due to changing wind and pressure systems.[3] Warm, humid climatic conditions have given rise to lush tropical rain forests teeming with a wide variety of plant and animal life.

In the more accessible areas, the forests have been cleared for intensive wet-rice cultivation, subsistence vegetable, fruit and fibre crops, and plantation plants, such as rubber, oil palm, tea, coffee, sugar, and pepper. The remaining forests continue to yield hardwoods, such as teak, ebony, aromatic woods, bamboo, rattan, and resins but not in the same quantities as before. These forested uplands continue to provide a livelihood for a number of indigenous semi-nomadic hunter-gatherers and slash-and-burn farmers. The coastal areas support stands of *nipa*, sago, and coconut palms, while the sea provides abundant fish and luxury items, such as shell, coral, and pearls. A complex geological structure has left parts of South-East Asia richly endowed with deposits of petroleum and tin. Manganese, nickel, copper, zinc, coal, iron, gold, and silver are also present in commercial quantities.

The Human Setting

Archaeological Background

Archaeological evidence suggests that mankind has been in South-East Asia since the dawn of history. Skeletal remains of *Pithecanthropus erectus* from the Solo River valley in Central Java, and other related finds, furnish us with evidence that prehistoric man lived in Indonesia nearly half a million years ago.[4] Tool and animal remains in the Cagayan valley of northern Luzon in the Philippines, dating back 200,000 years, indicate that this area was inhabited by a society supported by big-game hunting.[5] Further south, on the island of Palawan, archaeological finds from the Tabon Cave complex show that flourishing hunting and food-gathering communities were living on this site some 30,000 years ago.[6] This site is roughly contemporaneous with that of the Niah Caves of Sarawak, which is noted for its palaeolithic chopper tools and refined quadrangular adzes. This site also yielded the skull of a *Homo sapien* dating back 39,000 years, possibly the earliest definite representative of present-day man in insular South-East Asia.[7]

Sites on mainland South-East Asia have also provided us with tangible evidence of the activities of early man. Anyathian sites and the Padalin Caves in Burma have yielded a variety of chopper tools and other remains.[8] Various sites in the Malaysian states of Kedah, Kelantan, Perak, and Pahang reveal a variety of palaeolithic, mesolithic, and neolithic finds.[9]

Some of the most exciting finds in the post-war years have come from Thailand. In addition to stone and metal tools and pottery, there is evidence pertaining to farming activities. The fossilized remains of plants, such as candle nut, betel, pepper, bottle gourd, cucumber, water chestnut, and leguminous beans dating back to 12000 BC, have been discovered at the Spirit Cave near Mae Hong Son in northern Thailand. The Non Nok Tha site in Khon Kaen Province in the north-east, has yielded possible domestic animal remains, specimens of rice, and pieces of copper and bronze which date back to 3500 BC.[10]

The earliest firm evidence for the presence of textiles in South-East Asia comes from the 7,000-year-old funerary site of Ban Chiang in Udon Province in north-east Thailand. Remnants of natural coloured hemp fibres in a plain weave, dating to about 1000 BC, have been found attached to bronze bracelets. A silk thread was also found under pottery sherds near a skeleton.[11] Grooved earthenware rollers, which are thought to have been used to print patterns on textiles, have also been uncovered. These date from about 200 BC to AD 200.[12]

The earliest evidence to date for the presence of looms comes from the sites of Shizhaishan and Lizhaishan in the former Dian Kingdom in Yunnan province in south-east China, which borders South-East Asia. At both sites, bronze loom parts for a foot-braced body-tension loom have been recovered. A sculpture on the lid of a bronze cowrie container, depicting six women weaving, was also unearthed (Fig. 1). These finds may be dated to the Western Han Dynasty (206 BC–AD 8). The Dian, a non-Chinese people, are believed to be culturally related to the early peoples of South-East Asia. Their looms show strong affinities with those currently being used by some hill people in Taiwan and Hainan, and by the Nong Mnong and Maa peoples of central Vietnam.[13] Unfortunately, little is known about the dating and spread of this technology.[14]

Fig. 1. Sketch of a bronze figure of a weaver found on a cowrie container at Shizhaishan Tomb 1, Yunnan Province, China, Western Han Dynasty (206 BC–AD 8). (After Volmer.)

Archaeological evidence for the origins of some prevalent South-East Asian decorative motifs may be seen in beautifully ornamented ritual bronze artefacts, such as kettle drums, weapons, and bells, which have been uncovered from many sites throughout the region. The Bronze Age culture which flourished during the latter half of the first millennium BC, is generally referred to as the Dongson culture, after a village in Tongking, North Vietnam, where some of the earliest discoveries were made. The most spectacular examples are in the form of ritual bronze drums which have been uncovered in South China and in virtually every country of South-East Asia (Fig. 2). Typical decorative elements include curvilinear forms, such as spirals, hooks and meanders, and geometric figures, such as zig-zags, lozenges, and parallel lines, placed in a rigid symmetry across a patterned surface. Other motifs include stars, stylized reptiles, such as frogs, lizards, crocodiles and snakes, elephants, and anthropomorphs. These anthropomorphs are sometimes depicted in groups as dancing warriors, or as passengers on 'ships of the dead' (Fig. 3). These designs were widely disseminated throughout South-East Asia and appear on pottery, wood, metalwork, and textiles to this day.[15]

Some scholars are of the opinion that the highly stylized designs found on bronze objects from the

Fig. 2. Tympanum of a bronze drum depicting human figures, birds, and geometric designs. (After Wen Yu.)

Fig. 3. Patterns seen on bronze drums found in South China and South-East Asia. (After Wen Yu.)

late Chou period of China (eighth–third century BC) also had some influence on South-East Asian decorative motifs in terms of asymmetrical layout of design and the merging and modification of motifs to fit in with an overall schema. This influence is most evident on some textiles from Borneo and Sumatra.[16]

Migrations

Present-day South-East Asia is an anthropologist's delight. This corner of the world is remarkable for its incredible diversity of ethnic groups, each of which has its own language, customs, mores, and religious beliefs.

Historians and archaeologists are in general agreement that modern South-East Asia was *not* peopled by descendants of Java Man, a hominid species that ranged from Africa up to the glacial boundaries of Europe and all the way to China. It would appear that Java Man died after failing to adapt to a changing environment. It is also possible that he was pushed out or destroyed by a more advanced species.[17]

Scholars concur that the peopling of South-East Asia occurred through a series of southward migrations from the Asian mainland. The origins, the routes taken, the precise points in time of these various movements, and their extent and relationship to various archaeological remains of South-East Asia, have all stimulated lively debate.

South-East Asia today has a sprinkling of hunting-gathering Negrito people, such as the Aeta of the Philippines, the Semang of Malaysia, and the Orang Kubu of Sumatra. Living in family groups in isolated localities, these people are considered descendants of the earliest existing inhabitants of South-East Asia. Shunning contact with other groups, they sleep in tree or rock shelters and live off jungle fruits, roots, and game or fish caught with bows and arrows or ingenious traps.[18] Clothing is minimal and is made from bark, leaves, and grasses found in the forests. Weaving does not seem to be an important activity amongst these people.

Australoid peoples are also thought to have passed through South-East Asia during the early migration

period. Although no longer evident as a racial entity, traces of these people may be seen in the current racial stock of Timor and Flores. Vestiges of Melanesoid peoples have also been found amongst the Senoi in Malaysia, the Alfurs of the Moloccas, and on the Engano and Mentawi Islands off the south coast of Sumatra.[19]

The majority of the present-day indigenous population of South-East Asia are basically mongoloid, consisting of various peoples who migrated from South China and the Tibetan border areas at various times. The earliest arrivals, sometimes called Proto-Malay or Nesiot, are thought to have come around 2500–1500 BC in a series of drawn-out migrations. They were a Stone Age people who were hunters, cultivators, fishermen, and sailors. They built houses and boats. They were also skilled potters and adorned themselves with elaborate jewellery. As ancestor worshippers, they were preoccupied with notions of death, renewal, and fertility. The continued well-being of the community was all important. Megalithic monuments in the form of dolmens and menhirs remain as a testimony to their complex religious beliefs. In their art, images of ships, trees of life, ancestor figures, and sacred animals are commonly depicted.

Descendants of these people possibly include the Jakun of Malaysia, the Bataks of Sumatra, the Dayaks of Borneo, the Toradja of Sulawesi, the Igorot tribes of the northern Philippines, and some Montagnard groups of central Vietnam. Many of these people have a time-hallowed tradition of making cloth from bark. Some are also among the most outstanding weavers still at work in South-East Asia today. Using a simple body-tension loom and the warp *ikat* patterning technique, these people have created a remarkable range of textiles. The motifs offer a tantalizing glimpse into the world view of these ethnic groups. It is not known whether these people brought the art of weaving to South-East Asia, or whether they acquired it later through contact with more advanced cultures.

On arrival, these mongoloid people, in all probability, pushed the earlier Negrito into the interior uplands. In turn, they were forced by later invasions of more advanced peoples to vacate the lowland areas and seek their livelihood in the mountainous interiors and on the more remote islands, where they remain to this day. The later arrivals (sometimes called Deutero-Malay) include the present-day Malay population of Malaysia, Indonesia, and the Philippines, the Mons of Burma and Thailand, and the Khmers and Cham of lowland Indo-China. The Annamese of Vietnam, the Tibeto-Burmese peoples, and the numerous branches of the Thai race are thought to have arrived in South-East Asia in separate migrations at a later date.

The earliest arrivals of the Deutero-Malay group were neolithic peoples who had an advanced culture. They grew rice, used bamboo pipes for irrigation, and harnessed the buffalo to draw the plough. They were skilled woodworkers, had a rudimentary knowledge of metals, and were skilled navigators. They lived in matriarchal societies in organized village communities with simple laws.[20] They were animist in their beliefs and worshipped the ancestors. They certainly knew the art of weaving and could have passed it on to the less-advanced inhabitants of their new lands.

The Indianized Era: AD 100–1300

The earliest and most far-reaching influence which was to permeate South-East Asia was that of India. Attracted to South-East Asia's mineral wealth, Indian traders began visiting South-East Asia around the beginning of the Christian era. During these visits they came in contact with the Mon, Khmer, Malay, and Cham peoples. Early traders and adventurers were soon followed by priests and teachers who brought with them Indian religious ideas, customs, language, and literature.

Many local rulers adopted such Indianized ideas as a belief in a god-like king at the head of a hierarchical administration system, an ancestral genealogical pedigree system, the lunar-solar calendar, a system of astrology, and various Hindu rituals of worship.[21] The 'Laws of Manu' (part of the *Dharmasastras*) have provided the basis for codifying many local customs in South-East Asia.[22]

Traditional animist beliefs were overlaid with a veneer of popular Hinduism. Statues of Brahmanical deities have been found throughout mainland South-East Asia and Indonesia.

Buddhism was brought from India to South-East Asia by traders and missionaries at an early date.[23] The Mahayana sect initially was very popular, but was later superseded by the Hinayana sect which became deeply rooted in Laos, Cambodia, Thailand, and Burma. To this day, numerous temples stand as a moving testimony to the piety and technical skill of earlier generations of South-East Asians who were inspired by Indian religious beliefs. The prototypes for many of these temples may be found in India, but in design they have been subtly adapted to blend with indigenous architectural styles and religious beliefs.

The basic proportions and much of the iconography for South-East Asian religious art comes from Indian canons of design. Classical Indian art motifs, such as crowned snakes or *naga*, the *makara* water monster, the *kala* monster face mask, the *garuda* bird (the mount of the Hindu God Vishnu), peacocks, the lotus, the 'vase of plenty', and the 'wishing tree' have found their way into the design inventory of South-East Asia. Many of these motifs appear on South-East Asian textiles.

Local languages have been greatly enriched by loan words from Sanskrit, the ancient and sacred language of India. The *Pali* language of the Buddhist scriptures has also influenced local languages, though to a lesser extent. Indian writing systems were, at one time, widely used throughout South-East Asia. Today, only the Burmese, Shan, Thai, and Lao continue to use Indianized scripts. Indian epics, such as the *Mahabharata* and the *Ramayana*, along with the Buddhist *Jataka Tales*, form the basis of classical theatre in South-East Asia. The *Ramayana* is performed as a puppet play in southern Thailand, northern Malaysia, and Java, and as a dance drama in Burma, central Thailand, Laos, and Central Java. Diluted versions of the *Ramayana* survive in the southern Philippines in the form of recitations among the Muslim Maranao and Magindanao peoples.[24] Episodes from these stories also adorn temple frescoes and bas reliefs. Puppet figures from these epics (called *wayang*) appear on the textiles of Bali and Lombok in Indonesia.

The full flowering of Indian cultural influences culminated in some remarkable civilizations in South-East Asia. The earliest recorded is possibly Funan (*c.* AD 100–*c.* AD 500), which was located south-east of the Mekong Delta and may, at times, have extended along the Gulf of Thailand as far as southern Burma. In its heyday, it served as a halfway house on the sea route linking China with India.[25] Funan was superseded, in the sixth century, by Chenla, situated in the middle reaches of the Mekong.[26] These two civilizations are the forerunners to the magnificent ninth- to twelfth-century Khmer civilization centred on Siem Reap in Cambodia. Khmer artistic genius is beautifully expressed in the temples of Angkor Wat, Banteay Srei, and the Baphuon in Cambodia, and Phimai in north-east Thailand. These temples were embellished with exquisite Hindu and Buddhist statuary and lively bas reliefs. The Khmer civilization at its height controlled much of north-east and southern Thailand, southern Laos, and parts of Vietnam. Invasions from the Cham in the east, and later the Thai in the west, brought an end to that civilization.

The fall of Funan (*c.* AD 627) was followed by the rise of a new maritime empire and Buddhist centre in the south, called Sri Vijaya. Sri Vijaya was centred in Palembang on the south-east coast of Sumatra. By controlling the India–China sea route through the Straits of Malacca from the seventh to the thirteenth centuries, Sri Vijaya exerted considerable influence over much of South-East Asia. At the height of its power, Sri Vijaya had hegemony over the Malay Peninsula, southern Thailand, West Java, and parts of Borneo.[27]

In the middle of the eighth century, a new kingdom ruled by the Sailendra Dynasty (AD 778–864) arose in Central Java. Inspired by Mahayana beliefs, the magnificent Buddhist stupa of Borobudur and numerous monuments of the Kedu Plain were built under Sailendra patronage. The Sailendras vied with Sri Vijaya for hegemony in the region until the mid-ninth century, when a dynastic marriage was effected between the two royal houses. This eventually created a power vacuum in Java which was filled by a succession of dynasties based in East Java. The more notable were those of Mataram in the ninth century, Singosari in 1222, and Majapahit (1292–1520). These dynasties built a number of notable monuments in East Java incorporating both Indian and local styles.[28]

The mainland states of Burma and Thailand were settled in the southern areas by the Mon people who were pre-eminent from the sixth to the eleventh centuries. The Mons founded the kingdom of Dvaravati, which was located near Nakhon Pathom, west of Bangkok. From there, they spread as far north as Lamphun, just south of Chiangmai. They also founded settlements in Pegu and Thaton in Lower Burma. The Dvaravati period left temple remains, terra cotta figures, votive tablets, and some distinguished religious sculpture inspired by Indian Gupta and Pala artistic ideals.[29]

From the fifth to the ninth centuries, the Pyu, an early Tibeto-Burmese people, built a small but imposing kingdom at Mawza, a few kilometres out of Prome in central Burma. The remains of city walls, temples, Hindu and Buddhist statues, burial urns, jewellery, and silver coins have been recovered from these sites.[30]

Two centuries later, the Pyu were succeeded by the Pagan Dynasty in the dry zone of Burma. After the conquest of Thaton in 1056, the Burmese carried off the essence of the superior Mon civilization in the form of Theravada Buddhist texts, the king, his court, monks, scholars, scribes, and artisans.[31] So equipped, the Burmese in Pagan embarked on a 250-year orgy of temple building which resulted in the construction of over 5,000 Buddhist monuments embellished with *Jataka* plaques, gigantic Buddha images, and fine temple frescoes. The Pagan civilization came to an untimely end at the hands of the Mongols in 1287.

The decline of Pagan in the west and the Khmer in the east gave the Thais the opportunity to found kingdoms of their own in the Chao Phraya River valley. The Kingdom of Chiangmai was established in 1290–2, and was eventually overshadowed by the fast-expanding state of Sukhothai, which was founded in 1240. In 1438, nearly 200 years later, Sukhothai was absorbed by the stronger, more aggressive Thai Kingdom of Ayuthia.[32] Exquisite Buddhist statuary, temple remains, and ceramics from central Thailand, attest to the refinements of both these Thai kingdoms.

Our knowledge of textile production and use during the Indianized first millennium AD of South-East Asia's history, is extremely fragmentary.[33] Most of what we know has been gleaned from a variety of sources, such as stone inscriptions, local chronicles, contemporary writings (of which there are very few), and archaeological remains.

South-East Asia's warm, humid climate and abundant rodent and insect life have not been conducive to the preservation of fine fabrics. Frescoes at Pagan, in temples such as the Lokateikpan, Abeyadana, and Nandamanya, depict kings and worshippers in richly patterned clothes (Fig. 4).[34] The bas reliefs at Angkor Wat show heavenly maidens, called *apsara*, in elaborate sarongs patterned with small floral and geometric designs (Fig. 5).[35] Changes in clothing styles have been an important element in helping to assign sculpture to various periods of Khmer art.[36] Some thirteenth-century stone carvings from East Java show elaborate repetitive patterns on the clothing of some sculptures. These are identical to some of the patterns found on the walls of eighth-to-tenth-century temples, such as Prambanan and Candi Sewu in Central Java. Some patterns show strong affinities with woven and batik patterns seen on Indonesian textiles today.[37]

Chinese Interest in South-East Asia: AD 500–1600

From contemporary Chinese sources we have some fascinating accounts describing everyday life in the early Indianized states of South-East Asia. For example, we have this description from *A History of*

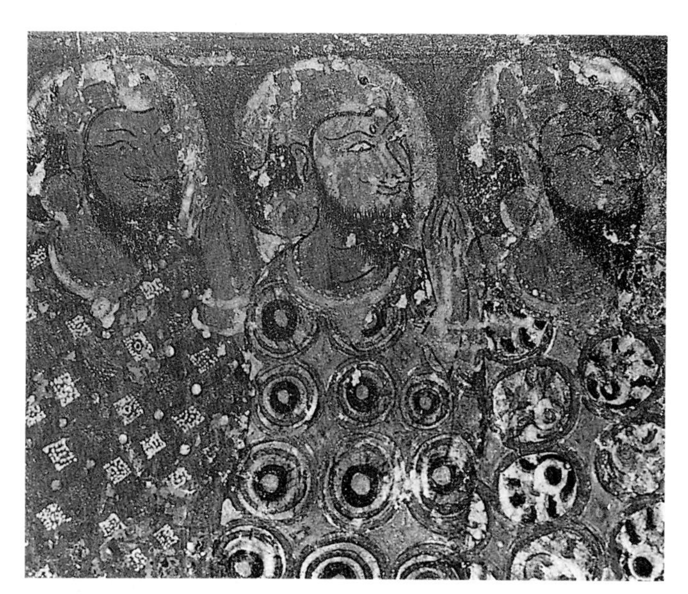

Fig. 4. Seated kings in gaily patterned garments paying homage to Lord Buddha, Abeyadana Temple, Pagan. Eleventh century.

Fig. 5. Bas relief of a pair of heavenly maidens, or *apsara*, wearing elaborate sarongs patterned with small floral and geometric designs, Angkor Wat. Twelfth century.

the Southern Chi, written about AD 400, of the clothing worn by the people of Funan:

The people of Funan are malicious and cunning.... As merchandise they have gold, silver and silks. The sons of great families cut brocade to make themselves sarongs; the women pass their heads [through some material to dress themselves]. The poor cover themselves with a piece of cloth.[38]

Chinese records from AD 518 mention the weaving of cotton on Sumatra and note that the king in the northern part of this island wore silk.[39]

Langkasuka, a kingdom in northern Malaysia–southern Thailand, was reported as establishing relations with China in 515. *The History of the Liang* has this to say about the local inhabitants:

Men and women let their hair hang loose and wear sleeveless garments of a material called *kan man* whose fibre is *ki-pe* cotton. The King and dignitaries of the kingdom add a piece of dawn red material over their dress which covers the upper part of the back between the shoulders. They gird their loins with a cord of gold and suspend gold rings from their ears. The women adorn themselves with beautiful scarves enriched with precious stones.[40]

A History of the Sui gives a description of the court in seventh-century Cambodia:

Every three days the King proceeds solemnly to the audience hall and sits on a couch made of five kinds of aromatic wood and decorated with seven precious things. Above the couch there rises a pavilion hung with magnificent fabrics.... The King wears a dawn red sash of *ki-pei* cotton that falls to his feet.... This robe is made of a very fine white fabric called *pe tie*.... The dress of great officials is similar to the King.[41]

The *Tang Shu*, the official chronicle of the Tang Dynasty (618–906), describes the dress of the ancient Pyu of Burma: 'For clothes they wear skirts made of cotton for they hold that silk should not be worn as it involves the taking of life. On the head they wear gold flowered hats with a blue net bag set in pearls.'[42]

Chao Ju Kua, a thirteenth-century salt tax inspector in Fukien Province, wrote in *Chu Fan Chi* (A Description of Barbarian Peoples) about the Philippines, as follows:

The country of Ma-i [Philippine archipelago] is to the north of Po-Ni [Borneo]. Over one hundred families are settled together along both sides of the creek [or gully]. The natives cover themselves with a sheet of cloth or hide the lower part of the body in a loin cloth.... The products of the country consist of yellow wax, cotton, pearls, tortoise shell, medicinal betel nuts, *yu-ta* cloth [abaca]. [The foreign] traders barter these for porcelain, trade gold, iron censers, lead, coloured glass beads and iron needles.[43]

Despite strong feelings of cultural superiority and economic self-sufficiency, the Chinese, like the Indians, had showed more than a passing interest in the mineral wealth and exotic tropical products of *Nanyang* (South-East Asia). Under the Tang Dynasty (618–906) there was a substantial trade in aromatic woods, such as camphor and sandalwood, kingfisher and bird of paradise feathers, and swallows' nests, in exchange for textiles, metalware, and porcelain. Trade with China was accompanied by a demand that the lesser states and kingdoms of South-East Asia acknowledge Chinese suzerainty by regularly accepting envoys and despatching missions to the Chinese capital with 'tribute' in the form of a sampling of their country's produce.[44]

The *Song Shi* and *Song Hui Yao* historical records list the various types of produce offered as trade tribute to the Chinese emperor during the Song Dynasty (960–1279) by some of the states of South-East Asia. Return gifts and payments made by China are also listed.[45] Tribute included a wide variety of items, ranging from turtle shell, ivory tusk, rhinoceros horn, lac, alloys of tin and lead, domesticated elephants, spices, such as cloves, nutmeg, aniseed, and cardamom, and aromatic wood, such as Borneo camphor, sandalwood, gharuwood, benzoin, and musk. There are also references to tribute in the form of textiles. For example, in 990, Jiao Zhi, in the northern part of Vietnam, was reported as offering (among other things), '10,000 rolls of silk, 1,000 rolls of cotton, 1,000 rolls of woven silk'. The kingdoms of Champa, Zhen Li Fu (both near modern Kampuchea) and Sri Viyaya, at different times sent samples of 'local' textiles.[46] In the year 1000, Ligor, a small state located near the present-day town of Nakhon Sri Thammarat in southern Thailand, offered dyestuffs in the form of 100 kati of madder for purple, 10,000 kati of sappan wood for red dye. One box of blankets and four pieces of floral textiles were also presented.[47]

In return, China usually presented such items as copper cash, silver, and numerous rolls of silk, all of which were forms of currency during Tang and Song times. Full sets of finely made garments, complete with accessories, such as boots, head-dresses, and gold and silver belts, were also offered as gifts to the chief emissaries. These tribute missions gave rise to great social occasions, accompanied by state banquets, sightseeing, and lavish entertainment.

Elaborate rules of protocol were established for presenting gifts and receiving tribute.[48]

During the first millennium, China did not exercise a tight rein over the states of South-East Asia. With the exception of North Vietnam, the states of South-East Asia were, by and large, left to their own devices and were free to develop their own institutions.[49] China did, however, keep an eye on developments in the region as various states rose and fell and vied with each other for hegemony. As overlord, China did not hesitate to bring a recalcitrant vassal to heel. During the Mongol Dynasty (1279–1368), the Khan sent his armies as far afield as Burma, Annam, and Java to enforce submission. China was also known to support a weaker or emerging state in the interests of maintaining a favourable balance of power in South-East Asia.[50]

During the early years of the Ming Dynasty (1368–1644), due to the aggressive efforts of Admiral Cheng Ho, China took a more active interest in South-East Asia.[51] From the thirteenth century onwards, there was a great expansion of trade with South-East Asia. Increased trade was accompanied by the founding of permanent settlements by Chinese traders in various well-situated coastal settlements in the Philippines, Indonesia, and possibly Malaysia. Many Chinese intermarried with local women and developed merchandising centres as middlemen, a role the Chinese have played for centuries in South-East Asia. In sixteenth-century Bantam, in West Java, Chinese merchants have been described as: 'The most influential.... [R]ows of Chinese shops with all their expensive goods: damask, velvet, satin, silk, gold thread, cloth of gold, porcelain, lacquered work, copperwork, woodwork, medicinal products and the like.'[52]

One account describes the cargo of two Chinese ships, seized in 1570 off the island of Mindoro in the Philippines, as being 'loaded with gold thread, silk and cotton cloth, gilded water jugs, porcelain vases, plates, bowls and some fine porcelain jars'.[53] In addition to documenting China's mercantile importance, these short descriptions reiterate the importance of China as a supplier of luxury textiles and fine porcelains to South-East Asia.

With the exception of Thailand, Cambodia, and Vietnam, the various peoples of South-East Asia produced only simple unglazed pottery. As a result, the porcelains and stonewares of China were in great demand, and a lively trade developed. Apart from utilitarian purposes, such as storing rice, water, oil, wine, condiments, medicines, and cosmetics, ceramic objects have also been used by many inhabitants of South-East Asia for ceremonial purposes, payment of fines, and as grave goods.

Aesthetically, these ceramics were admired for their handsome, robust shapes and lively motifs. Chinese ceramic shapes have been imitated in brassware by the Maranao people of the southern Philippines, while some ceramic motifs have been widely copied on South-East Asian textiles. Images of dragons from stoneware jars may occasionally be seen on warp *ikat* cloths from Sumba.[54] More stylized versions may be seen on Maranao woven *langkit* decorative strips and on Iban blankets. In Bali and Thailand, the *banji*, or swastika, motif is a pattern regularly encountered on silk weft *ikat*. It may also be seen in a supplementary weft on Lao, Shan, and Kachin textiles.

The influx of Chinese settlers led to the erection of Chinese temples throughout South-East Asia. Sumptuous embroidered altar cloths and brightly coloured banners, adorned with figures and emblems from Buddhist and Taoist mythology, were imported from China to furnish these new edifices. Chinese motifs, such as the 'eight precious objects', the Twelve Lohans, the Eight Taoist Immortals, the Gods of Wealth, Long Life and Numerous Progeny, the phoenix, bats, peonies, bamboo, pine, plum blossom, cloud, and rock motifs, and auspicious characters depicted on these cloths, eventually found their way on to Javanese batik and various local embroideries. Embroidery stitches, such as the long and short satin-stitch and the Peking knot, were also incorporated into local needlework.

The Coming of Islam to South-East Asia

Arab traders came to South-East Asia as early as the fourth century AD. Following the death of the Prophet Mohammed in the seventh century, Islam swept across the Middle East and reached northern India in the eighth century. By the ninth century, there were small communities of Muslim merchants in several ports along the route to China.[55] The Gujerati port of Cambay, in north-east India, fell into Muslim hands in 1298. This event was to have great repercussions for South-East Asia, for Cambay was an important emporium and rendezvous point for trade between China, South-East Asia, the Persian Gulf, and Europe. Piece goods bound for South-East Asia were taken on here in exchange for spices, aromatic woods, musk, and porcelains.[56]

Many local Indian merchants who had business dealings with insular South-East Asia, fervently embraced the Muslim faith.[57] With their knowledge of commerce and the outside world, Muslim traders in many cases became widely respected by local rulers. These merchants sold much of their produce to the rulers at wholesale prices and, in return, received the basic commodities and supplies they needed, thereby forming mutually profitable partnerships. Some merchants rose to the position of *shahbandar*, or 'director of trade', at various ports of Malaysia and insular South-East Asia. Whetted by a desire for increased trade, wealth, and power, many local rulers saw conversion to Islam as a key to furthering their ambitions. In anticipation of subsequent events, many Muslim merchants brought their mullahs and holy men with them. On the conversion of a ruler, the population under his jurisdiction also became Muslim. Islam was also spread through marriage, for the religion required that a non-Muslim spouse convert to the faith prior to the wedding ceremony.

The rise of Malacca in the early fifteenth century to a position of commercial pre-eminence, coupled with the conversion of its first ruler, Parameswara, through his marriage to a daughter of the Sultan of Pasai in Sumatra, gave great impetus to the spread of the Muslim faith throughout Malaysia and the Indonesian Archipelago.[58] By 1477, much of the Malay Peninsula, present-day southern Thailand, and the east coast of Sumatra had embraced Islam. Muslim missionaries were despatched along various trade routes as far as the Philippines. Islam reached the Moluccas around 1475, and the ruler of Brunei in Borneo was converted in 1500, as were the Minangkabau in West Sumatra. Through trading relations, Islam also began to penetrate Java. By the end of the fifteenth century, much of Java was Muslim and the coastal peoples of the southern Philippines were being drawn into the Islamic fold. The Bugis of Sulawesi converted the islands of Sumbawa and Lombok early in the seventeenth century.[59] Only Bali was to resist attempts at conversion and has remained Hindu to this day. Flores and Timor, because of their relative remoteness, were less affected. On the larger islands, Islam was largely confined to the coastal areas. The interior peoples retained their animist beliefs.

The most sought-after goods that the Muslim traders brought to South-East Asia were Indian silk and cotton textiles from Gujerat and the Coromandel coast. Most famous was the soft, loosely woven double *ikat*-patterned *patola* cloth, sometimes called *cinde* (Plate 1).[60] At a very early date, Indian textile artisans had mastered the art of making brilliant colours in natural dyes and fixing them with mordants made from alum and iron.[61] These cloths were primarily in glowing reds, with motifs outlined in white, complimented by brown and indigo, and enlivened with splashes of yellow and green. They completely captivated the South-East Asians who were used to textiles of more sombre colours.[62]

From 1450 onwards, Indian cloth was first brought to ports on the main trade routes in coastal Sumatra, Bantam in Java, and Pegu in Lower Burma, and then traded widely throughout South-East Asia. In the early sixteenth century, a Portuguese, Duarte Barbosa, noted that into Pegu:

> Hither come every year many moorish ships to trade and bring an abundance of printed Cambaya (and Paleacate) cloths both cotton and silk, which they call patolas. These are coloured with great skill and are here worth a lot of money.

He continues:

> In Ambon [in the Moluccas] Cambaya cloths are held in great value here, and every man toils to hold so great a pile of them that when they are folded and lain on the ground one on the other they form a pile as high as himself. Who so possesses this holds himself to be free and alive, for if he is taken captive, he cannot be ransomed save for so great a pile of cloth.[63]

It is also reported 'that in 1603 an average piece of imported cloth was worth forty pounds of nutmeg on the island of Banda'.[64] It appears that the sale of imported Indian cloth provided the lion's share of the capital outlay involved in purchasing the spices sought by Europe to preserve food during the winter months.

Because of its tremendous popularity, *patola* cloth had a major influence on local textile development in South-East Asia, both in terms of design motifs and pattern layout. *Patola* cloth is characterized by a large centre field filled with finely patterned repetitive motifs. The selvages are usually bordered by narrow stripes, which may enclose smaller *ikat* figures. The warp edges are framed by two to three bands enclosing rows of simple vegetal/floral elements, diamonds or heart-shaped leaves, and are finished by a line of triangles in a saw-tooth pattern. Some *patola* textiles may resemble an oriental carpet, having wide distinctive borders in a darker colour on all sides of the cloth.

In Indonesia, *patola*-inspired pattern layout can be seen on weft *ikat* cloths from Singaraja in Bali, the nearby island of Nusa Penida, coastal East Sumatra, and on warp *ikat* from East Flores and the island of Roti. The supplementary weft-patterned cloth of Bali, Palembang, and northern Malaysia has also been greatly influenced by *patola* design arrangement. The royal weft *ikat* cloth of Thailand and Cambodia, although differing in some design elements, shows strong affinities with *patola* cloth in pattern alignment, as do the weft *ikat* cloths from Inle Lake in Burma. Some sarongs, such as the *malong andon* of the southern Philippines, also show clear *patola*-like pattern arrangements.

One of the most prevalent design elements seen on imported *patola* cloth is an eight-rayed star-like rosette set in squares, circles, or lozenges which touch but do not overlap. Called *jelamprang* in Indonesia, this motif has been widely imitated on warp *ikat* from Nusa Tenggara, and on Javanese batik. It may appear as an overall pattern in the central field, as on Lio *ikat* in Flores, or it may be seen in bands, as in nearby Sikka and on the island of Savu. On Roti, this motif appears in a variety of original layouts in the centre field.

Geometric forms, such as rhombs and squares, are also important design elements, both in the centre field and along the warp ends of *patola* cloth. They may be very simple and merely form a row or trellis-like framework for another motif, or they may be arranged in complex clusters filled with stars, hooks, crosses, small floral patterns, and subtle splashes of colours of varying intensity. Such patterns may be seen on the double *ikat* cloths of Tenganan and the weft *ikat* cloths of Singaraja in Bali and Nusa Penida. These designs also appear on the centre bands of blankets made for royalty on the island of Sumba. Small heart-shaped motifs, a popular design element on *patola* cloths, may be seen on warp *ikat* blankets from Lomblen Island, north-east of Flores (Fig. 6).

Bands of isosceles triangles along the narrow ends of the cloth are referred to as *tumpul* in Malay. They may be plain in a solid colour or filled with very elaborate floral elements. The sides may have decoration along the edges, while some may be in the form of curling leaves or feathers. As a decorative item, the *tumpul* has been used in South-East Asia since neolithic times.[65] However, its placement along the lateral or warp edges of the cloth, and its design elaboration, are due to *patola* influence. The *tumpul* pattern is widely seen on cloth made by Muslim weavers in Malaysia and Indonesia; it appears to a lesser extent on cloth made by Muslims in both Thailand and the Philippines.[66]

Because of their rare and exotic beauty, *patola* textiles became symbols of wealth and prestige. Many were kept as treasured heirlooms to be used on special occasions. The courts of Central Java reserved the wearing of *patola* cloth to rulers and those of high nobility. The textiles were eagerly sought as marriage gifts, and were sometimes part of the wedding attire for Javanese and Savunese royalty. They sheltered the bier of deceased royalty on the island of Roti, and in the southern Philippines they were piled on a corpse as it lay in state. Magical powers attributed to them led to their use with ceremonial objects such as the *kris* (Malay dagger) and *wayang* puppet figures.[67] In an effort to repel evil spirits on the island of Bali, *patola* cloths were prominently displayed as banners or used as altar cloths during religious ceremonies and cremations.

Fig. 6. A double *ikat patola* cloth featuring diamonds, heart shapes, and floral patterns. Arts of Asia Gallery, Denpasar.

Fig. 7. An antique Indian resist-patterned textile. Arts of Asia Gallery, Denpasar.

Numerous resist-patterned cotton textiles were imported into South-East Asia from the Coromandel coast of India (Fig. 7). The range of colours and intricacy of design on surviving examples are extraordinary. Unfortunately, there has been little research on the influence of these imported designs on South-East Asian traditions. Like the *patola*, they were considered family heirlooms, but they do not seem to have acquired quite the same exhalted status as *patola* cloth.[68]

Muslim traders are also credited with introducing plaid-patterned textiles to South-East Asia. Along with stripes, the plaid is the most widespread pattern on textiles today throughout the area.[69] Woven both in cotton and silk, plaids are particularly favoured by men, both for everyday wear and for formal occasions. Plaids vary from large, brightly coloured checks in vibrant silks, as worn in South Sulawesi by the Bugis people, through gay tartans in green, yellow, and red, as seen in Thailand, to small checks in conservative blue, beige, brown, and olive green, as worn by males in Burma.

The *ikat* patterning technique of dyeing the weft threads prior to weaving was probably introduced from India via Muslim traders.[70] Weft *ikat* is a prevalent patterning technique throughout insular South-East Asia where Muslim influence remains strong. It is also an important decorative technique amongst Buddhist communities on mainland South-East Asia.

Muslim traders also brought gold and silver thread to South-East Asia. Inspired by the sumptuous gold weavings of northern India, Iran, and North Africa, the use of metallic threads may have been introduced as a supplementary weft to create patterns. Except in Bali, the art of weaving with metallic threads has traditionally been limited to Muslim weavers in South-East Asia.[71] Some embroidery techniques using gold yarns, sequins, and pieces of glass may also be attributed to Muslim influence.

Because of a religious proscription against the representation of living things in a realistic manner, design elements in strongly Islamic areas make extensive use of geometric and floral forms. The traditional South-East Asian design register owes much of its inspiration to the region's abundant plant and animal life. The artist or craftsman could not completely abandon his heritage nor ignore his environment in meeting the tenets of his new Muslim faith, so he modified various design elements to make sure that flora and fauna was not expressed in a naturalistic way. He took one or more key features of a plant or an animal, such as a bud, a leaf, a tail, a feather, or a wing and, through elaborate embellishment, transformed it into a new motif. With the passage of time, these motifs became more stylized and the original inspiration was forgotten. Sometimes, names of motifs continue to provide a clue. Despite the proscription, animal and human figures continue to appear surrepticiously in border patterns on textiles from staunchly Muslim areas, such as Palembang and Sumbawa.

European Influence

News of the whereabouts of the fabulous Spice Islands was brought to Europe in the fourteenth century by Marco Polo. Prior to this, Europeans had little idea of exactly where the spices so necessary to flavour their preserved meat came from. Transport of these spices, first by sea, and then through politically unstable Asia Minor over to Venice, was hazardous; by the time the spices reached Venice for European distribution, prices were exorbitant. Something so expensive and exotic in taste undoubtedly came from a distant, mysterious land.

The Renaissance in Europe had opened man's eyes to a wider world. Combined with improvements in navigation and ship building, and a desire to propagate the Christian faith, this expanded world-view made Europeans determined to find a sea route for the lucrative spice trade. Thus was 'The Age of Discovery' inaugurated.

The Portuguese were first off the mark, followed by Spain. Vasco da Gama rounded the Cape of Good Hope in 1498, and the Portuguese soon established trading posts along the coast of India. They seized Malacca in 1511, and from this base despatched missions to Siam, Burma, Cambodia, and later China and Japan. They soon established a network of fortified trading posts along the main sea lanes of the Indonesian Archipelago, as far as Timor and the Moluccas. Local rulers were prevailed upon to pay tribute and to grant the Portuguese a monopoly on their spices and other products in return for the much sought-after textiles acquired from their trading posts in India.[72] The Portuguese also traded vast quantities of Chinese textiles in Europe and Japan.[73]

Magellan's circumnavigation of the globe in 1521 led Spain to take an active interest in the spice trade of the Moluccas. After some initial jostling with the Portuguese for hegemony, Spain eventually redirected its interest to another of Magellan's discoveries—the Philippines. Manila, on the island of Luzon, was made the capital in 1571, and Spanish systems of administration and feudal estates were established. The majority of the population were converted to the Roman Catholic faith, and lowland Filipinos adopted Spanish-style dress. The Philippines became an outlying appendage of the Spanish Empire which was centred on Central and South America. Yearly galleons plying between the Philippines and Acupulco carried shawls, silks, and porcelains from China, screens from Japan, Filipino chests, wood carvings, ivory icons, and Indonesian spices, in return for Mexican silver dollars, which became an accepted international currency in the ports of China and South-East Asia.[74]

During the seventeenth century, maritime supremacy passed from the Iberian Peninsula to north-west Europe. England, Holland, and France all set up trading companies within the first five years of the seventeenth century, with the express purpose of seeking direct access to Eastern markets.

Between 1595 and 1601, some sixty-five Dutch ships sailed to Indonesia and successfully purchased pepper from Bantam and spices from the Moluccas. By 1616, they had established fifteen trading posts in the Indonesian Archipelago. Over a similar time-frame, but on a smaller scale, the British East India Company established agents at Aceh in North Sumatra, Makassar in Sulawesi, and Patani and Ayuthia in Siam. They also made determined efforts to enter the spice trade but were thwarted by the Dutch at the Massacre of Amboyna in 1623, an event which caused the English to retreat and redirect their attention to India. With the capture of Malacca in 1641, and the extension of their control over Ceylon between 1640 and 1650, the Dutch supplanted the Portuguese as the pre-eminent power in the Indonesian Archipelago.[75]

The Dutch East India Company, or VOC (Vereenigne Oostindisch Compagnie), from its base in Batavia (Jakarta), developed into a vast inter-Asian trading concern. Items of trade included Javanese rice, Timor sandalwood, cloves and nutmeg from the Moluccas, pepper from Borneo and Sumatra, cinnamon from Ceylon, copper from Japan, gems from Persia, silk and porcelains from China, and cloth from India.

As far as imported Indian textiles were concerned, the Dutch picked up the trade where the Muslims and Portuguese left off. In Nusa Tenggara (eastern Indonesia), trade was conducted with local rulers who were expected to provide tribute in items such as slaves, wax, and foodstuffs, in return for muskets, alcohol, and *patola* cloth. The Dutch awarded *patola* cloth only to the most important potentates with whom they traded, and the right to wear this cloth became the exclusive prerogative of the leading nobility on each island.[76]

Over the years, various merchants catered to the personal preferences of individual areas throughout South-East Asia. It has been reported that over forty-three types of imported Indian textiles were handled by Dutch traders on the Coromandel coast.[77] The VOC apparently conducted a very profitable trade with the Royal House of Ayuthia, Siam, in imported textiles. Siamese sumptuary laws decreed that court costumes had to be superior to, and different from, those of commoners. Imported metallic patterned brocades (called *zarabaft*—a golden brocade 'atlas' satin with a gold and silver design, and *kimbhab*, a silk fabric with gold or silver thread in a striped design) were bought from Surat and Ahmedabad in Gujerat and Benares in eastern India. Made to order resist-patterned cotton with Thai patterns, such as praying devas called *thepanom* and *kranok* flame foliage, were created in white against a deep maroon or blackish indigo background (Fig. 8).[78] So great was this trade in luxury fabrics that by 1690 the King of Thailand tried to staunch the flow from the royal treasury by enlisting the assistance of the VOC in acquiring cotton seeds and dyes, along with Indian dyers and weavers, to begin a local industry. The venture does not appear to have been successful.[79]

While the Thai were most impressed with beautiful Indian fabrics, it appears that the French court of Louis XIV was no less enchanted by the beauty of the Thai *ikat* silk cloth worn by the ambassadors

Fig. 8. A resist-patterned textile featuring *thepanom* figures surrounded by *kranok* flame designs. This fabric was made in India for the Thai Court. National Museum, Bangkok.

from Siam. It is reported that their garments created quite a stir amongst French silk manufacturers who made imitations in cotton and silk called *Siamoise*.[80]

With the decline of the spice trade in the early eighteenth century because of improved farming practices in Europe, the Dutch increasingly turned their attention to the production of tropical commodities which were needed in Europe. Beginning in 1830, a system of forced deliveries produced sugar, coffee, indigo, and pepper for export. The profits of these exports enabled the Dutch to build a vast merchant navy and to establish important manufacturing industries in Holland. Machine-printed imitations of Toradja and Javanese batik cloths were also made for sale in Indonesia.

From the seventeenth through the nineteenth centuries, Anglo-French rivalry manifested itself in South-East Asia.[81] The French were driven from India in 1757, and to protect India, Penang was acquired by the British in 1786. Raffles acquired Singapore in 1819. By 1886, all of Burma was under British control. In 1888, Sarawak, Brunei, and North Borneo (now Sabah) came under British protection. Beginning in 1874, the British gradually extended their control over the Malay Peninsula.

In response, the French turned their attention to Indo-China. In 1787, they acquired Pulo Condore; within the following century they managed to add Cochin China, Annam, and Tongking, Cambodia, and Laos to the Union of Indo-China, which was established in 1887.

Following the defeat of the Spanish in the Spanish–American War in 1898, the Philippines became an American colony. By 1900, with the exception of Thailand, which served as a buffer between French and British interests, all of South-East Asia was under Western rule and was to remain that way until Japanese occupation during World War II.

Although the period of European occupation was relatively short, the changes made through colonization were far-reaching. Prior to the 'Age of Discovery', South-East Asia was primarily concerned with local events and, to a lesser extent, the affairs of India and China. As a result of European contacts, South-East Asian countries changed from self-sufficient economies to ones which produced crops and minerals for export. Under colonial rule, South-East Asia provided much of the world's rubber, cinchona, rice, copra, palm oil, and tin. In return, South-East Asia took vast quantities of manufactured goods from the West. This latter development had a devastating effect on local handicrafts, including weaving. With increased trade, communications developed and urban centres came into being, complete with factories, administrative and educational institutions, and attendant problems, such as slums and overcrowding. There began a drift from the rural areas to the towns. These economic changes also brought a great influx of Chinese and Indians, who came as indentured labourers, coolies, or small businessmen. Their arrival added to the pressures and tensions of an already complicated racial and religious mix in the region.

Subsequently, the spread of Western hygiene, improved sanitation, and health care have led to a decreased death rate, especially among infants. The result has been a phenomenal increase in the population. While education and the spread of Western ideas and technology have been responsible for much material progress, they have contributed to a cleavage between new and traditional ways of life.

Like Indian, Chinese, and Muslim contacts, interaction with European culture added new motifs to the South-East Asian textile design register. European floral motifs, in particular, have been widely incorporated into South-East Asian designs. English garden flowers, such as roses, daisies, and primroses may be seen on cloth from Arakan and Amarapura in Burma. Small rosettes and floral sprigs, possibly adopted from French lace patterns, sometimes appear in supplementary weft patterns on women's textiles from the Vientiane–Luang Prabang area of Laos. Floral bouquets are very popular on north coast Javanese batik. The warp *ikat* of Savu, Roti, Flores and, to a lesser extent, Timor, abound with European-style motifs set side by side with traditional designs.

Dogs, teapots, and automobiles may surreptitiously appear along with 'trees of life' and other traditional motifs in warp *ikat* made by the Lio in Flores. World War I tanks and Communist hammers and sickles have appeared on Javanese batik. Helicopters have recently made their appearance on warp *ikat* from Borneo and central Vietnam. European fairytale characters, such as Little Red Riding Hood and Cinderella, have appeared on batik from the north coast of Java. Cherubs and European dancing ladies may be seen on modern warp *ikat* from Timor. Not surprisingly, Christian crosses and chalices have appeared on religious vestments and altar cloths from the Philippines. Heraldic motifs, such as rampant lions flanking shields or bearing sceptres (taken from old Dutch coins) may be seen on warp *ikat* blankets from Sumba. Former Dutch royalty and coat of arms motifs may be seen on textiles from Sumba and Savu.

The Current Setting

For centuries, the production of textiles for ritual and daily use has been the work of women. Along with other household tasks, the weaving of cloth has been integrated into the pattern of the women's everyday lives, which very much revolve around the agricultural cycle and childbirth. Only a few noblewomen, such as those on the island of Sumba, or in the royal courts of Java, were freed from the usual household and agricultural chores, and could devote themselves full-time to the art of producing cloth.

Traditionally, women were responsible for the entire process. They planted and harvested the cotton (or raised their own silkworms). Under the shelter of the house or an overhanging roof, fibres were carded, twisted, and spun, warp threads were carefully counted and measured before being placed on the loom, and weft threads were wound onto bobbins. Materials for natural dyes were gathered and laboriously prepared so as to get the deep rich tones desired. Textiles were created with the aid of simple looms and other implements made by the menfolk from locally available wood and bamboo.[82]

The art of weaving was a craft traditionally handed down from mother to daughter. At age seven or eight, a little girl would begin to assist her mother with the carding, spinning, and winding of thread. She would watch her mother or grandmother set up the loom and offer assistance where possible. About the age of fifteen, she would begin to learn weaving. In the case of a body-tension loom, a girl would not begin weaving until she was strong enough to control the tension of the warp yarns. As a beginner, a young girl would start with small items, such as the backing for belts and small bags in a plain weave. Gradually she would progress to more complicated patterns in various techniques using the motifs common to the society in which she lived. By sixteen or seventeen, the average girl was a competent weaver, happy to be seen by all industriously seated at her loom weaving garments for her trousseau. Mastery of the art of weaving was a yardstick by which a young girl was judged 'ready', in the practical sense, for marriage. The loom and spinning wheel were popular rendezvous places for young people of both sexes to gather in the cool of the evening. South-East Asian literature abounds with romantic poems and songs associated with weaving.

Prospective in-laws were known to look closely at a girl's handiwork to determine whether she would make a suitable wife for an eligible son. In some societies, it is believed that a girl's nature is revealed in her weaving and stitchery. For example, in Mien (Yao) society, a girl's choice of colour combinations in embroidery supposedly reveals her nature, while the quality of her workmanship shows her to be fastidious or careless, hardworking or lazy, imaginative, docile, or quick tempered.[83] In Hmong (Meo) society, skill at weaving and making batik commands a higher bride price, and enhances a girl's attractiveness to men and her status amongst women.[84]

Such a situation, where weaving is done entirely by family members to meet domestic and ritual needs, using home-grown, hand-processed yarns coloured with natural dyes, is now limited to a few remote islands of Indonesia and isolated upland areas of South-East Asia. The day when it was common to see a loom under every house has gone. With Westernization, greater economic opportunities have beckoned in the towns. Many rural folk have abandoned their traditional self-sufficient way of life to move to urban areas, to take up labouring and factory employment. On relocating, women no longer weave. They prefer to spend part of their hard-earned wages on colourful, ready-made Western-style clothing purchased from roadside stalls and markets.

For those who continue to weave in the less remote rural areas and small villages, store-purchased threads and chemical dyes are readily available, as are mechanically-operated devices for spinning, warping, and preparing the weft. Hand-operated flying shuttles have been attached to floor looms to speed up the weaving process. Women, in many cases, no longer weave purely for domestic needs. They may be organized into small cottage industries and go to weave for a set number of hours at the back of the factory owner's house. Some women in rural areas may be contracted to do piecework at home. They receive materials from an entrepreneur or craft organization. On completion, the weaver turns in the cloth she has woven and is paid according to the amount woven and the quality of workmanship.

Thus, we have seen that a study of textiles serves as a useful lens through which to view a number of important social and cultural developments in South-East Asia. The textile arts of South-East Asia have been intricately interwoven into the history of the region. South-East Asia's historic position at the crossroads of inter-ocean trade has exposed the region to a steady flow of ideas, skills, and materials from diverse sources. Indian, Chinese, Muslim, and European influences have all, at one time or another, helped shape the textile traditions of South-East Asia. The diverse inhabitants of the region have shown a remarkable propensity to absorb and digest a wide range of cultural influences. These they have taken and selectively blended with older indigenous elements to produce new, more intricate textile forms, each unique in some way to their particular ethnic group.

While much has been written on the importance of foreign influences on South-East Asian fabrics, the influence of South-East Asia textiles on other textile traditions has, to date, been little studied. We do know that textiles have always been important items of trade within the region. People with a weaving tradition traded textiles with those who did not. Particular cloths, noted for their fine craftsmanship or supernatural powers, were eagerly traded with people of neighbouring islands.

Looking a little further afield, we do know that the Chinese were most interested in receiving cotton textiles from South-East Asia as part of an individual country's tribute. Dye-stuffs, too, were listed by the Chinese as suitable tributary items.[85] The art of *hanaori*, a supplementary weft patterning technique which is practised to this day by the weavers of Yomitan, Okinawa, is believed to have come from South-East Asia a few hundred years ago, when Yomitan was an important port with trading links in South-East Asia.[86] Much more research into the historic trade routes in and around the region needs to be done before it is possible to accurately begin to gauge the extent and importance of South-East Asian textile influences.

As previously noted, Thai silk patterns were imitated by the French during the time of Louis XIV. Indonesian batik designs have been copied on European cloth, and special batiks in subdued greens, blues, and yellow were, at one time, made for the Arab market.[87] Contemporary fabric designers, such as Jack Lenore Larsen of New York and Inger McCabe Elliot of China Seas Inc., draw inspiration from South-East Asian motifs for some of their distinctive modern textiles.[88] While South-East Asia has continued to absorb selective elements from a wide range of cultural influences, other textile traditions have, in turn, drawn inspiration from South-East Asia's dazzling array of finely crafted fabrics and distinctive motifs.

1. Charles A. Fisher, *South East Asia*, Methuen, London, 1964, pp. 15–17.
2. Gabriel Casal *et al.*, *People and Art of the Philippines*, Museum of Cultural History, University of California, Los Angeles, 1981, p. 11.
3. Fisher, op. cit., pp. 21–6.
4. Ibid., p. 64.
5. Casal *et al.*, op. cit., p. 13.
6. Ibid., pp. 13 and 23.
7. Lucas Chin, *Cultural Heritage of Sarawak*, Sarawak Museum, Kuching, 1980, pp. 5–7.
8. Aung Thaw, 'Neolithic Culture of the Padalin Caves', *Journal of the Burma Research Society*, Vol. 52, No. 1, 1969, p. 10.
9. M. W. F. Tweedie, 'Prehistory in Malaya', *Journal of the Royal Asiatic Society*, Vol. 26, No. 2, 1942, pp. 1–13.
10. Pisit Charoenwongsa and M. C. Subhadradis Diskul, *Archeologia Mvndi: Thailand*, Nagel, Switzerland, 1978, pp. 39–43.
11. Chiraporn Aranyanak, 'Ancient Fragments from Ban Chiang', *Muang Boran Journal*, Vol. 2, No. 1, 1985, pp. 83–4; and Chira Chongkol, 'Textiles and Costume in Thailand', *Arts of Asia*, November–December 1982, p. 124.
12. Robert S. Griffin, 'Thailand's Ban Chiang: The Birth Place of a Civilisation?', *Arts of Asia*, November–December 1973, pp. 32–3.
13. John Vollmer, 'Archaeological Evidence for Looms from Yunnan', in Irene Emery and Patricia Fiske (eds.), *Looms and their Products: Irene Emery Roundtable on Museum Textiles, 1977 Proceedings*, Textile Museum, Washington, DC, 1979, pp. 78–80. This loom is also known by the Angami Naga of Assam and Melanesian groups of north-west Irian.
14. Bronwen Solyom and Garrett Solyom, *Fabric Traditions of Indonesia*, exhibition catalogue, Museum of Art, Washington State University and Washington State University Press, Pullman, 1984, p. 1.
15. P. V. van Stein Callenfels, 'The Age of Bronze Kettle Drums', *Bulletin Raffles Museum*, Series B, Vol. I, No. 3, 1937, pp. 150–3; and John Lowenstein and G. de G. Sieveking, 'The Origin of the Malay Metal Age', *Journal of the Malayan Branch of the Royal Asiatic Society*, Vol. 24, Pt. 2, 1956, pp. 13–42.
16. Nigel Bullough, *Woven Treasures of Insular South East Asia*, Auckland Institute and Museum, Auckland, 1981, p. 5.
17. Bill Dalton, *Indonesia Handbook*, 2nd edn., Moon Publications, Hong Kong, 1980, p. 4.
18. Fisher, op. cit., p. 65. Their origin is a mystery. There are some scholars who believe that they possibly came to South-East Asia via India some 30,000 years ago.
19. Ibid., p. 66.
20. George Coedès, *The Indianized States of Southeast Asia*, 3rd edn., East–West Center Press, Honolulu, 1971, pp. 8–9. The present writer has relied very heavily on this book for much of the information pertaining to the Indianized period.
21. The courts of the former kings of Burma, Cambodia, and Laos, until their demise, employed court brahmins. The Royal Thai Court continues to use the services of resident brahmins to predict auspicious days and to perform in annual rites, such as at the Ploughing Ceremony, and at royal rites of passage.
22. Coedès, op. cit., pp. 254–5.

23. Local South-East Asian Buddhist literature has numerous legends pertaining to the early introduction of Buddhism to respective countries. Some legends date the introduction of Buddhism contemporary with that of the life of the Buddha, while others date the introduction of Buddhism to the time of King Ashoka (*c*.269–*c*.237 BC). The earliest archaeological remains include a Buddha in Amaravati style from Sulawesi dating from AD third–fifth centuries. For a photograph, see Philip Rawson, *The Art of Southeast Asia*, Thames and Hudson, London, 1967, p. 209.

24. Perala Ratnam, *Laos and Its Culture*, Tulsi Publishing House, New Delhi, 1982, p. 13.

25. Rawson, op. cit., p. 18.

26. Ibid., pp. 24–5.

27. Assigning Palembang as the site of the capital of Sri Vijaya is controversial. Due to the fact that few remains have been found in the vicinity of Palembang, there are scholars who are inclined to place the capital somewhere in Thailand. See M. C. Subhadradis Diskul, *Art in Thailand, A Brief History*, Amarin Press, Bangkok, 1969, pp. 12–13; and Michael Wright, 'Where Was Sri Vijaya? Another Approach', *Siam Society Newsletter*, Vol. 1, No. 1, 1985, pp. 4–12.

28. Rawson, op. cit., pp. 207–8.

29. Diskul, op. cit., pp. 4–9.

30. Aung Thaw, *Historical Sites in Burma*, Ministry of Union Culture, Burma, 1972, pp. 1–33.

31. Maung Htin Aung, *A History of Burma*, Columbia University Press, New York, 1967, p. 33.

32. D. G. E. Hall, *A History of South-East Asia*, 2nd edn., Macmillan, New York, 1966, p. 160.

33. Textile fragments from a temple banner dating back to the Pagan period were discovered by UNESCO archaeologist Dr Pierre Prichard at Temple No. 315 at Pagan.

34. Ono Toru and Inoue Takao, *Pagan Mural Paintings of the Buddhist Temples of Burma*, Kodansha, Tokyo, 1979; see photographs on pp. 29, 44, 68 and 69.

35. For photographs of *apsara*, see K. M. Srivastava, 'Apsaras at Angkor Wat', *Arts of Asia*, May–June 1984, pp. 57–67.

36. Lecture given on Khmer Art by M. C. Subhadradis Diskul at the National Museum, Bangkok, April 1981.

37. A. N. J. Th. a. Th. Van Der Hoop, *Indonesian Ornamental Design*, Bandung, 1949, pp. 81, 85 and 89.

38. Coedès, op. cit., p. 58.

39. Mattiebelle Gittinger, *Splendid Symbols, Textiles and Tradition in Indonesia*, Textile Museum, Washington, DC, 1979, p. 14.

40. Coedès, op. cit., p. 51.

41. Ibid., p. 74.

42. Reginald Le May, *The Culture of South-East Asia*, George Allen and Unwin Ltd., London, 1954, p. 45.

43. Harry J. Benda and John A. Larkin, *The World of Southeast Asia: Selected Historical Readings*, Harper & Row, New York, 1967, pp. 7–8.

44. Fisher, op, cit., p. 90.

45. Grace Wong, 'Tributary Trade between China and Southeast Asia in the Sung Dynasty', in *Chinese Celadons and Other Related Wares in Southeast Asia*, compiled by the Southeast Asia Ceramic Society, Ars Orientalis, Singapore, 1979, pp. 74–87.

46. Unfortunately, the 'local' textiles have not been described.

47. Wong, op. cit., p. 79.

48. Ibid., p. 89.

49. China ruled over Annam and Tongking from 111 BC to AD 939 imposing Chinese administration and institutions. See Hall, op. cit., p. 184.

50. For example, China supported the emerging Thai Kingdom of Sukhothai in the thirteenth century. Later China was to encourage the growth of Malacca as a counterweight to the growing power of the Thais in the Malay Peninsula.

51. This policy of active engagement in South-East Asia was abandoned in the mid-fifteenth century due to Ming preoccupation with having to defend the kingdom from attacks by nomadic herdsmen along the north-west border.

52. Hall, op. cit., p. 256.

53. Robert Fox, 'The Philippines Since the Beginning of Time', in *Glimpses of Philippine Culture*, National Museum, Manila, 1967, p. 34.

54. Bullough, op. cit., p. 7.

55. Hall, op. cit., p. 190.

56. Ibid., pp. 191 and 197.

57. Islam, which stressed a simple direct personal relationship between man and God regardless of social status, and encouraged the virtues of prosperity and hard work, along with allowing for high individual initiative and freedom of movement, was very attractive to traders. See Dalton, op. cit., pp. 6–7.

58. Richard O. Winstedt, *The Malays, A Cultural History*, Routledge Kegan Paul, 1947, revised by Than Seong Chee, Graham Brash, Singapore, 1981, p. 34.

59. See map, Hall, op. cit., p. 192.

60. Alfred Buhler, 'Patola Influences in Southeast Asia', *Journal of Indian Textile History*, No. 4, 1959, pp. 1–4.

61. Mattiebelle Gittinger, 'Master Dyers to the World: Early Indian Dyed Cotton Textiles', *Orientations*, Vol. 14, No. 2, 1983, p. 12.

62. Wanda Warming and Michael Gaworski, *The World of Indonesian Textiles*, Kodansha, Tokyo, 1981, p. 103.

63. Buhler, op. cit., p. 2.

64. Gittinger, *Splendid Symbols*, p. 15.

65. Van der Hoop, op. cit., pp. 24–5.

66. Figurative elements, such as elephants, lions, birds, and human figures may also be seen on imported *patola* cloths. These elements have appeared on South-East Asian textiles since earliest times, so it is difficult to say whether they have been imitated from *patola* cloths. Some scholars are of the opinion that figurative elements, such as elephants, seen on cloth from Flores and Sumba could have been influenced by *patola* cloth. See Warming and Gaworski, op. cit., p. 106; and Buhler, op. cit., p. 5.

67. Gittinger, 'Master Dyers', p. 20.

68. Ibid., p. 18.

69. Gittinger, *Splendid Symbols*, pp. 15–16.

70. Bullough, op. cit., p. 6.

71. John R. Maxwell and Robyn J. Maxwell, *Textiles of Indonesia: An Introductory Handbook*, exhibition catalogue, Indonesian Arts Society and National Gallery of Victoria, Australia, 1976, p. 6.

72. Fisher, op. cit., pp. 127–30.

73. R. Soame Jenyns, *Chinese Art III*, rev. edn., Rizzoli, New York, 1985, p. 28.

74. Casal *et al.*, op. cit., p. 92.

75. Fisher, op. cit., pp. 134–6. The Portuguese only managed to retain East Timor.

76. James J. Fox, 'Roti, Ndau and Savu', in Mary Hunt

Kahlenburg (ed.), *Textile Traditions of Indonesia*, exhibition catalogue, Los Angeles County Museum of Art, Los Angeles, 1977, p. 98.

77. Gittinger, 'Master Dyers', p. 18.

78. Chira Chongkol, op. cit., p. 129.

79. Gittinger, 'Master Dyers', p. 20.

80. Charles F. Ikle, 'The Ikat Technique and Dutch East Indian Ikats', *Bulletin of the Needle and Bobbin Club*, Vol. 15, Nos. 1 and 2, 1931, 1934 reprint, p. 50.

81. Much of the information pertaining to colonial expansion has been taken from Hall, op. cit., Pt. 3.

82. Some of these weaving implements are objects of great beauty, lovingly crafted by men for their wives and sweethearts. Examples may be seen in provincial museums throughout South-East Asia.

83. Margaret Campbell, *From the Hands of the Hills*, Media Transasia, Bangkok, 1978, 2nd edn., 1981, p. 106.

84. Ibid., p. 132.

85. Wong, op. cit., pp. 74–87.

86. Hisao Suzuki, *Living Crafts of Okinawa*, Weatherhill, Tokyo, New York, 1973, p. 42.

87. The batik collection of Mrs Eiko Kusuma of Jakarta has some samples of European imitations and examples of textiles for the Arab market.

88. Jack Lenor Larsen, 'Evolution of Thai Silk: Homage to Jim Thompson', *Arts of Asia*, May–June 1978, p. 77.

CHAPTER TWO

Getting Ready to Weave

Yarns

Cotton

COTTON, a soft fibrous substance surrounding the seeds of the plant *Gossypium*, is the predominant textile yarn of South-East Asia.[1] Although possibly not native to South-East Asia, the cotton plant has been widely cultivated throughout the region for at least 2,000 years. Chinese records make reference to the weaving of cotton on Sumatra as early as the sixth century AD.[2] Gifts of cotton textiles, as tribute from South-East Asia between the seventh and fifteenth centuries, have been noted by Chinese historians. Upon arriving in South-East Asia, the Portuguese found the production of cotton well-established in the Indonesian Archipelago.[3] Up until the nineteenth century, South-East Asia produced sufficient cotton to supply domestic needs, and Burma and Thailand were reported as having small surpluses for export.[4]

The import of cheap European cotton goods at the beginning in the nineteenth century, and the displacement of subsistence crops in favour of export crops, caused a serious decline in cotton production throughout South-East Asia. Today, most countries in the region need to import cotton yarn from India and the United States to meet their domestic requirements.

A number of different strains of cotton are cultivated in South-East Asia. The most common is *Gossypium herbaceum*, a short-staple cotton which, when woven, produces a thick, warm, nubbly cloth (Fig. 9). The plant is hardy and fairly disease resistant. The fibres hold dye colours well. It is widely cultivated in the uplands of mainland South-East Asia and on the outer islands of Indonesia. Attempts have been made to cultivate longer-staple cottons, but success has been limited. *Gossypium nanking*, *Gossypium paniculatum*, and local hybrids with a longer staple, may be grown in the more accessible areas but they require more care than the *Gossypium herbaceum*. *Gossypium arboretum*, a tall tree-like perennial, is also grown for cotton in some parts of South-East Asia. Cotton is widely cultivated as a subsistence crop amongst the hill tribe people of South-East Asia and on the islands of Nusa Tenggara in Indonesia. It may be grown in a corner of a rice field or vegetable garden, or be planted in alternate rows between other crops, such as maize and tapioca. In some areas, it grows wild on abandoned swiddens. In other places, large acreages may be planted with cotton as the principal crop. Cotton requires a six-month growing season and does best in areas which have some moisture during growth, followed by a pronounced dry spell when the plant reaches maturity.

Fig. 9. The cotton plant (*Gossypium herbaceum*) which is widely grown throughout South-East Asia.

In Burma, cotton is grown on a large scale in the Mandalay and Sagaing areas, and in the Pakokku district of central Burma. During the eighteenth century, cotton from these areas was an important export crop to China.[5] Cotton is also grown as a subsistence crop in the Chin, Kachin, and Shan States of Burma,[6] and is an important subsidiary crop in north-east Thailand. The Nan and Loei Provinces in the north produce cotton on a commercial scale. It is also widely grown by the Karen, Lawa, and Akha hill tribe peoples who inhabit the border areas of Thailand. The hill tribe peoples of Vietnam and Laos also cultivate cotton. During colonial times, cotton was an important crop in the Kompong Cham, Kandal, and Takeo Provinces of Cambodia, but production has declined since World War II.[7] Because of its equatorial climate, with no marked dry season, Malaysia has not traditionally been a great producer of cotton. The Dayaks and Dusun peoples of Malaysian Borneo, however, cultivate cotton to meet local needs. Cotton was once widely cultivated in the Philippines by the mountain people of northern Luzon, and was a major crop in the Ilocos region along the north-west coast of Luzon. It was also cultivated on the Visayan Islands in the central Philippines, but does not seem to have been cultivated on the island of Mindanao.[8] Cotton has been widely cultivated throughout Indonesia as a subsistence crop. Over recent years, the Indonesian Government has been making quite an effort to make the country self-sufficient in its cotton requirements. Cotton is now grown on a commercial basis in government-sponsored projects in North Sumatra, Lombok, Flores, and Sumbawa.

The preparation of hand-processed cotton, while subject to a few local variations, is largely the same throughout South-East Asia. Traditional weaving implements in museums and private collections attest to the similarity of the processes throughout the region (Fig. 10).

Fig. 10. Traditional weaving implements.

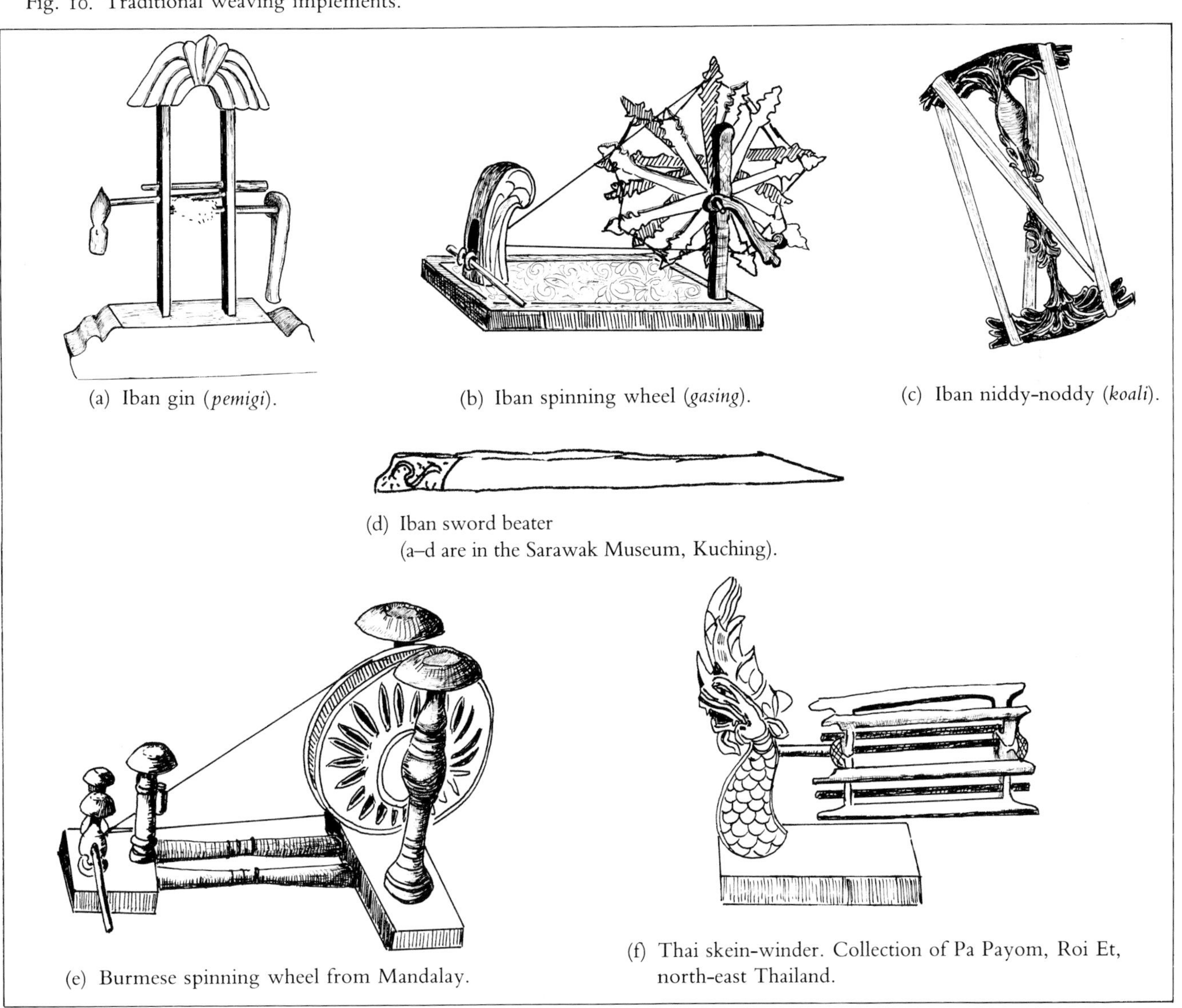

(a) Iban gin (*pemigi*).

(b) Iban spinning wheel (*gasing*).

(c) Iban niddy-noddy (*koali*).

(d) Iban sword beater
(a–d are in the Sarawak Museum, Kuching).

(e) Burmese spinning wheel from Mandalay.

(f) Thai skein-winder. Collection of Pa Payom, Roi Et, north-east Thailand.

After the cotton has been picked and the remains of the pericarp removed, it is dried in the sun for four to five days. The raw cotton is then put through a gin to separate the seeds from the fibres (Fig. 11). A cotton gin consists of an upright wooden frame set into a sturdy wooden base. In the centre of the frame are two rollers which revolve in opposite directions when turned by a projecting handle at the side. Additional pieces of wood (tightened by side wedges), may be added above and below the rollers to allow pressure to be adjusted during operation.

To make cotton fibre easier to spin, it is pulled apart in the hands before being fluffed up by a simple but ingenious instrument—a bow made from pliant wood fixed with a taut string of rattan or twine (Fig. 12). As the string is plucked across a handful of fibres, the sharp vibrations stir up the cotton, causing it to become light and airy and easier to spin. In some parts of the northern Philippines and Indonesia (such as Java and Sumbawa), the cotton may be beaten with rattan to fluff it up.[9] The cotton is then rolled into small cylinders or tufts for spinning.

Spinning wheels are widely used throughout South-East Asia to twist cotton fibres into yarn (Fig. 13). Spinning wheels consist of a spoked or drum wheel which is turned by a handle to rotate a mounted spindle by means of a cord. Sitting on the floor with the spinning wheel usually to the right, the spinner first attaches a few fibres from a tuft of cotton to the spindle. She turns the wheel to rotate the spindle while simultaneously paying out fibres from the wad of cotton in the other hand. Some weavers today use a pedal-operated spinning wheel which leaves both of the spinner's hands free to spin thread faster. In some of the more remote areas of Nusa Tenggara and mainland South-East Asia, the ancient drop spindle is still used to spin yarn (Fig. 14). Weighted with a wooden or clay whorl, the spindle is set in motion and allowed to fall with a few fibres attached. As it drops, the spindle sets up a circular action, thereby drawing and twisting the hand-held fibres into yarn. With practice, women can become very adept at using this basic spindle and may be seen plying it whenever their hands are free, be it on the way to market or returning home from the fields. A skilled spinner is able to produce thread that is as even as manufactured yarn. If a two-ply thread (commonly used throughout South-East Asia) is required, the yarn is respun with two threads being twisted together. This task is often done by the weaver's children.[10]

In preparation for weaving, the spun yarn is removed from the spindle and looped around a wooden 'H'-shaped frame, called a 'niddy-noddy', to form skeins (Fig. 15). For ease of handling, the skeins may be stiffened by being dipped into a starch solution made from rice, cassava, or corn water. These skeins are then transferred onto a 'swift', an implement with rotating arms set on a stand from which the weaver may draw off the thread as required. Sometimes the spinning wheel and swift may be used

Fig. 11. A young girl removing seeds from the cotton with a wooden gin. Maumere, Flores.

Fig. 12. Carding and fluffing up the cotton with a wooden bow prior to spinning. Maumere, Flores.

Fig. 13. Spinning cotton from yarn with a spinning wheel. Maumere, Flores.

together to wind thread onto a spool attached to the spindle. Thread may be wound around a variety of devices, such as frames, drums, basket-like rattan spools, and four-pronged stands, in readiness for warp and weft preparation. There are many local variations in spinning wheels and swifts, but the basic operating principle is the same.

In many small textile workshops, spinning and skeining are now done using machines. Machine-processed cotton yarn is readily available throughout South-East Asia.

Bast and Leaf Fibres

Hemp (*Cannabis sativa*) from the marijuana family could well have been the original textile fibre of South-East Asia (Fig. 16). In Thailand, it rivals silk in antiquity and was probably once widely used for everyday clothing. Today, its use in South-East Asia is confined to the hill tribes of Thailand and the Shan States of Burma. In Thailand, hemp is grown from seed in small hillside plots by the Hmong and Akha people. On reaching maturity, after about three months, the stems are cut down and shorn of leaves before being dried in the sun for three to five days. Using a long thumb nail especially grown for the purpose, the women split the dry stalks into very fine strips. The fibres are then frayed at the ends and folded three at a time into a continuous chain before being bleached by submersion for a day in boiling water mixed with ashes. The hemp is first twisted, then woven into heavy, serviceable cloth. Because of its greater weight and bounce, hemp continues to be the fibre preferred by the Hmong women to weave their beautiful batik-decorated accordian-pleated skirts. The Lisu at one time wove all their clothes and blankets from hemp. The Akha people make net bags for carrying fish and the strings of their cross-bows from hemp.[11]

Ramie (*Boehmeria nivea*), a 0.9–2.4 m high shrub, which may be grown from seeds, cuttings, or layers, yields a tough, white, lustrous stem fibre 10–180 cm in length. This plant is native to East Asia, including the Philippines. On reaching maturity, the stems are cut and the outer layers stripped off in ribbons. The fibre is degummed by soaking in a caustic soda solution, and may be further softened by passing through fluted rollers. When woven it resembles linen but it is not as elastic or as resistant to abrasion. To increase its versatility, ramie is often mixed with other fibres.[12] Ramie is grown in the central Philippines and is woven on a floor loom to produce fabric for clothing and table linen. The weaving of ramie in the Philippines today is quite commercialized. The Hre people of South Vietnam occasionally use ramie to produce cloth, while in Cambodia it has been traditionally grown to make fishing nets.[13] The Indonesian Government has been experimenting with growing ramie in North Sumatra.[14]

Abaca (*Musa textilis*) is a type of banana plantain which grows wild in central and southern Philippines and on the islands of Sangir and Talaud north

Fig. 14. This woman from Kapan in Central Timor is using the ancient drop spindle to twist cotton thread.

Fig. 15. This young lady is using a niddy-noddy to wind the cotton yarn into skeins prior to warping. Maumere.

Fig. 16. The leaves and stems of the hemp plant (*Cannabis savita*), northern Thailand.

Fig. 17. A grove of abaca trees (*Musa textilis*) growing wild in the central and southern Philippines. Legazpi, southern Luzon.

Fig. 18. Stripping the fibre-bearing layers of the abaca trunk. Legazpi, southern Luzon.

Fig. 19. Long silken abaca threads are hung on lines to be combed and bleached by the sun. Legazpi, southern Luzon.

of Sulawesi, where it is called *koffo* (Fig. 17). The coarse 150–300 cm white fibres produced from the leaf stems are strong and durable. The plant is a perennial with numerous suckers that grow from the root forming a cluster of ten to twenty-five overlapping leaf stalks which end in typical banana leaves. On reaching maturity (at about eighteen months to two years), the stalk is cut down and the outer fibre-bearing layer of each successive leaf is stripped away in ribbons (Fig. 18). It is then passed through a toothed stripper to remove the pulp and other waste. What remains are long silken threads which are hung on clothes lines to be combed and bleached by the sun until they are pliant and flaxen (Fig. 19). The fibres are knotted end to end and stretched on a bamboo frame prior to weaving. Abaca fibres are not as soft as cotton and feel slippery and greasy when wet. They accept dyes well and through polishing, by rubbing with ashes and a shell, they assume a beautiful sheen.[15] The upland people of Mindanao make clothing and blankets from abaca, while the people of Sangir and Talaud use it to make curtains, screens, and room dividers.

The central Philippines also produces a fine lustrous fibre from the 100–200 cm long, sword-shaped leaves of the pineapple plant (*Ananas comosus*). The 90–120 cm long fibres are obtained by retting the leaves in water and scraping to remove the outer skin and pulp. Like abaca, the yarn is formed by knotting the fibre strands together. A delicate semi-diaphanous but durable, off-white coloured fabric called *piña* is woven from the fibres. The finest *piña* cloth uses single fibres, and is woven into small objects, such as handkerchiefs and altar cloths, which are frequently adorned with exquisite embroidery (Fig. 20). For coarser quality cloth, several fibres are joined together prior to weaving. The Toradja people of South Sulawesi also once wove pineapple fibre for making ritual clothing and screens to partition sleeping areas for guests.[16] The Philippines also produces a fine sheer fabric called *jusi*, which is made from a pineapple or hemp warp and a silk, cotton, or fine abaca weft. This material, produced in a plain weave, is made into shirts and dresses. It is often embroidered with floral designs.[17]

The Bahau people in the Mahakam area of south-west Kalimantan and the Dusun of Sabah use a fibre made from the leaves of the lemba plant (*Curculigo latifolia*) for weaving. The leaves are picked and soaked in water to separate the fibrous leaf veins from the pulp. The fibres are then hung in bunches on poles to dry in the sun for several days. Each strand is then rolled against the arm and twisted before being knotted to form a yarn which resembles rough string.[18]

Palm leaves such as *nipah*, reeds such as *bemban*, and spiky plants such as *pandanus*, have long been used in South-East Asia to weave baskets, mats, and hats. Similar fibres have also been used to weave textiles. Palm leaf fibres are prepared by beating to

Fig. 20. An embroidered handkerchief from the central Philippines made of *piña* cloth woven from pineapple leaf fibres.

Fig. 21. *Lontar* palm (*Borassus flabelliformis*), called *gewang*, which was at one time used by the Toradja of Sulawesi, the Rotinese, and the people of southern Moluccas to make cloth.

Fig. 22. Trays of silkworms feeding on mulberry leaves, north-east Thailand.

loosen the fibre from the woody pulp, which is then scraped off. The fibres are split with a comb and dried in the sun before being knotted end to end to make yarn. *Corypha elata Roxb*, called *buri* in the Philippines, produces a raffia-type fibre, and was at one time woven in several communities in the central Philippines where it was used for clothing, sails, mosquito nets, and for packaging tobacco.[19] The Rotinese and the Toradja of Indonesia at one time used the leaves of this palm (which they call *gewang*) to make cloth for everyday work clothes. The Rotinese fashion their distinctive sixteenth-century Portuguese-style hats from the leaves of the *lontar* palm (*Borassus flabelliformis*) (Fig. 21). On the island of Tanimbar in the Moluccas, the fibres of that same plant may be mixed with cotton to weave a distinctive sarong. The Toradja also make use of a rush (*Fimbristylis globus* Kath) to produce cloth.[20]

Silk and Other Yarns

Silk is a very strong lustrous filament secreted by the caterpillar of the moth (*Bombyx mori*) to cushion its cocoon when it passes into the pupa stage of its life cycle. The female moth lives three to four days and lays 200–500 eggs. After ten to twelve days, the eggs hatch into 2–3 mm long caterpillars. They feed continuously on mulberry leaves, shedding their skin three or four times before they become ready to spin their cocoons (Fig. 22).

To obtain the silk, the cocoons are immersed in water heated to 60 °C. The hot water kills the pupa and loosens the silk filaments. Suspended over the water container is a crossbar, above which is mounted a small reel. The person reeling the silk fishes in the water with a double-pronged bamboo fork for the ends of loosened filaments which are lifted, and ten to twelve are then twisted together to form a single thread (Fig. 23). This thread is passed through a hole in the crossbar and drawn over the reel and into a basket to await being wound into skeins with yarns of uniform thickness. The silk, which contains 20–30 per cent seracin, or gum, is then boiled in a solution of water, soap, and soda ash to remove the gum, leaving hanks of softened, whitened yarns (Fig. 24). The yarns are then twisted on a spinning wheel into various plys depending on the type of material being woven. South-East Asian silk yarn is generally rougher than Chinese or Japanese silk, because it has a higher seracin content. This extra roughness makes the silk of South-East Asia a little more difficult to weave. In most places where silk weaving has become commercialized, the use of local silk is largely confined to the weft, while Chinese or Japanese silk is used for the warp threads.[21]

Silk, which is second to cotton in importance as a textile fibre in South-East Asia, has long been prized for making luxurious fabrics and furnishings. The accounts of early travellers to Asia contain numerous

Fig. 23. A woman, with the aid of a bamboo fork, reels and twists filaments of silk from the cocoons placed in a pot of hot water, north-east Thailand.

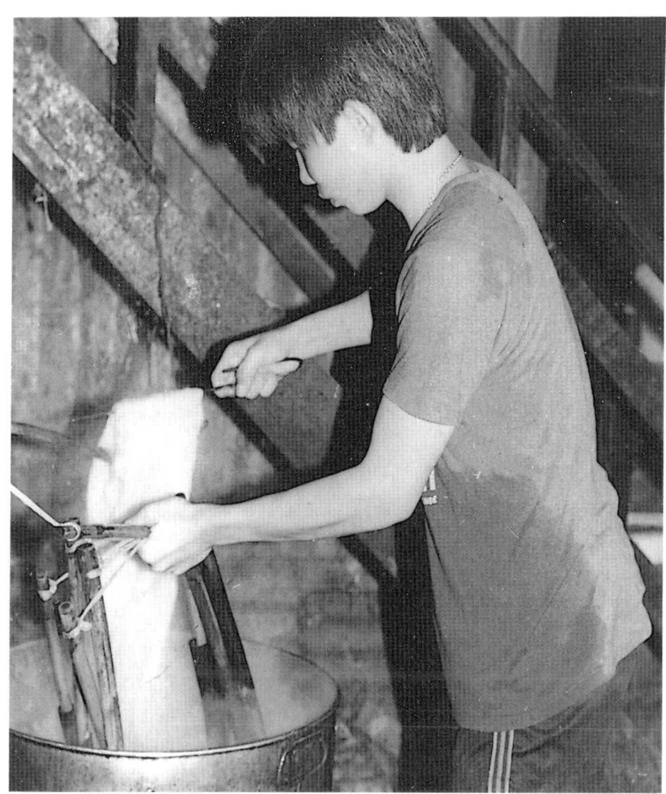

Fig. 24. Silk is degummed in a solution of boiling water, soap, and soda ash. Sai Mai Silk Weaving Factory, Bangkok.

references to local monarchy and nobility being splendidly attired in colourful silk fabrics.

Although sericulture was introduced from China to South-East Asia at an early date, with the exception of Thailand, silk cultivation (unlike cotton), has not been widespread in the region. South-East Asia has traditionally relied heavily on imports of silk thread from China and woven silk fabrics from India.

Thailand, the greatest producer of silk in South-East Asia, has historically traded in silk with its Kampuchean and Burmese neighbours.[22] Modern Thai silk production is centred in the north-east where some 330,000 households produce 500 000 kg of silk a year.[23] A little silk is also produced in northern Thailand in the Chiangmai area. Despite this domestic production, Thailand imports silk from China, Japan, and other countries for use as warp threads. The Kampuchean and Lao peoples have also traditionally raised silkworms for domestic use. Burma has not produced a great deal of silk, in part because of the Buddhist prohibition against taking life. The Burmese textile producing area of Shwedaung, near Prome, has traditionally relied on the Yabein, an animist slash-and-burn people living on the nearby hillsides, to grow the silkworm for coarse silk yarns used to produce everyday clothing, while another animist tribe, the Riang (sometimes called Yang), have traditionally grown silk for the Buddhist Shans of Burma.[24] The best quality silk has always been imported overland from China. Today, a little silk is produced in the hill station of Maymyo north of Mandalay, and in the Chin Hills in north-west Burma.[25] Malaysia and the Philippines do not have a history of silk cultivation. Silk was cultivated intermittently on the island of Sumatra from about the beginning of the Sri Vijayan Empire (*c.* AD 1000).[26] Today, production has ceased except for a small experimental project at Paya Kumbuh some 36 km west of Bukit Tinggi in West Sumatra.[27] The only other silk-producing area in Indonesia is at Tajuncu in South Sulawesi, which supplies the Bugis silk industry at Sengkang with much of its thread.[28]

Wool is not produced in South-East Asia. It is, however, used as a weft yarn by hill tribe peoples where extra warmth is required. Wool tassels, pom-poms, and wool yarns for embroidery are also popular for decorating hill tribe clothing. Imported wool yarns are obtained from weekly markets held in the more remote areas of South-East Asia. Heavier materials, such as imported velvets and flannel, are also popular for jackets and trim amongst some hill tribe people. Goat's hair has been known to decorate Kachin bags, while in Surakarta, in Java, human hair has been used to weave special sashes and waistbands.[29]

Synthetic yarns, such as rayon, acetate, acrylic, and polyester are well known in South-East Asia and are becoming more readily available, even to weavers in the more remote areas. These fibres are easy to weave, being less susceptible to climatic variations than natural fibres. They do not break easily and readily absorb dye substances. The increasing demand for crease-resistant, wash-and-wear-type fabrics has increased the popularity of synthetic fibres. In recent years, brightly coloured acrylic yarns have virtually replaced wool in most areas. Artificial fibres may be interwoven with cotton and silk yarns.

Metallic threads, especially gold and silver, are very popular for use in textiles in many parts of South-East Asia. Although these precious metals are mined in a number of areas, most metallic yarns used today are imported. At one time, some metallic yarns were made from coins melted down by village craftsmen. The metal was drawn through a perforated plate to form wire, and then hammered into fine ribbons before being tightly wound around a strand of cotton or silk.[30] In Burma, metallic gold or silver filaments used to highlight special silk fabrics were wound around a very fine vine tendril to create a thread.[31] According to weavers, the best metallic threads today come from Japan and France, but they are expensive and not always obtainable. Cheaper Indian metallic threads are readily available, but they do not last as well. In Palembang, Sumatra, weavers are known to unravel gold yarns from older cloths for new, specially commissioned pieces.[32]

Dyes

South-East Asia has traditionally made use of a large number of plants to produce a wide range of rich, beautiful dyes. One report, compiled by the British Government in 1880, lists over thirty local plants that were used in southern Burma to obtain dyes.[33] Researchers of Indonesian and Filipino textiles in the early twentieth century have commented on the large number of natural dyes known to weavers, and have marvelled at the dyers' ingenuity in making creative use of resources available in the environment.[34]

Dyeing yarns using vegetable substances is a complicated and arduous process. The basic raw materials have to be cultivated or gathered from the forest, and processed by chopping, pounding with a pestle and mortar, soaking, squeezing, boiling, and evaporating in wide-mouthed earthernware jars and pots. Auxiliary agents, such as mordants, essential to fix the dyes and to improve tone and lustre, need also to be found in the environment and added to the dye solution. These may include lime, salt, urine, leaves, and various substances containing alum and iron compounds. Some esoteric matter may also be added at the dyer's whim to improve the dyes and help secure a successful outcome. Dyeing practices vary from village to village. Traditionally, each dyer had her own special recipes, the details of which were a closely guarded secret. At one time it was often possible to tell the provenance of a textile by closely examining the colours.

No two natural dye shades are exactly the same, and the results are not always to the dyer's satisfaction. The dyeing process is fraught with pitfalls and frustrations stemming from the availability of raw materials, and the fact that the chemical content will vary depending on the soil, climate, age, and condition of the plant. As a consequence of this unpredictability, a whole folklore of beliefs and taboos has evolved around the dyer's art.

The dyeing process has traditionally been the preserve of women. Up until recently it was usually done in an isolated area well away from prying eyes. In places such as Sumba (an island in eastern Indonesia) and in north-east Thailand, dyers worked in a walled or fenced-in enclosure.[35] Men, particularly monks and shamans, were excluded from the dyeing area, as were pregnant and menstruating women, whose presence it was felt might impair success.[36] On the island of Roti, not far from Sumba, it was feared that evil spirits might tamper with the dyes and ruin them. To counteract this possibility, talismans of thatched hen feathers were put over the dye pots to protect them.[37] Prior to embarking on an important dyeing project, the Iban women of Borneo traditionally prayed to various goddesses in their pantheon who were credited with instructing the Iban the art of dyeing. When attempting to concoct a particularly difficult dye, such as red, the dyer would pray and bite on a piece of steel to 'strengthen' her soul to help procure a successful outcome.[38] The Karen women of Thailand also associated dyeing with spirit worship and believed that they would anger the deities if they divulged any information pertaining to dyeing. They also observed taboos, such as refraining from speaking or touching oil while dyeing to ensure that the colour would turn out even.[39] When dyeing, Javanese batik dyers (who in this case were usually men), took care to avoid domestic quarrels on the eve of an important dyeing project.[40] Women skilled in the art of dyeing have always been highly regarded in their respective communities. On the islands of Sumba and nearby Savu, dyeing was at one time limited to those of noble birth. Master dyers in Iban society might be recompensed in goods or small precious objects for their assistance.[41]

Indigo

The most important and well-known dye throughout South-East Asia is indigo, which is derived from a 1 m high shrub with light, feathery leaves (Fig. 25). This dye has been used in China since ancient times. In South-East Asia, it has long been an important export item, at one time ranking immediately behind spices and aromatic woods in importance. Throughout the seventeenth and eighteenth centuries, it was Indonesian indigo which coloured the clothes worn by the Calvinist Dutch Burghers. The British, at the same time, carried on a very profitable trade in indigo from their Far Eastern possessions.[42]

Fig. 25. A clump of indigo plants (*Indigofera anil*) in the Sikka district of East Flores.

Although the making of indigo dye varies in each area in the number of days for steeping and the quantities of specific additives, the processes and key ingredients are essentially the same. The most important indigo-producing species are *Indigofera tinctoria* and *Indigofera anil*, which may be cultivated in small plots or grown wild.[43] Three months after planting, the leaves are collected, pounded, placed in an earthenware jar, and steeped in water for three to ten days until they disintegrate through fermentation (Fig. 26). To render the dye soluble in water, slaked lime, in the approximate proportion of 1 : 5 in weight, is added to the solution. A lye made from wood ashes is also added. Various substances, such as palm sugar, molasses, and rice whisky, may be introduced in small quantities to aid fermentation. Local ingredients, such as a dash of kerosene, banana, cassava, betel, or shredded chicken meat have been known to have sometimes been included in an effort to procure a better dye colour.

Fig. 26. Making indigo dye in an earthenware pot. Leaves of indigo are placed with lime (in the coconut shell) to ferment. A few wood ashes (in the bucket beside the woman) are also added. Maumere.

When the solution is ready, the cloth or yarn is completely submerged in the dye overnight, and then hung out to dry in the sun the following day. This drying process, after dipping, is necessary to oxidize the dye so as to form the indigo colour on the fibre. To get the characteristic blue-black shade, the cloth must be dipped and dried, up to thirty times and rinsed occasionally in fresh water. The process may take a few weeks to complete, depending on the strength of the dye solution. With the exception of Indonesia, mordants do not seem to be widely used for indigo, so the colour tends to bleed a little in the first few washes until it becomes fixed.[44] In some areas where the indigo plant is scarce, a dye made from leaves mixed with lime is evaporated to leave a solid blue substance, which may be saved and activated by mixing with water and a lye of ashes.[45]

Despite the fact that chemical substitutes, such as aniline dyes derived from coal tar, and other hydrocarbon sources are readily available, the art of indigo dyeing using leaves is still quite widespread. To the skilled dyer, a number of shades, ranging from blue-black and royal blue through to light shades of grey, lavender, and pink, are possible from indigo.[46] The women of Savu and Roti are famous for their distinctive bluish-black indigos, while Sumba colours its famous *ikat* with an arresting lighter blue. Indigo is the most important colour to the hill tribes of South-East Asia. It is the only dye used by the Hmong, Mien, and Akha, who consider it a most fitting ground colour for their distinctive batik, embroidery, and appliqué decoration. Indigo-dyed clothing is also very popular for everyday wear throughout rural Thailand. In South-East Asia, indigo is always the first colour to be applied to the yarns.

Red

Red is second in importance to indigo as a colorant and is generally a much more difficult dye to make. A large number of processes and pigments are known for creating red dyes. Depending on the dye substance used, the fibres may require extensive pre-treatment to prepare them to accept the dye.

The root bast of the morinda tree (*Morinda citrifolia*), which contains an alizarin substance, is widely used throughout South-East Asia as a source of red dye capable of producing tones ranging from vibrant red through to deep maroon and chocolate (Fig. 27).[47] Known by various names, such as 'turkey red' and *mengkudu* and *kombu* in Indonesia, and *koh* by the hill tribes of Thailand, the knowledge of this dye substance is thought to have originated in the Middle East and spread to India and South-East Asia through Indian and Arab contacts.

The morinda roots are collected, crushed, pounded, and mixed with water. After filtering, a mordant made from the crushed leaves of alumina-bearing plants, such as the *Symplocos* genus, are added, along with a lye of ashes to improve the solubility of the pigment. The dye is usually boiled in a pot over an open fire for a period of time. So that the dye colour will adhere, the fibres are repeatedly pre-treated with an oily substance mixed with a lye of ashes. Many types of oil are used. The Balinese use coconut, while the people of Nusa Tenggara favour the *kemiri* nut (candlenut). The Batak people of Sumatra use buffalo lard. The Iban are known to pre-dye their threads with yellow so as to get a particular shade of red. The pre-treatment process may be almost as lengthy as the dyeing process.

The actual dyeing may take many months until the desired shade is achieved. For an ordinary red, the yarn is soaked overnight and dried during the day at least ten times. To achieve a darker hue, the yarn might be dipped and dried over thirty times at three-to-fifteen-day intervals. In the village of Tenganan in Bali, it can take up to six years to achieve a particular highly prized shade of red for ritual cloths.[48] In the central and southern Philippines, dyes are applied to the yarns by a steaming process using two earthenware pots, rather like a double boiler.[49]

Dyers may be hampered by the availability of raw materials. In the drier areas of Nusa Tenggara where

Fig. 27. The morinda tree (*Morinda citrifolia*), the roots of which provide a red dye commonly used throughout South-East Asia. Rende, East Sumba.

the morinda tree is not prolific, dyers may have to wait until the roots sufficiently regenerate before they can make more dye.[50] On the island of Flores in eastern Indonesia, it may take up to 27 kg of morinda root to produce enough dye for one sarong.[51]

When boiled with water, the wood shavings of sappan, *Caesalpinia sappan L.*, a low, thorny bush with yellow flowers, produce a red dye which is widely known throughout the area. In the Philippines, after dyeing with sappan, the threads may be cooled in river clay to fix the colour. The seeds of the anatto (*Bixa orellana*), a South American tree widely grown in the tropics, are dipped in water to produce an orange-red dye which is used in the Philippines, Thailand, and Kampuchea.[52] The weavers of north-east Malaysia at one time made a red dye for silk from the fruit of *asam gelugor* (*Garcinia atroviridis*) mixed with tamarind and a pinch of alum.[53] The Iban and Dusun of Borneo make a dye from the resin-coated scales which cover the developing fruit of a species of rattan (*Damonorhops didymophyllus*) found in the Malaysian and Sumatran forests. In Sabah, the Dusun collect the fruit when ripe and shake the juice and resin into a basket. This liquid is boiled until it evaporates into a paste which adheres to a stick. This stick is saved and reheated when the dye is required.[54] A similarly coloured substance secreted by the lac insect (*Lucifer lacca Ker*) is used in Thailand and Burma for making a red dye. The resin-coated twigs are collected, dried in the sun, and then pounded to a powder, which is mixed with an acidic tamarind paste and water. Cotton and silk yarns are submerged in the solution for thirty minutes before being hung to dry.[55] In addition to those mentioned, every country has a number of specific indigenous plants which are known to produce a variety of red dye substances.

Other Colours

A large number of plants are known to give yellow dyes. The most important is the turmeric rhizome (*Curcuma longa L.*), a member of the ginger family which is widely grown throughout South-East Asia. The root is ground to a coarse powder and mixed with water. After filtering, it is ready for use. This dye is readily absorbed by fibres, so a mordant is not always used.[56] Specific substances may be added to alter the characteristic mustard shade. The addition of slaked lime produces an orange hue, while a squeeze of an acidic juice, such as citrus or tamarind, can render the solution bright yellow.[57] The heartwood of *Cudrana Javanensis*, a spiny shrub, is used in both Thailand and Indonesia to yield a yellow dye. The dye is obtained by soaking chips of heartwood overnight and boiling the following day. After stirring, a mordant is added and the yarns are boiled in the dye.[58]

The wood of the jackfruit tree (*Artocarpus integrifolia L.*) was at one time used exclusively to dye the robes of Buddhist monks in Thailand, Kampuchea, and Burma a particular sober shade of yellow, as prescribed in the *Pali Vinaya*, a book of rules pertaining to monks. Prior to dyeing, the material for the robes was put in a vat containing a solution prepared from cow dung, red earth, or selected parts of plants, such as rhizomes, roots, bark, wood, flowers, or fruits. The heartwood of the jackfruit was cut into slices or fragments and boiled until the liquid was dark brown in colour. The wet robe was steeped in the solution until a dull yellow was obtained.[59] The Iban also used the jackfruit wood, mixed with an alkali, as a dye. It was also once used with alum as a dye for Javanese batik.[60] The wood of the jackfruit is capable of producing a wide variety of shades, ranging from muddy yellow through to apricot and tan.

The people of Nusa Tenggara are also known to use the bark of the mango tree (*Mangifera laurinia*) to produce a yellow dye.[61] The safflower (*Carthamus tinctorius*), which has had a long history as a dye in the ancient civilizations of Asia and Africa, probably came to South-East Asia via India. The crushed petals, after being kneaded and stirred in cold water, first yield a yellow dye, which is poured off. The base of the petals produces an acidic red dye, which is neutralized by adding lime as a mordant prior to use.[62] Safflower is used as a dye in Indonesia and Thailand. Some dyers in Indonesia also use the dried imported flowers of *Saphora japonica* to produce a yellow dye. In Burma, the dried stigma of *Crocus sativus* was at one time used to produce a strong yellow shade.

There are a number of trees which yield brown dyes. The best known in Indonesia is the bark of the *soga* tree (*Pelthophorium ferrungineum*), which gives dyes ranging from yellowish tan to deep chocolate brown. It is used to dye Central Javanese batik its soft brown colour. Shavings of bark are boiled in water to produce the dye. As with morinda, the cloth or yarn must be pre-treated with oil prior to dipping many times in a lukewarm dye solution. After dyeing, the fibres are submerged in a lime solution and then in a fixing bath. The latter may include one or more of the following: borax, alum, sugar, and lemon juice. *Soga* may be mixed with other dyes, such as morinda, to get different shades.[63]

The leaves of the teak tree (*Tectona grandis*) are used to produce a brown-to-red dye in Sumbawa and Sulawesi.[64] The Philippines also has a number of trees which produce strong brown dyes, such as the *talisay* tree (*Terminalia catapa*) and the *katuray* (*Sesbania grandifolia*).[65] A dull brown is obtained from the wood of the ubiquitous betel-nut tree (*Areca catechu*). The wood is cut into chips and boiled in water to release the dye. The yarn is immersed in the hot liquid for about half an hour for a lighter shade and overnight for a darker hue. This dye is used in Kalimantan and Burma. The coastal communities of South-East Asia have, in the past, used the bark and leaves of the mangrove (*Rhizophora mucronata*) to get brown-to-black shades. The Iban use its bark and that of the *samak* tree to get a rich reddish brown colour to dye their distinctive warp *ikat*.[66]

Green shades are usually obtained by overdyeing a yellow, such as jackfruit or turmeric, with a weak solution of indigo. In Thailand, the outer pulp of pineapple leaves and the leguminous leaves of the *Sesbania grandifolia* have been used to produce green dyes.[67] The Sikka district of Flores has been known to use the leaves of the cotton plant and, central Timor, the pods of beans to produce green.[68] In Sabah, the Bajau have used the leaves of the red pepper (*Capsium annum*) to obtain a green colorant.[69]

In some areas of Indonesia, such as Bali and Kalimantan, a purplish black colour was traditionally obtained by overdyeing indigo with a red-brown dye. A number of trees are known to produce black dyes. In north-east Thailand, the berries of the *krajai* tree are pulverized and soaked in water for thirty minutes, while at Baan Na Mun Sri in southern Thailand dyers at one time obtained a black dye from the hairy skins of the rambutan (*Nephelium lappaceum*).[70] The Iban are known to have used the large, juicy leaves of the *Medinillopsis beccariana* to obtain black.[71] After being boiled and pounded, the green fruit of the ebony (*Diospyros ebenum*) also produces a black dye.[72] In Amarasi, in West Timor, the women boil the bark of the wild mango tree along with the leaves of another local plant. The threads are soaked in this solution overnight before being buried in the mud of a nearby lake. The threads take on a deep black hue due to the iron salt which acts as a mordant.[73] The Toradja are known to mix the crushed leaves of the *bilante* (*Homolanthus populneas*) with mud to create a black dye for their mourning clothes.[74] The Iban and the hill people of the Philippines also use mud to darken and fix their dyes. Soot from lampblack mixed with resin is used to dye cloth black in Kalimantan and in Tanimbar. In Sulawesi and Timor, soot may be painted on indigo cloth to produce a black shade.[75]

With the possible exception of indigo, chemical dyes, which are faster and easier to prepare and handle, have largely replaced natural dyes throughout South-East Asia. Even in the more remote areas, chemical dyes are widely known. Many of these dyes come complete with instructions in the vernacular printed on the packet, which the weavers (or their children), with a rudimentary education, can read. The steeping time has been reduced from days to hours and minutes. Often, only a quick dipping is required. Chemical dyes produce more predictable results and offer better resistance to sunlight and frequent washing. With the advent of chemical dyes, the task of dyeing, in many cases, has passed from women to men. In small establishments, it is the men, clad in shorts and knee-high rubber boots, who may be seen hard at work bending over vats, dipping and lifting sodden yarns to and from racks and pulleys.

Being relatively new to chemical dyes, many dyers in South-East Asia have not yet acquired consummate skill in mixing and blending these powerful agents to get a range of rich subtle shades. The results are often harsh and jarring, which has caused connoisseurs of fine traditional textiles to view the spread of chemical dyes with regret.[76] Much of the appeal of many antique South-East Asian textiles lies in their rich blue and red ground colours, which are enhanced by adding natural fibre tones and subtle touches of subdued greens and yellows and, occasionally, a deep, muted purple. Throughout South-East Asia, natural dyes have traditionally been combined with exquisite taste on the part of the weaver to create textiles of great visual beauty.

Looms

The process of weaving (Fig. 28) consists of interlacing one series of parallel longitudinal threads, called the WARP, at right angles with another series of lateral threads, called the WEFT. This process forms a continuous WEB, or cloth. Most societies make use of a special device, called a loom, to facilitate the weaving process. The warp threads extend in parallel lines between two lateral beams, called a WARP BEAM and a BREAST or CLOTH BEAM, which is closer to the weaver. During the process of

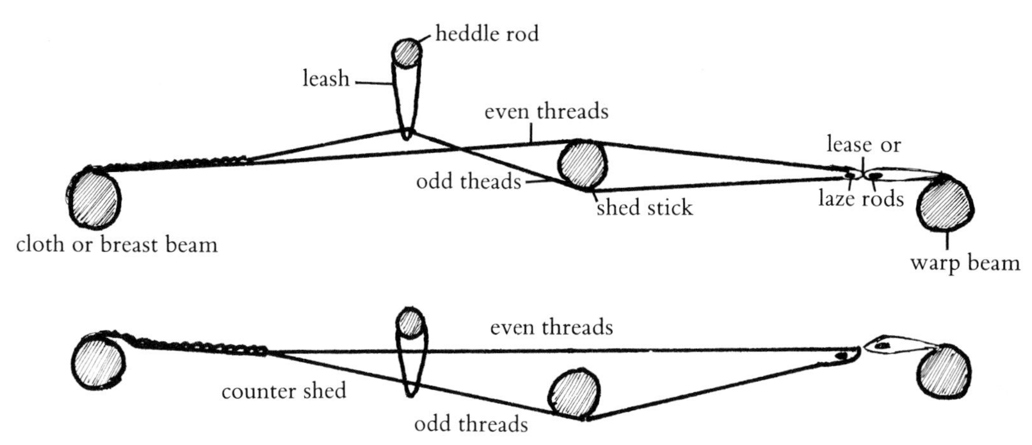

Fig. 28. Principles of weaving.

weaving, the weaver alternately raises and lowers the odd and even warp threads creating a space, or SHED, between the two set of threads. The weft thread is passed through each shed forming a PICK, or SHOT, which is firmly pressed into the already woven cloth with a SWORD STICK, a long knife-like baton of wood.

To make the process of weaving easier and faster, a HEDDLE stick, or HEALD ROD, to which the odd threads are lightly attached by a series of cotton loops, or LEASHES, is used. When the heddle is raised, all the odd threads are lifted together to form a shed. After the weft pick has been made, the heddle rod is lowered and the weight of a large cylindrical stick set behind the heddle, called the SHED STICK, forces the odd threads below the even threads to create a COUNTER SHED. The weaver turns the sword stick on its side to enlarge the shed to make it easier for the weft thread to pass through the even threads. As it is woven, the cloth is rolled onto the breast or cloth beam.

To prevent the warp threads from becoming entangled, the threads may be held in position behind the heddle and shed stick by crossing individual threads alternately over and under a pair of sticks called LEASE or LAZE RODS. Sometimes, instead of using laze rods, individual threads may be wound around a single COIL ROD. Various movable devices with dents or teeth, such as WARP SPACERS, REEDS, or BEATERS, may also be placed in the warp in front of the heddle to maintain the correct order and spacing of warp yarns. These attachments may be used to push each pick into the already woven cloth. To keep the width of the fabric even, a flat narrow stick, called a TEMPLE, with two to three small teeth at each end, may be inserted into each selvage edge of the newly woven cloth, directly in front of the breast beam. The temple is adjusted as weaving progresses.

To facilitate the passage of the weft thread through the warp, a number of devices may be used. For the simplest looms, the weft thread is wound lengthwise onto a stick which serves as a SPOOL (Fig. 29). An amount of thread is released in advance by the

Fig. 29. Spools and shuttles.

(a) This type of spool, where the thread is wound lengthwise, is often used with a body-tension loom.

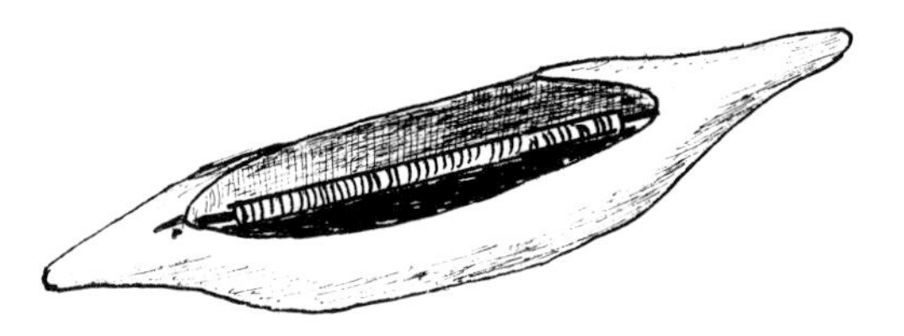

(b) This boat-shaped shuttle is often used with a frame loom, such as by the Thai, Burmese, and Akha.

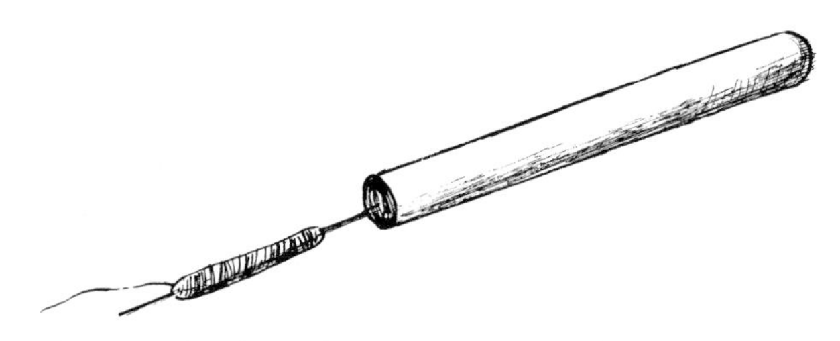

(c) This cylindrical shuttle is used with a Malay loom.

(d) This type of spool is used for weaving supplementary weft threads in Brunei and Bali.

weaver prior to passing the spool through the shed. Thread may also be wound on at right angles around bobbins and placed in bamboo or light metal cylindrical or boat-shaped wooden SHUTTLES. The thread is wound onto bobbins in such a way that it is immediately released as the shuttle passes through the shed.[77]

A number of different types of looms are known in South-East Asia. They range from a very simple body-tension (sometimes called backstrap) loom, through belt and card looms, to more sophisticated hand-operated frame looms capable of doing a number of different weaves.

Body-tension Looms

The oldest and simplest type of loom, and one which is still widely used throughout South-East Asia, is a body-tension loom (Fig. 30). It is so called because the pressure of the body is used to keep the warp threads taut. This is made possible by means of a backstrap consisting of a pad of leather, woven matting, or wood attached by a string or rope to the breast beam. This attachment passes around the back of the weaver's waist when she is seated on the ground before the loom with feet outstretched. By shifting her weight backwards and forwards, the weaver controls the tension of the warp yarns extended before her. As weaving progresses, the woven cloth is rolled under or around the breast beam, which rests on the weaver's lap. The weaver usually sits on a woven mat surrounded by baskets of freshly spun yarn and spools of thread. Small coconut-fibre brushes or corn husks are usually close at hand along with a bowl of rice water. The weaver may periodically brush the warp yarns with this

Fig. 30. Simple body-tension loom as used by the Montagnard people of Vietnam.

solution to keep them stiff while weaving. She may also have a small, sharp knife to trim stray pieces of yarn and a pick to help align threads and remove any impurities from the cotton yarns.

The body-tension loom is generally as wide as is manageable and its width usually falls within the arm span of the weaver. In the Philippines, some body-tension looms may be as wide as 80 cm. More common, however, is a smaller loom with a width of around 60 cm. The length of the cloth is determined by the length of the warp threads. Due to its width limitations, lengths of cloth woven on a body-tension loom are not generally very large. Some items, such as blankets and sarongs, may be made by joining two or three separately woven pieces. Every country in South-East Asia has groups of people who use a version of this loom for their weaving. Although a very basic implement, this loom is capable of weaving the finest *ikat* and very sophisticated supplementary warp and weft textiles.

The simplest and most primitive body-tension loom found in South-East Asia has a CONTINUOUS WARP (Fig. 31). That is to say, the warp forms a circle around the breast beam, resting in the weaver's lap, and the warp beam which is supported by the weaver's feet. Cloth woven on this loom is of cotton or bast fibre and is not more than about 160 cm long, because the length cannot exceed more than twice the distance from the waist to the feet of the weaver. This loom is operated by the alternate raising of the heddle and shed stick to create a passage for the weft thread, which is wound lengthwise around a spool stick. The sword stick serves to enlarge the sheds and to beat the weft into place. Since the warp is not very long, laze rods and warp spacers are not generally used. As weaving progresses, the weaver shifts the newly woven cloth under the breast beam, at the same time drawing new warp threads into her lap. At the conclusion of weaving, the cylindrical-shaped piece of cloth is removed from the loom. If a rectangular length of cloth is required, the cloth is cut along a small unwoven section. The unwoven warp threads then serve as fringes. The use of this loom today seems to be confined to a few Mon–Khmer minority groups in Vietnam, such as the Mnong, Maa, Stieng, Bahnar, and Jeh.[78]

A more widely used body-tension loom is one with a continuous warp, where the warp beam is secured to a house pillar, tree trunk, or other convenient support. The weaver sits on the ground or on a low stool, her feet braced against a firmly anchored piece of wood, with the warp threads

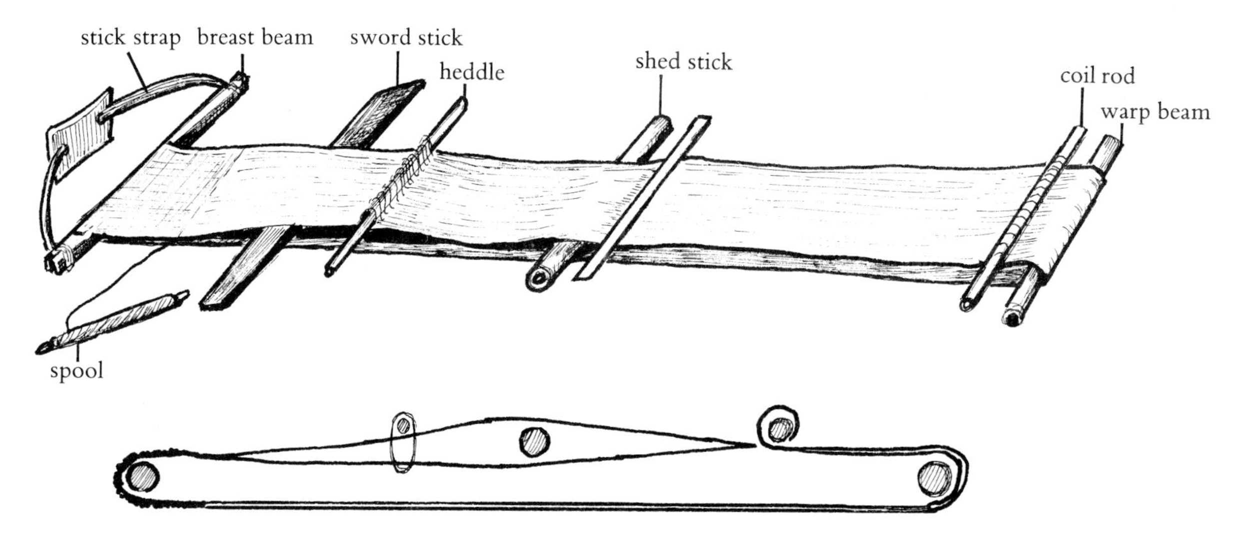

Fig. 31. Body-tension loom with a continuous warp. (After Dournes and Roth.)

inclined at an angle towards her. The warp threads average 2–3 m in length but may be up to 9 m long in parts of the Philippines. A coil rod or laze rods are placed in the warp to maintain the correct ordering of yarns. The breast beam varies on such looms. It may be a simple, round bamboo pole; alternatively, it may consist of two pieces of wood firmly hafted, or dovetailed, together to act as a clamp to hold the newly woven cloth firmly between them. This beam can be loosened to shift the newly woven cloth under it. Some weavers use a warp spacer or reed with this type of loom. While weaving, the woman strikes her sword stick against the back of the reed to push the newly woven pick into the cloth. Like the earlier loom, the weft thread is wound onto a spool. Sword sticks may vary in shape, from a long knife to wedge-shaped implements. Cotton, silk, and bast fibres are woven on this loom.

Another widely used body-tension loom is one where the warp beam is slotted into a sturdy upright wooden frame rather than being attached to a post. The warp on this loom is NOT CONTINUOUS (Fig. 32). It consists of parallel threads carefully wound around a plank-like warp beam which is held

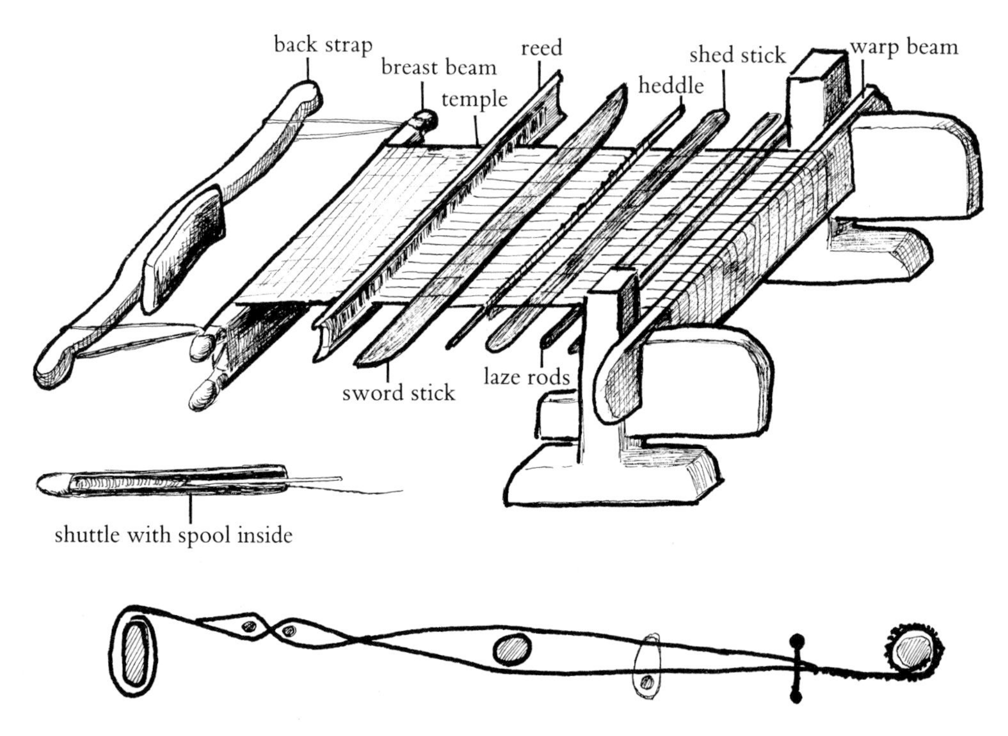

Fig. 32. Body-tension loom with a discontinuous warp. (After Hitchcock and Roth.)

vertically in the frame. As weaving progresses, the weaver releases the warp threads stored on the warp beam. It is possible to weave fairly long lengths of cloth on this loom. Since the excess warp threads can be rolled up, this loom does not take up as much space as the previously mentioned loom. The weaver is also able to brace her feet against the frame supporting the warp beam. Laze and coil rods, reeds, and temples may be used to align the yarns correctly. In many cases, the weft thread is placed in a bullet- or boat-shaped shuttle containing a spool. Being more streamlined, the shuttle is able to move through the warp faster than a spool. In addition to cotton textiles, this loom is used to produce the beautiful silk and gold-patterned textiles found in Bali, Lombok, Sumbawa, and Sulawesi.[79]

There are a number of variations to the body-tension loom. Some weavers in Palembang and Lombok have their looms set in a wooden frame on the floor. The weaver sits between two parallel extensions of wood which support the brace holding the warp beam.[80] Weavers of traditional cloth patterned with a supplementary weft in Gianyar, Bali, sit on benches in small workshops with the warp beam of their body-tension looms slotted into bedstead-type frames. If the weaver wishes to take her weaving home, she can easily lift out the warp beam, roll up the warp, and reset her loom in her house. All that is required is a simple brace to support the warp beam.

The Hmong people of northern Thailand use a body-tension loom which combines some features of the frame loom (Fig. 33). The warp threads are wound around a flat warp beam which is supported in an upright frame. The weaver sits at a bench with the backstrap fastened around her waist and the cloth beam resting in her lap. A pair of levers attached to the frame support a heddle bar with string heddles. The heddle bar is also connected to a rope near the weaver's feet. By pulling on this rope with her feet, the weaver raises the heddle rod to create a shed. Releasing the rope lowers the shed stick to create a counter shed. This loom is used to weave cotton and hemp yarns for clothing.[81]

Body-tension looms vary greatly in size and shape. Smaller versions of this loom may be used to weave narrow bands of cloth. For example, the Maranao of the southern Philippines use a miniature body-tension loom to weave their distinctive *langkit* strips which are attached to the sarong.[82] The women of Sumba also use a smaller version of their body-tension loom to create attractive bands at the ends of their blankets (see Fig. 245). These looms have a

Fig. 33. The Hmong loom, which combines some features of the frame loom. (After Lewis.)

simple bamboo rod as a warp beam. The weft thread is usually inserted with the fingers and beaten down with a sword stick.

Band and Card Looms

Narrow body-tension looms can also be considered band looms, of which several are known in South-East Asia. The Buginese people of South Sulawesi, for example, use a simple loom to make intricately patterned belts (Fig. 34). The end warps are tied to a tree or post. A pair of heddle rods, with string heddles, are suspended from a bar above the warp. This bar is attached to some fixed object, such as the branch of a tree or a pole. Below the heddles are loops of string, weighted with pieces of wood which serve to steady the loom and act as make-shift treadles. The upward and downward tilting of the heddles creates a shed for the weft thread. The weft is inserted with the fingers.[83] The Balinese were also known at one time to use a simple warp-weighted loom to weave narrow bands. The warp was suspended between two frames. An overhead pole parallel to the warp held the shedding mechanism, along with the treadles, which were moved forward as weaving progressed.[84]

Card or tablet weaving was at one time practised on Central Sulawesi and parts of Java and Burma (Fig. 35). The shedding device for a card loom

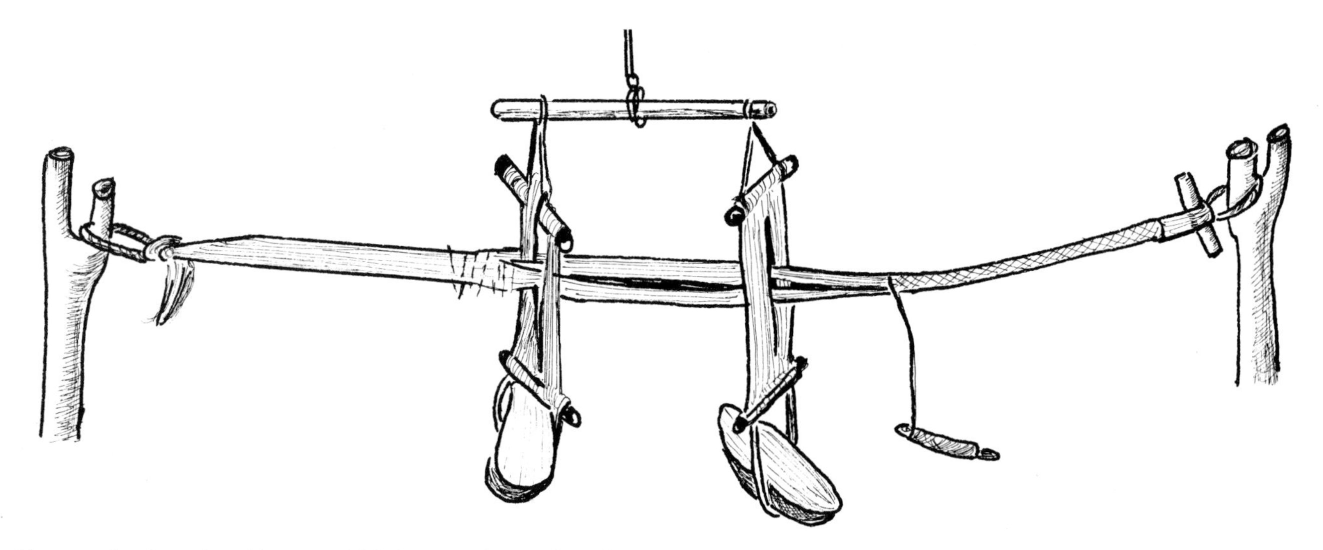

Fig. 34. Buginese band loom, which is secured at each end to a tree or post. The heddles operate as levers of a scale, tipping up and down. (After Newman.)

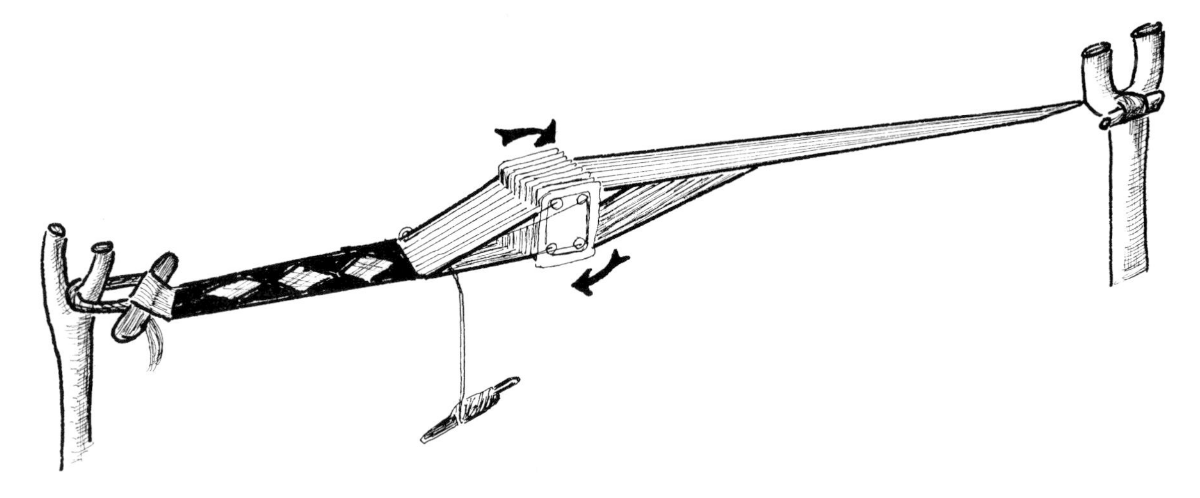

Fig. 35. Card or tablet weaving, in which the cards are rotated to form sheds. (After Bolland.)

consists of a series of 50–200 thin, square, tortoiseshell cards with a hole in each corner, through which are threaded the warp threads. The warp ends are anchored to a post or hook to keep them taut. The shed is created with each turn or rotation of the cards, which move along the warp in the course of the weaving. The cards may be raised, turned, and twisted to create a variety of tightly woven patterns. Numerous colour combinations and textures are also possible through changing the colour of the weft threads and altering the direction of the card rotation.[85]

Frame Looms

Several types of simple frame looms (sometimes called floor and shaft looms) are found in South-East Asia. They differ from the body-tension loom in that, instead of the weaver's body supporting the warp and maintaining an even tension, a frame is used. The warp threads, which are considerably longer than those on a body-tension loom, are wound around a warp beam which is set into a support towards the end of the frame. On some looms there may be a seat for the weaver built into the frame. The newly woven fabric is rolled around a cloth beam which, like the warp beam, is set into some type of support attached to the frame.

The shedding system is different on a frame loom. All warp yarns pass through heddles, or healds, consisting of numerous cotton leashes with loops or eyes for holding individual threads. The leashes are mounted top and bottom to a pair of wooden rods clamped tightly together. The heddles are supported by two harnesses joined to a pulley and bar system suspended from the frame at right angles above the warp. Below the warp, the heddles are attached by string or rope to foot treadles. By working the treadles, the weaver alternately raises and lowers each heddle to create the shed and counter shed for the passage of the shuttle bearing the weft threads. A large reed and beater with bamboo or metal teeth is hung from the top of the frame, and is set in the warp to serve both as a warp spacer and beater. After each weft pick has been made, the reed is pulled forward by the weaver to beat the pick into the cloth. On a frame loom, a sword stick is not required to beat in the weft; however, it may be used to assist with counting and lifting threads for certain patterns. A temple may be inserted during the course of weaving to keep the selvages even. A cylindrical or boat-shaped shuttle is used to hold the bobbin containing the weft thread.

The Akha of north-west Thailand use a very rudimentary frame loom consisting of four 2 m high posts sunk into the ground (Fig. 36). Supporting cross poles are either tied or fitted into holes in the uprights. The warp ends are tied to a pole at the extreme end of the loom and the woven cloth is wound around a cloth bar. A pair of string heddles are suspended from a cylindrical stick balanced across the top of the frame. Below the warp, the heddles are attached with string or rope to bamboo treadles. While weaving, the weaver moves the shedding device along the length of the warp. The weft is passed through the warp in a large boat-shaped wooden shuttle. Because of the simple harness suspension system and absence of a warp beam, it is not possible to weave wide widths of cloth on this loom. Cloth woven on the loom averages about 20 cm in width.[86] This type of loom is sometimes used by other mainland South-East Asian groups, such as the Lawa and the Lahu. The Lahu of Burma (sometimes called Muhso or Muser) are known to have a slightly more advanced version of the Akha loom. The weaver sits at a crossbar behind the cloth beam to operate the treadles. She moves the stick anchoring the ends of the warp threads towards her as weaving progresses.[87] One advantage of these rudimentary looms is that they are easily assembled and dismantled. The loom is usually set up at the beginning of the dry season, after harvesting in late November, when the climate is drier and there is more time for weaving. With the onset of the first rains, which herald the beginning of the planting season, the loom is dismantled. Besides being too busy, the women do not like to weave during the rainy season because the cotton absorbs the moisture, causing the threads on the loom to be less taut.

Fig. 36. Rudimentary Akha frame loom. (After Lewis.)

One loom which is widely used in Malaysia, southern Thailand, the coastal towns of Borneo, Central Sumatra, and certain areas of Java is the Malay loom (Fig. 37). It is used mainly for weaving supplementary weft-patterned textiles called *kain songket*. The loom consists of a 1.2 m high frame which supports a flat warp beam fixed on its edge into slotted wooden guides suspended from a crossbar at the top of the frame. Another crossbar provides support for a reed and a pair of harnesses held by strings passing through two bamboo tubes, which support a pair of heddles made from two pieces of wood tightly clamped together and mounted with fine cords for holding individual warp threads. Below the warp, the heddles are connected by strings to a pair of transverse treadles suspended about 18 cm above the ground. The weaver sits at a wooden bench built into the loom, with the cloth bar before her. On some Malay looms, the warp threads may be inclined downwards away from the weaver. As the weaver moves the heddles by pushing the treadles with her feet, she passes the bobbin of weft thread inside a cylindrical bamboo or aluminium shuttle through the shed created in the warp.[88]

Temple mural paintings in both Thailand and Burma attest to the fact that frame looms have been in existence in these countries for at least 150 years. The Burmese/Thai frame loom consists of a cloth beam, temple, reed/beater, a pair of heddles, and laze rods (Fig. 38). There is no warp beam. Prior to being woven, the warp is passed over the top of the loom and tied above the weaver's head. When more warp is required, the weaver is able to release warp threads without having to leave her seat. The heddles consist of cotton leashes clamped onto sticks hung

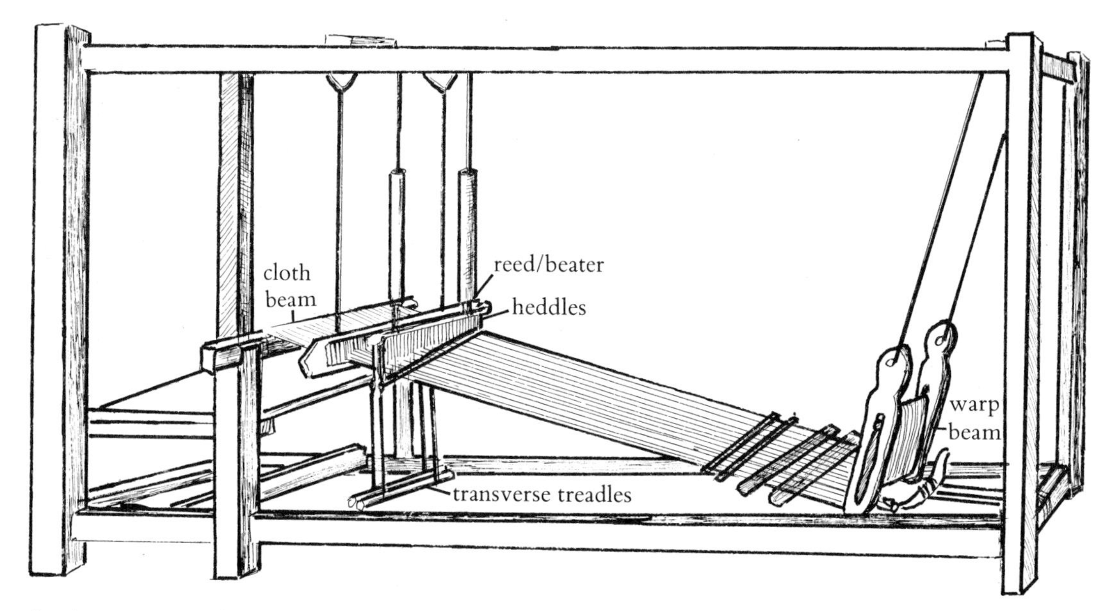

Fig. 37. Malay loom, used mainly for weaving *kain songket*.

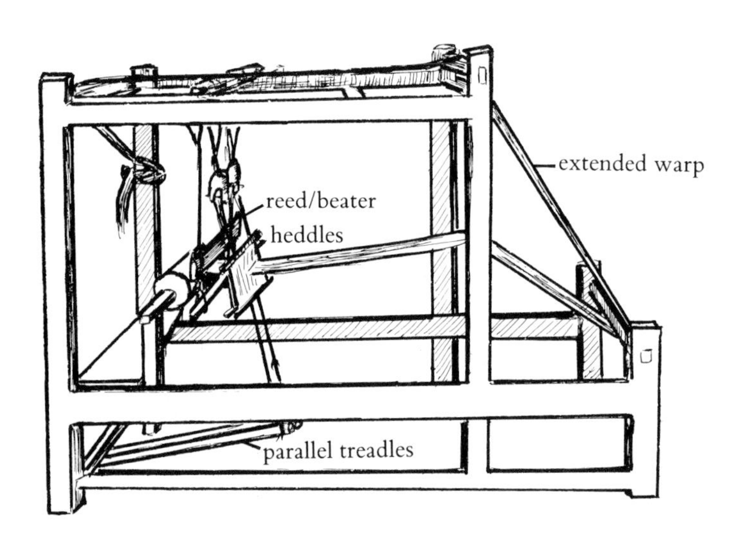

Fig. 38. Burmese/Thai frame loom.

Fig. 39. Loom pulleys for the traditional Burmese frame loom. Collection of U Myint Sein.

from cords passing through wooden pulleys. The latter may be carved in the shape of a bird (Fig. 39). The entire arrangement rests on a transverse bar supported by the top of the loom frame. Under the warp, the heddles are connected to a pair of central cords which support a pair of parallel pedals. On some early Burmese looms, discs made from half a coconut shell were used instead of pedals. Resting her heel on the ground with her toes curled around the cord supporting the shells, the Burmese weaver moved the heddles by pressing up and down on the coconut discs. The traditional Burmese loom is also characterized by an exceptionally large, flamboyant-looking hardwood reed/beater. In some cases, these were elaborately carved into minor works of art. This loom (minus the elaborate beater) is still used in some areas of the Shan States for weaving small objects, such as sashes and bags. In northern Thailand, it is still used by a few weavers in Nan Province and in the Chom Thong district of Chiangmai Province. In north-east Thailand, this loom has been modified to include a warp beam to store the warp threads. The shedding device on the north-east loom is supported below the warp by cords attached to the treadles. Directly above the warp, the heddles are held in place by loop cords attached to a transverse bar across the top of the frame.[89]

The Cambodians also have an interesting traditional loom with an extraordinary long warp length of over 5 m (Fig. 40). It has a flat, horizontally suspended warp beam. The shedding device is similar to that of a Thai loom, in that the heddles are

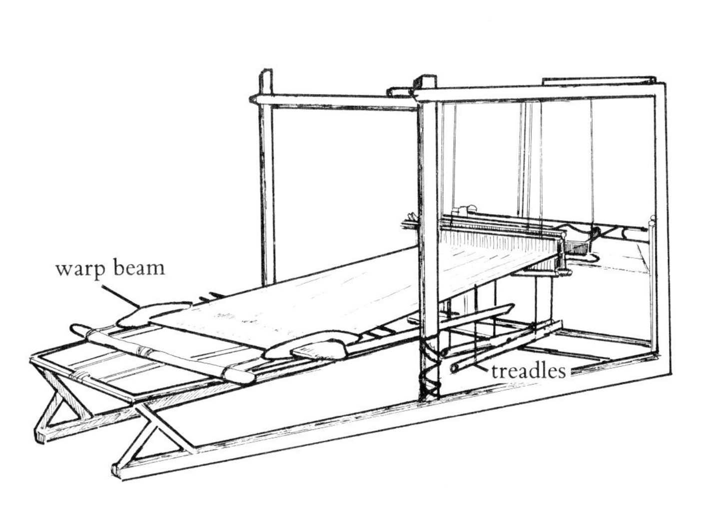

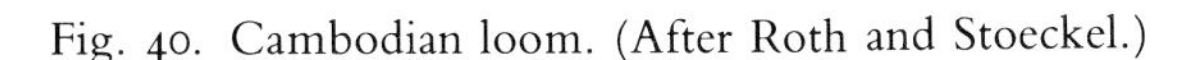
Fig. 40. Cambodian loom. (After Roth and Stoeckel.)

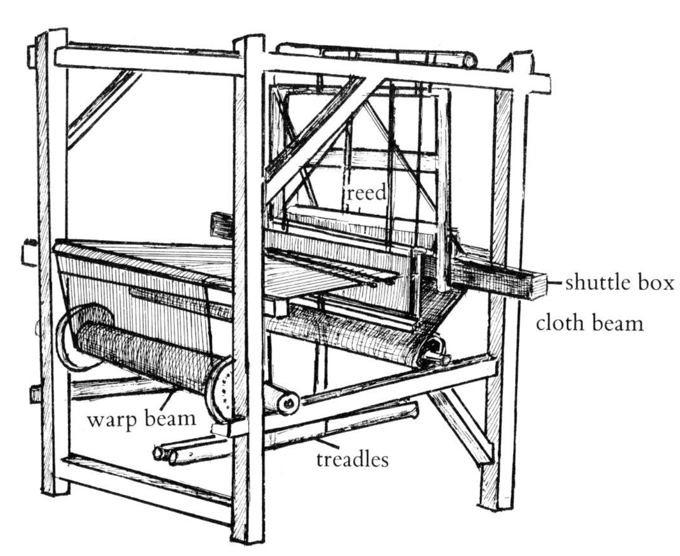

Fig. 41. 'Siamese' loom, an updated version of the traditional Thai loom.

lowered and raised by parallel foot treadles. There may be an extra transversal treadle, which is used to create interesting patterns and to do more complex weaves. On this loom, it is possible to 'hook up' different pattern heddles to this extra treadle as they are needed to create a variety of designs.[90] This loom is used to weave both cotton and silk in the Phnom Penh area. A number of Thai weavers in the Surin area, close to the Kampuchean border, use this loom to weave silk *ikat*.

Where weaving is done as a cottage industry in South-East Asia, the previously described frame looms have been largely superseded by what is popularly called the 'Siamese' loom, which is an updated version of the traditional Thai loom (Fig. 41). There are a number of local variations, but basically the loom consists of a circular warp beam suspended above or below the extended warp at the end of the frame. It is connected to a notched device which slowly rotates the warp beam causing it to gradually release warp threads in the course of weaving. The cloth beam in front of the weaver also has a similar device to aid in rolling up the newly woven cloth. The shedding system is similar to the frame looms previously described. Metal or cotton leashes set in a wooden frame are attached by rope to a transversal crossbar towards the top of the loom. Cords attached to the lower part of the heddles below the warp support two or more wooden or bamboo treadles aligned parallel to the warp. In front of the heddles, set at right angles to the warp, is a movable wooden beater with bamboo or metal teeth in the centre. A box-like attachment on both sides of this frame holds the shuttle. A cord linked to the shuttle box is suspended above the weaver. As the heddles are alternately raised and lowered by pushing on the foot treadles, the weaver tugs repeatedly on the shuttle cord to activate a simple mechanism inside the shuttle box, which sends the shuttle 'flying' back and forth across the warp. At the same time, the weaver rhythmically moves the beater backwards and forwards to push the newly woven weft threads tightly into the web. This loom is very fast and two or three sarong lengths can be woven in a day. Cloth woven on this loom is much wider and longer than that woven on other looms. Both men and women may be seen operating this loom, which is widely used in Burma, Thailand, central Philippines, Java, Sumatra, and Sulawesi, to produce cloth for commercial purposes.

1. For general definitions pertaining to fibres the author has used material from various editions of *Encyclopaedia Britannica* and Dr Isobel Wingate's *Fairchild's Dictionary of Textiles*, Fairchild Publications, Inc., New York, 1974.

2. Mattiebelle Gittinger, *Splendid Symbols, Textiles and Tradition in Indonesia*, Textile Museum, Washington, DC, 1979, p. 14.

3. Michael Hitchcock, *Indonesian Textile Techniques*, Shire Publications, Ayelesbury, England, 1985, p. 27.

4. Charles A. Fisher, *South East Asia*, Methuen, London, 1964, pp. 441 and 493.

5. Michael Symes, *An Account of an Embassy to the Kingdom of Ava, Sent by the Governor-General of India, 1795*, Nicol and Wright, London, 1800, p. 325.

6. Information pertaining to the growing of cotton in Burma is from U Than Aung, Chairman, Mandalay Division Cooperative Syndicate.

7. Thomas Fitzsimmons (ed.), *Cambodia*, Country Survey Series, Human Relations Area Files Press, New Haven, 1957, p. 198.

8. Gabriel Casal *et al.*, *People and Art of the Philippines*, Museum of Cultural History, University of California, Los Angeles, 1981, p. 131.

9. Fay-Cooper Cole, *The Tinguian, Social, Religious and Economic Life of a Philippine Tribe*, Publication 209, Anthropological Series, Field Museum of Natural History, Chicago, 1922, p. 417; and Hitchcock, op. cit., p. 28.

10. Wanda Warming and Michael Gaworski, *The World of Indonesian Textiles*, Kodansha, Tokyo, 1981, p. 58.

11. Margaret Campbell, *From the Hands of the Hills*, Media Transasia, Bangkok, 1978, 2nd edn., 1981, p. 88.

12. Wingate, op. cit., p. 470.

13. Robert Mole, *The Montagnards of South Vietnam*, Tuttle, Tokyo, 1970, p. 207; and Fitzsimmons, op. cit., p. 199.

14. Fisher, op. cit., p. 324.

15. 'T'boli Arts and Crafts: T'nalak', pamphlet, Santa Cruz Mission, Lake Sebu, South Cotobato; Wingate, op. cit., p. 2; and Casal *et al.*, op. cit., p. 132.

16. Gittinger, *Splendid Symbols*, p. 206; and Hetty Nooy Palm, 'Dress and Adornment of the Sa'adan Toradja', *Tropical Man*, Vol. 2, pp. 168–9.

17. Wingate, op. cit., p. 307.

18. John Alman and Elizabeth Alman, *Handcrafts in Sabah*, Borneo Literature Bureau, Kuching, 1968, reprinted 1973, p. 63; and Gittinger, *Splendid Symbols*, p. 217.

19. Luther Parker, 'Primitive Looms and Weaving in the Philippines', *The Philippine Craftsman*, Vol. 2, No. 6, 1913, p. 376.

20. Gittinger, *Splendid Symbols*, pp. 189, 195 and 206.

21. Much of this information was obtained during a visit to U Piankusol Silk Weaving establishment, Chiangmai, 8 September 1985.

22. Stephen Browne, 'Origins of Thai Silk', *Arts of Asia*, September–October 1979, p. 93.

23. Ibid., p. 100.

24. H. R. Spearman, *British Burma Gazetter*, Vol. 1, Government Press, Rangoon, 1880, p. 412; and Henny Harald Hansen, 'Some Costumes of Highland Burma at the Ethnographical Museum of Gothenburg', *Etnologiska Studier*, Vol. 24, Goteborg, 1960, p. 13.

25. Information from U Than Aung, Chairman, Mandalay Division Cooperative Syndicate.

26. Gittinger, *Splendid Symbols*, pp. 14 and 102.

27. Information from Nyonya H. Sannar Ramli of Pusako Weaving House, Pandai Sikat, Sumatra.

28. Warming and Gaworski, op. cit., p. 114.

29. R. A. Innes, *Costumes of Upper Burma and the Shan States in the Collections of Bankfield Museum*, Halifax, 1957, p. 41; and Bronwen Solyom and Garrett Solyom, 'Notes and Observations on Indonesian Textiles', in Joseph Fischer (ed.), *Threads of Tradition, Textiles of Indonesia and Sarawak*, exhibition catalogue, Lowie Museum of Anthropology and the University Art Museum, Berkeley, California, 1979, p. 15.

30. Hitchcock, op. cit., pp. 31–2.

31. Information given to the author by Mrs Jane Terry Bailey of Denison University, Granville, Ohio.

32. Gittinger, *Splendid Symbols*, p. 102.

33. Spearman, op. cit., pp. 136–8.

34. Early researchers of Indonesian textiles include G. P. Rouffaer, H. H. Juynboll, J. E. Jasper, and Mas Pirgandie. Those for the Philippines include Fay-Cooper Cole, Alfred Kroeber, and Alejandro Livioko.

35. Jack Lenor Larsen *et al.*, *The Dyer's Art: Ikat, Batik, Plangi*, Van Nostrand Reinhold, New York, 1967, p. 150.

36. Vilmophan Peetathawatchai, *Esarn Cloth Design*, Faculty of Education, Khon Kaen University, Khon Kaen, Thailand, 1973, p. 48.

37. Larsen, op. cit., p. 164.

38. Alfred Haddon and Laura E. Start, *Iban or Sea Dayak Fabrics and Their Patterns*, Cambridge University Press, 1936, reprinted Ruth Bean, Bedford, 1982, p. 22.

39. Campbell, op. cit., pp. 133 and 140.

40. Tassilo Adam, 'The Art of Batik in Java', *Bulletin of the Needle and Bobbin Club*, Vol. 8, No. 2, 1934, p. 12.

41. Haddon and Start, op. cit., pp. 21–2.

42. Hitchcock, op. cit., p. 32.

43. Thailand also obtains indigo from *I. Sumatrana* and *I. arecta.* See Sananikone Thao Peng, 'Further Notes—Dyeing', *Brooklyn Botanic Garden*, 1964, p. 46. The Philippines also gets indigo from *I. suffructicosa Mill* and *I. hursuta L.*—see Casal *et al.*, op. cit., p. 132.

44. In Indonesia, a mordant such as the bark of the *Tinggi* tree (*Ceriops candolleana*) is sometimes added.

45. Warming and Gaworski, op. cit., p. 67–8.

46. Cynthia Hyatt, 'A Dyeing Art, or an Introduction to Natural Dyes', unpublished typescript, Thailand, 1985.

47. In Burma and Thailand, the species used is *Morinda tinctoria*, while in the Philippines *Morinda bractaeta Roxb.* is the local equivalent.

48. Alfred Buhler, 'Turkey Red Dyeing in South and Southeast Asia', *Ciba Review*, Vol. 39, 1941, pp. 1423–6.

49. 'T'boli Arts and Crafts: T'nalak', op. cit.

50. Warming and Gaworski, op. cit., p. 70.

51. Kent Watters, 'Flores', in Mary Hunt Kahlenberg (ed.), *Textile Traditions of Indonesia*, exhibition catalogue, Los Angeles County Museum of Art, Los Angeles, 1977, p. 88.

52. Casal *et al.*, op. cit., p. 132.

53. W. W. Skeat, 'Silk and Cotton Dyeing by Malays', *Journal of the Straits Branch of the Royal Asiatic Society*, No. 38, 1902, p. 123.

54. Alman and Alman, op. cit., pp. 52–3; and Haddon and Start, op. cit., p. 20.

55. Peetathawatchai, *Esarn Cloth Design*, pp. 46–7.

56. Warming and Gaworski, op. cit., p. 71.

57. Hyatt, op. cit.

58. Hitchcock, op. cit., p. 33.

59. Suvatabandhu Kasin, 'Buddhist Rules Prescribe Dyes for Monks' Robes', *Brooklyn Botanic Garden*, 1964, p. 45.

60, Haddon and Start, op. cit.

61. Warming and Gaworski, op. cit.

62. Hitchcock, op. cit., p. 33.

63. Adam, op. cit., p. 14.

64. Hitchcock, op. cit., p. 35.

65. Casal *et al.*, op. cit., p. 132.

66. Haddon and Start, op. cit., p. 19.

67. Suvatabandhu, op. cit., p. 46.

68. Warming and Gaworski, op. cit., p. 72.

69. John H. Alman, 'Bajau Weaving', *Sarawak Museum*

Journal, Vol. 9, Nos. 15–16, 1960, p. 604.

70. Peetathawatchai, *Esarn Cloth Design*, p. 46; and also by the same author, *Folkcrafts of the South*, Housewives' Voluntary Foundation Committee, Bangkok, 1976, p. 124.

71. Haddon and Start, op. cit., p. 20.

72. Hyatt, op. cit.

73. Warming and Gaworski, op. cit., p. 72.

74. Eric Crystal, 'Mountain Ikats and Coastal Silks: Traditional Textiles in South Sulawesi', in Fischer (ed.), *Threads of Tradition*, p. 61.

75. Hitchcock, op. cit., p. 35.

76. Alfred Buhler, 'Dyes and Dyeing Methods for Ikat Threads', *Ciba Review*, No. 44, 1942, p. 1602.

77. Henry Ling Roth, *Studies in Primitive Looms*, 1918, 3rd edn., Bankfield Museum, Halifax, 1950, pp. 2–6.

78. Mattiebelle Gittinger, 'An Introduction to the Body-Tension Looms and Simple Frame Looms of Southeast Asia', in Irene Emery and Patricia Fiske (eds.), *Looms and their Products: Irene Emery Roundtable on Museum Textiles, 1977 Proceedings*, Textile Museum, Washington, DC, 1979, p. 54.

79. Ibid., p. 55.

80. Warming and Gaworski, op. cit., p. 130; and Rita Bolland and A. Polak, 'Manufacture and Use of Some Sacred Woven Fabrics in a North Lombok Community', *Tropical Man*, Vol. 4, 1971, p. 154 (line drawing).

81. Gittinger, 'Introduction to the Body-Tension Looms', p. 56; and Campbell, op. cit., p. 89.

82. Thelma R. Newman, *Contemporary Southeast Asian Arts and Crafts*, Crown Publishers Inc., New York, 1977, p. 66.

83. Ibid., p. 61.

84. See photograph, Gittinger, 'Introduction to the Body-Tension Looms', p. 63.

85. Candace Crockett, 'Card Weaving', in Emery and Fiske (eds.), *Looms and their Products*, pp. 54–68; and Rita Bolland, 'Three Looms for Tablet Weaving', *Tropical Man*, No. 3, 1970, pp. 160–89.

86. Gittinger, 'Introduction to the Body-Tension Looms', p. 58; and Campbell, op. cit., p. 89.

87. Roth, op. cit., p. 92; and Innes, op. cit., p. 42.

88. Gittinger, 'Introduction to the Body-Tension Looms', p. 58; and *Serian Songkit* (in Bahasa Malaysia), Perbadanan Kemajuan Kraftangan, Malaysia, Kuala Lumpur, n.d., p. 36.

89. Peetathawatchai, *Esarn Cloth Design*. See diagram in front of book.

90. Gittinger, 'Introduction to the Body-Tension Looms', p. 59; and Jean Stoeckel, 'Etude sur le Tissage au Cambodge', *Arts et Archeologie Khmers*, Vol. 1, No. 4, p. 338. In Roth, op. cit., p. 92, there is a drawing of a Cambodian loom which has transversal treadles rather than parallel treadles.

CHAPTER THREE

Textile Techniques

Decorative Dyeing Techniques

Ikat

THE best known and most widespread technique of patterning cloth in South-East Asia is that of *ikat*, a process by which designs are dyed onto the threads prior to being woven. The word '*ikat*' comes from the Malay word *mengikat*, which means 'to bind, tie, or wind around'.[1] In this process, the parts of the yarn which are to remain undyed are reserved by binding them with a material that resists the penetration of the dye. The *ikat* technique may be applied to either the warp or the weft threads; more rarely, it is applied to both warp and weft yarns. Patterns of staggering complexity in colour, design, and ingenuity may be created by the skilled use of this technique. Due to a slight bleeding of the dyes around the resist bindings, patterns created by this method are characterized by a softness or blurriness which blends in well with the texture of the fabric.

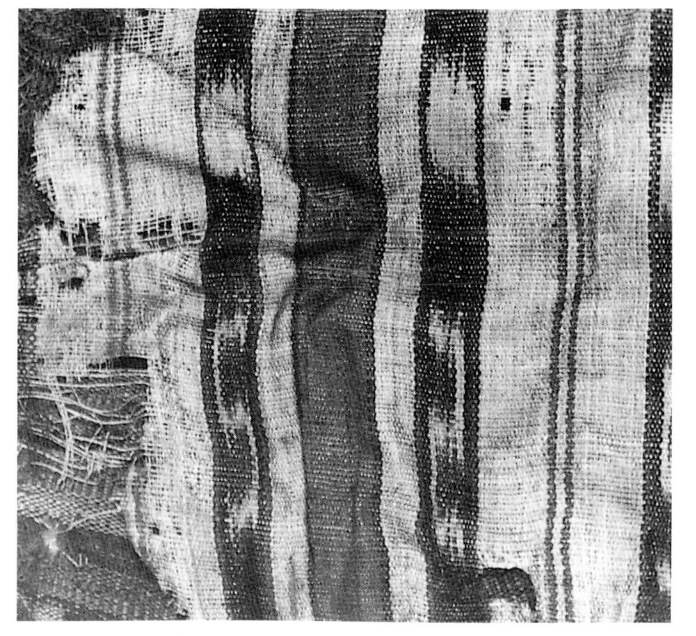

Fig. 42. Earliest example of warp *ikat* in South-East Asia from Banton Island in the Philippines. Photograph courtesy of the Regents of the University of California.

Warp Ikat

The oldest method of resist dyeing in South-East Asia is that of warp *ikat*, which is thought to have been used almost as long as weaving itself.[2] Whether this ancient method of decorating textiles was introduced from outside the region, or was developed independently in South-East Asia, is a matter of conjecture. The earliest example of warp *ikat* is from Banton Island in the Philippines, where two pieces of cotton burial cloth patterned with red stripes and parallel bands of warp *ikat* in white on a black ground, were found inside secondary coffin burials (Fig. 42). They were found in association with frontally reformed skulls and blue and white ceramics dating from the fourteenth and fifteenth centuries.[3]

Warp *ikat* remains a prevalent decorative technique (Fig. 43) amongst descendants of the earliest inhabitants of South-East Asia, such as the Batak of Sumatra, the Dayaks of Borneo, the Toradja of Sulawesi, the peoples of East Nusa Tenggara in Indonesia, and the T'boli, Mandaya, and Ifugao peoples of the Philippines. Warp *ikat*-patterned textiles have traditionally served as important ritual accoutrements in the social and religious lives of these peoples, a fact which points to considerable antiquity for both warp *ikat* and weaving in South-East Asia.[4]

Warp *ikat* patterning is applied primarily to cotton or bast fibres. Only rarely has it been applied to silk.[5] It is usually woven on a simple body-tension loom using a continuous warp structure. The process of preparing the threads for warp *ikat* patterning begins with warping. The warp threads have to be stretched between two adjustable slats or rollers, which will later be replaced by the warp and breast beams, respectively, on the loom. The distance between the two rollers is set at approximately half the length of the intended textile (Fig. 44). Two women sit in front of the frame and continuously pass between them two balls of thread placed in con-

Fig. 43. *Ikat* techniques.

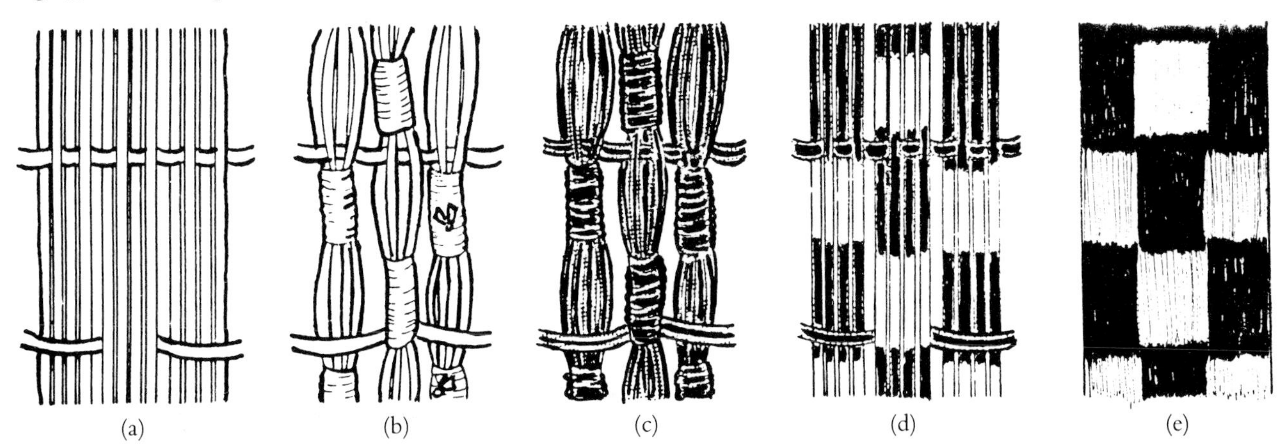

(a) Bundles of warp or weft threads are first separated.
(b) Bindings of bast or plastic string are then applied to the areas that are *not* to be coloured by the dye.
(c) The complete unwoven warp or weft threads are then steeped in dye.
(d) The resist bindings are removed to show the protected areas in the original colour of the thread.
(e) The *ikat* pattern emerges as the cloth is woven. (After Peacock.)

tainers, such as coconut shells or aluminium bowls, to prevent tangling (Fig. 45). The thread is passed over the warping frame, around each pole, and back under the frame. At the same time, the women divide the odd and even threads into an upper and lower layer by means of two lease cords placed a few centimetres apart at right angles to the warp yarns. This division into odd and even threads makes possible the creation of sheds during weaving. It also helps prevent the threads from becoming entangled during the binding and dyeing processes.[6]

As warping takes place, the threads are carefully counted and marked off in readiness for binding, according to the intended design. If the design is to be in parallel bands separated by plain areas, as is common in Savu and in parts of Flores, hanks of threads which are to have the same design are arranged together and bound with a twist of bast fibre to keep them separate from yarns which are to bear other patterns. If the design is to consist of repetitive motifs in mirror images, as on the island of Sumba, the weaver reduces the dyeing and tying time by doubling and quadrupling the number of threads that will need to be bound for each section.[7] A weaver may often prepare the threads for more than one identical cloth at the same time. This is often necessary in many cases, for narrow widths of identical cloth have to be joined together to make a large textile. In some areas, two identical textiles may be required to complete an outfit, as in Sumba, where men wear a pair of matching blankets as a ceremonial costume.

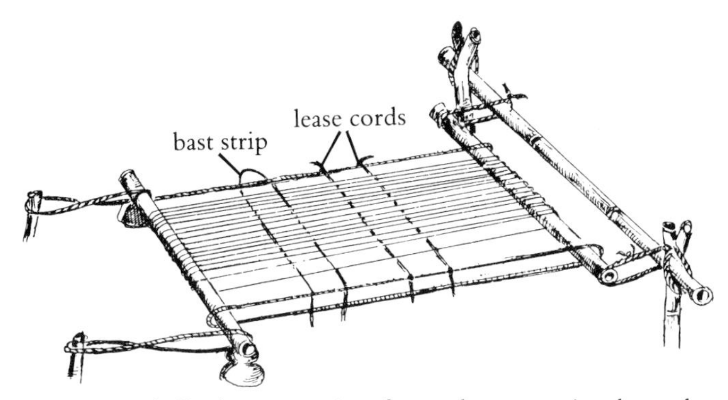

Fig. 44. A Rotinese warping frame demonstrating how the yarn is mounted as a continuous warp around a pair of rollers which will later be replaced by the breast and warp beams when set on the loom. The yarn is passed alternately over and under a pair of lease cords to divide the threads into odd and even sets to create sheds for weaving. A bast strip separates groups of thread in readiness for binding. (After Gill.)

Fig. 45. Two women on the island of Sumba work together to set up the warp in readiness for making warp *ikat*.

Once the threads have been divided into pre-arranged design sections, they are ready to be bound. A variety of frames are used to hold the threads taut while tying takes place. For traditional designs which are well known to the weaver, no pattern guides are necessary. If a new design is attempted, the weaver may refer to a paper sketch or to a sample textile (Fig. 46). Prior to tying a design, the dyer may block in the main elements with charcoal. Distances are measured with a piece of binding fibre or with the length and span of the weaver's fingers.

To create a resist pattern, small areas along groups of three to ten warp threads are tightly wrapped, prior to dyeing, with bast fibre from the *gewang* palm, coconut leaves, dried banana stems, or various grasses. The fibre is tightly wound two to three times around a small group of threads before being secured with a double knot. The excess fibre is trimmed with a knife. In other places, such as Roti, Borneo, and the southern Philippines, the bindings may be coated with bee's wax to increase resistance to dyes. In some places, such as on Roti, small differences in the knotting system are used to indicate colour changes in the course of the dyeing.[8] The tying and dyeing process is extremely time-consuming. A warp *ikat* textile on the islands of Flores and Sumba may have a warp of 3,000–4,000 threads. It could take up to six months to complete such an *ikat* with natural dyes, especially if red is the dominant colour.

Warp *ikat* in South-East Asia is usually first dyed blue, then red. Secondary colours (such as green and yellow) which appear as highlights in the main design, are usually dyed last. The parts of the pattern which are to retain the natural colour of the yarn are the first to be tied, followed by those areas which are to be dyed red. The bound threads are submerged in an indigo bath and then dried in the sun. If natural dyes are used, there may be several immersions and subsequent dryings until the desired shade is reached. The threads which are to be dyed red then have their bindings removed, while the areas to remain blue are protected by wrapping the appropriate yarns. This process is repeated for every additional colour until all colours have been applied. On completion, all remaining bindings are cut away.

The dyed threads are carefully arranged into their respective positions on a binding frame to be stretched for a few days (Fig. 47). They are then starched before being transferred to a body-tension loom. Alternate threads are threaded through string loops attached to the heddle rod. Weft threads are normally the same hue as the predominant warp colour. So that the warp-patterned design will show clearly on the finished textile, there are more warp than weft threads in the weave. Warp *ikat* is usually woven in a plain weave.

Fig. 46. A woman from Central Timor, with the help of a sketch, is tying in a resist pattern onto warp threads held taut in a frame.

Fig. 47. After dyeing, the warp threads are starched and stretched on a frame prior to being transferred to a loom for weaving. Prailiu, Sumba.

Weft Ikat

It is generally believed that the technique of tying and dyeing weft threads to create textile patterns was introduced into South-East Asia via Indian and Arab traders around the fourteenth or fifteenth century. In insular South-East Asia, the technique of weft *ikat* is widely practised amongst staunchly Muslim peoples, such as the Maranao and Maginda-nao in the Philippines, the Bugis of Sulawesi, and the Malay populations of coastal Sumatra and north-east Java. It was also at one time an important patterning technique in north-east Malaysia. The Hindu island of Bali also makes wide use of this patterning technique. On mainland South-East Asia, it is a particularly important decorative technique amongst the Buddhist Thai people, including the Shans of Burma and the Lao of Thailand and Laos. The Khmer people of Kampuchea are also leading exponents of this art and are credited with introducing weft *ikat* to Burma and Malaysia.

Weft *ikat* traditionally uses silk rather than cotton yarn. Many commercialized workshops now produce a wide range of weft *ikat* in both cotton and rayon. However, for fine quality cloth, silk continues to be the preferred yarn. Weft *ikat*, like warp *ikat*, was at one time produced solely by women working at home on body-tension or simple frame looms. Today, frame looms with hand-operated flying shuttles do much of the weaving. Chemical dyes are also widely used. In many areas, the making of weft *ikat* has been organized as a cottage industry, where people come to the home of the owner of the workshop to work. Men may be employed to assist with the tying and dyeing of the thread. Weaving, though, is still very much the preserve of women. Weaving may also be contracted out for women to do at home in their spare time. Yarns and dyes are supplied by the owner of the workshop, who also specifies the designs to be woven.

The process begins with the weft threads being wound onto a frame which is approximately the same width as the finished cloth. This is often done by hand, as with warp *ikat*. In many areas today, however, the weft threads are quickly wrapped around a revolving frame by drawing threads from a rack which holds twenty to thirty bobbins (Fig. 48). Sufficient threads to make more than one identical sarong may be drawn off at a time.

As with warp *ikat*, threads that carry an identical motif are grouped together and placed on a tying frame. Plastic string is used to bind the resist patterns (Fig. 49). A woman working at home on a traditional design will bind the pattern from memory. She may also follow a tying and dyeing process similar to that of warp *ikat*. In a workshop, where much of the tying and dyeing is done by men, guide-lines and even templates may be used to block out the pattern on the weft threads (Fig. 50).

Where weft *ikat* is commercially produced, a number of short cuts in the dyeing process may be used to save time. To create the gay, multicoloured patterns seen on much of modern *ikat*, a reverse *ikat* process, called *cetak*, is used. To dye small design

Fig. 48. The weft threads are wound from a rack of bobbins onto a revolving frame. Cili Weft Ikat Factory, Bali.

Fig. 49. Tying in a traditional weft *ikat* resist pattern using plastic string. Pak Thong Chai, north-east Thailand.

Fig. 50. Making an outline of the design on the weft threads using cardboard templates. Troso, Java.

Fig. 51. Applying a dye using the *cetak* method of spooning dye onto small exposed areas of weft threads. San Khan, Burma.

areas, the outline of a particular motif is bound around the outside and the dye colour is carefully spooned inside the enclosed outline (Fig. 51). Once the excess moisture in the dye has evaporated, the newly coloured area is bound over with plastic string to protect it from subsequent dyeing. This process is repeated for all colours except the ground colour. After all the minor colours have been applied and sealed with plastic string, the weft threads are submerged in a vat of dye for the ground colour.[9]

In Thailand, a few factories produce weft *ikat* using a 'discharge' process, which makes possible a greater number of colours and also enables the free use of dark colours early in the dyeing process. An area is dyed in a particular hue and covered with bindings where the colour is to remain. The excess colour is discharged, or removed, from the weft by the application of chlorine or some other bleaching agent. The remainder of the weft can then be over-dyed in another shade.[10]

After the last dye bath, when all the threads are dry, the *ikat* bindings are removed and the ends of individual threads are fitted into grooves of a winding wheel to be wound onto bobbins. In weft *ikat*, where repetitive geometric and floral designs tend to predominate, each bobbin contains sufficient thread to weave a complete repetition of a particular motif. The bobbins are carefully numbered so that they will be used in the correct sequence.

For weaving weft *ikat*, a plain-coloured warp thread is set up in the loom. It may be the same ground shade as the weft, or a contrasting colour to create a striking 'shot' effect when woven in silk. While weaving, the weaver must take care to see that the designs are correctly aligned. A small pick may be used to urge recalcitrant threads slightly to the left or right. On some weft *ikat* cloth, the selvage areas might be a little rough, due to the difficulties in precisely matching elements of a particular design. Failure to align the weft thread correctly will result in a blurred design. To make the design stand out clearly on the finished cloth, the weft is packed down tightly after each pick. In weft *ikat* there are more weft than warp threads in the weave.

Double Ikat

The process of making double *ikat* by dyeing both the warp and weft threads is known in only one place in South-East Asia, the small, remote village of Tenganan Pageringsingan in East Bali. This village is inhabited by the Bali Aga who claim their descent from the pre-Hindu people of Bali. The Bali Aga make a special cloth, called *geringsing*, by dyeing both the warp and weft threads. This complicated technique is known in only a few parts of the world, including India and Japan. How or from where the Bali Aga acquired this technique is not known. Since the people of Tenganan have generally kept aloof from their neighbours, it is quite possible that they developed the technique independent of outside influences. The process of spinning, dyeing, and weaving for *geringsing*, complicated by strict ritual observances regarding its manufacture, is so time-consuming that it can take eight to ten years to make a single cloth. The cloth is of cotton and is woven on a small body-tension loom with a continuous warp structure (Fig. 52).

For double *ikat*, the threads are prepared in the usual way. The yarns are alternately soaked overnight in a solution of *kemiri* nut oil, wood ashes, and water and then dried in the sun for five to twelve days to prepare them for the red *mengkudu* dye, which is made from morinda. The warp and weft threads are carefully measured, laid out, and bound on separate frames. The weft frame is adjustable to the desired width of the weaving. Since the cloth is small, and to save time, the weaver may dye threads for a number of cloths at the same time. An actual sample or charcoal guide-lines may be used in the initial stages of tying the bast fibres. The bindings of both the warp and the weft designs are carefully measured so that the pattern will be correctly integrated during the weaving process. Once the initial binding is finished, the threads are sent to the nearby town of Bug Bug to be dyed blue, for the women of Tenganan do not dye with indigo. On completion of the first dyeing, the cloth is returned to Tenganan Pageringsingan and the bindings for the areas to be dyed red are removed. Some of the now blue areas, which are to become a rusty purplish brown, are left to be overdyed in red. The parts to remain blue are bound. The red *mengkudu* dye, which comes from the nearby offshore island of Nusa Penida, is mixed with an aluminium mordant obtained from the bark of the *symplocos* tree. The yarns are steeped in this solution over a number of nights, alternated with drying during the day, until the desired shade is reached. To obtain the characteristic rich red shade may take many months.[11]

The circular warp threads are unbound and stiffened with a starch solution before being laid out on a body-tension loom. As they are untied, the weft threads are wound onto bobbins. So that the threads show equally on the woven fabric, there are equal numbers of warp and weft threads to the square centimetre. The weave is rather loose. Great care is taken throughout the weaving process to maintain an even tension so that the emerging patterns match exactly. As she beats the weft yarns into place, the weaver may use a small pick to help align the threads correctly in the design. At the time of this writing, only four women were known to be practising the art of weaving traditional *geringsing* cloth by the time-hallowed double *ikat* method.

Fig. 52. A young woman at the village of Tenganan Pageringsingan, Bali, weaving a double *ikat geringsing* cloth.

Batik, Tritik, and Pelangi

Batik, *tritik*, and *pelangi* are largely outside the scope of this book in that these resist patterning techniques are applied to previously woven cloth, which today is generally machine-produced, rather than handwoven. The batik technique, which uses a wax compound to cover those parts of the cloth that are to resist a particular colour during the dyeing process, is still applied to handwoven cloth in a few villages, such as Kerek, Dongkol, and Bejagung near the town of Tuban in north-east Java. In these villages, a number of women grow and weave a coarse cotton cloth and pattern it with nineteenth-century style, Cirebon-inspired batik designs featuring swooping phoenixes amidst spiky foliage in indigo, *soga* brown, and morinda natural dyes. Telago Bira, a shipbuilding community on the island of Madura, just off the north-east coast of Java, is also reported to produce its own cotton cloth to make a local style of batik.[12] While tie-dye *pelangi* and thread-resist *tritik* are rarely applied to handwoven cloth today, there are some exquisite examples of fine handwoven cloth in a number of South-East Asian and Western museums.

Decorative Weaving Techniques

Setting up the Loom

Before weaving can begin, the warp threads must be placed on the loom. To do this, the weaver must decide on the width of the cloth to be woven, and carefully lay out the warp threads in a process called WARPING UP. To make the threads easier to handle, they are coated with a starch solution made from a gruel of rice, maize, tapioca, or sago. In the simplest method of warping up, two poles representing the warp and cloth beams are set up some distance from one another (Fig. 53). The weaver walks back and forth looping the yarns from a spool in her hand in a weaver's cross, over and around the poles. Additional poles may be added in the line of the warp to keep it taut. These supplementary poles assist in dividing the warp into odd and even threads as an aid to placing leashes and heddle rods later. Some of the hill tribes of mainland South-East Asia and the Yakan of Basilan Island use this method to prepare the warp.[13]

For greater convenience, a warping frame may be used. It may vary from a rudimentary implement consisting of a pair of adjustable parallel bars connected by rough pieces of wood (Fig. 54), to a large, rectangular horizontal (Fig. 55) or vertical (Fig. 56) frame with a series of pegs arranged diagonally to each other along two opposite sides. Drawing threads from a rack with numerous bobbins, and looping the threads around the pegs, it is possible to warp up long yardages at one time while taking up little space. Frames with numerous pegs are widely used in areas where weaving has become more commercialized, particularly by establishments specializing in making weft *ikat*. Warp threads for up to thirty sarongs may be drawn off at one time.

The warp then has to be placed on the loom. For a continuous warp textile, placement on a body-tension loom is a relatively simple operation. The warp and breast beams are substituted for the poles or bars supporting the warp threads at each end. For a discontinuous warp, such as is used on a frame loom, the warp threads are removed from the frame, extended for a considerable length, and held taut while a mechanical device, such as a raddle, is used to help spread the warp threads evenly as they are wound onto the warp beam (Fig. 57). Heddles and a reed/beater need to be threaded through the warp yarns. Prior to weaving a piece of cloth on a frame loom, the new warp ends are tied onto the remains of a previous warp near the cloth beam. At the same time, weft threads are wound onto spools and bobbins. This may either be done by hand, using simple spinning and reeling devices, or by machine. The latter is common in the larger workshops.

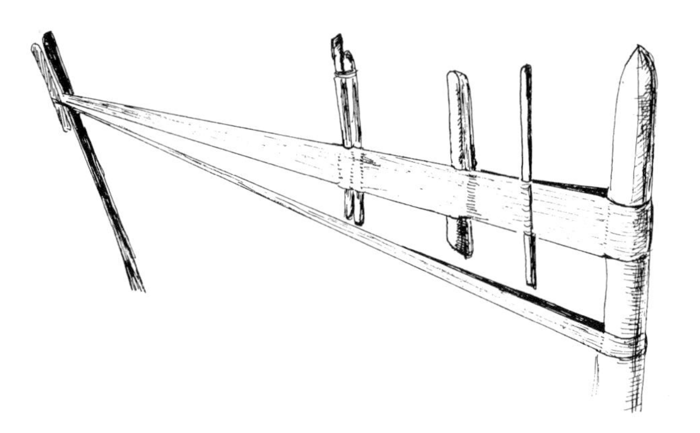

Fig. 53. Simplest method of warping up between two poles, which is widely used by many hill tribe peoples.

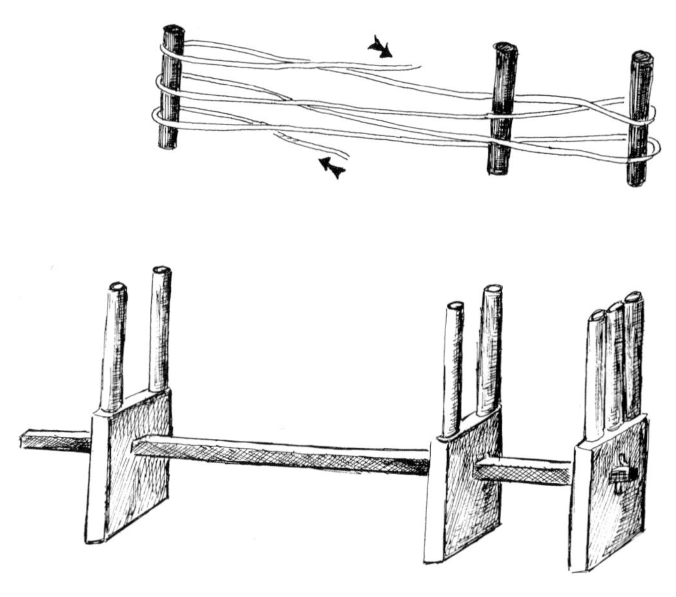

Fig. 54. Simple warping frame as used by many peoples of Indonesia. (After Bolland.)

Fig. 55. Warping up using a large horizontal warping frame. Surin, north-east Thailand.

Fig. 57. Winding the warp threads onto the warp beam. Ko Yor, southern Thailand.

Fig. 56. Warping up using a vertical frame. Nongkhai area, north-east Thailand.

Plain Weaves

The interlacing of individual warp and weft threads produces what is called a PLAIN or TABBY WEAVE (Fig. 58). This is the oldest, simplest, and most widely used of all weaves. The woven cloth appears the same on both sides. If the warp and weft threads are equally spaced and are equal in size and flexibility, the cloth may be described as having a BALANCED weave. If there are more warp than weft threads per square centimetre (as in warp *ikat*), the weave is described as WARP FACED. This means that after weaving the warp threads will be more noticeable while the weft yarns will be largely concealed. Conversely, if there are more weft threads per centimetre than warp (as in weft *ikat*), the weft will be visually predominant and the weave will be called WEFT FACED.[14]

The majority of textiles in South-East Asia are woven in a plain weave. Natural undyed and indigo-coloured cloth, fashioned into simple, tailored hill tribe and lowland farmer's clothing, is usually woven in a plain weave, as are the ubiquitous stripes and plaid sarongs seen throughout South-East Asia. The stripes on these textiles are made by setting up bands of various colours in the warp, and weaving them together using a single-coloured weft thread. Plaids are created by using bands of different-coloured threads for both the warp and the weft. For added interest, warp or weft threads may be composed of two different strands of colour twisted together.

Fig. 58. Plain weaves.

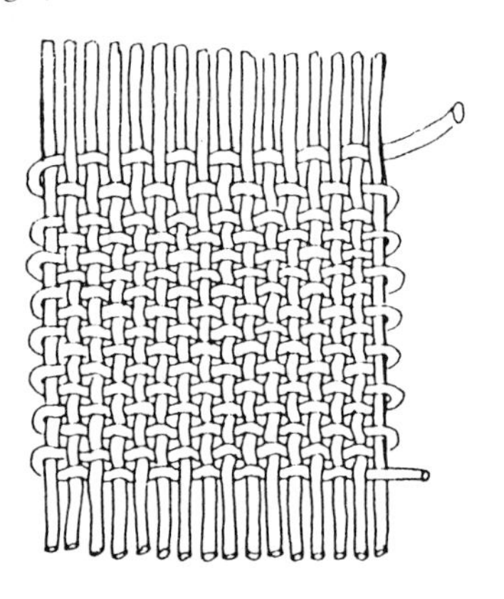

(a) Balanced (tabby) weave.

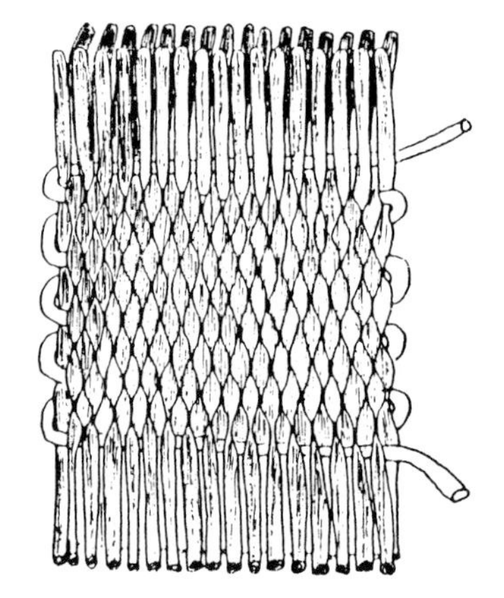

(b) Warp-faced weave.

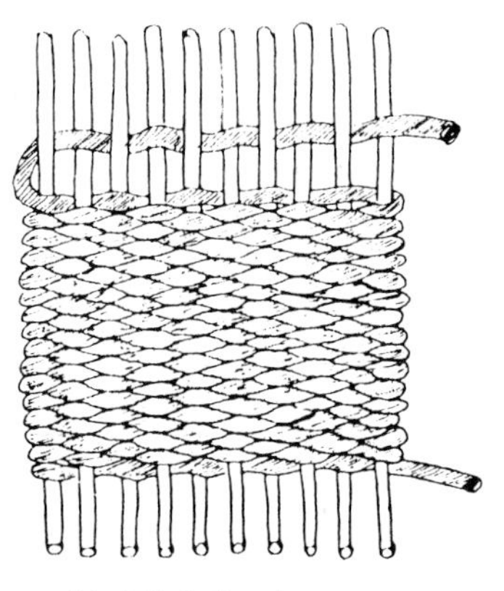

(c) Weft-faced weave.

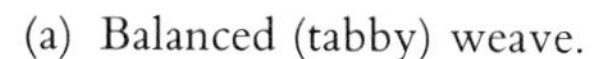

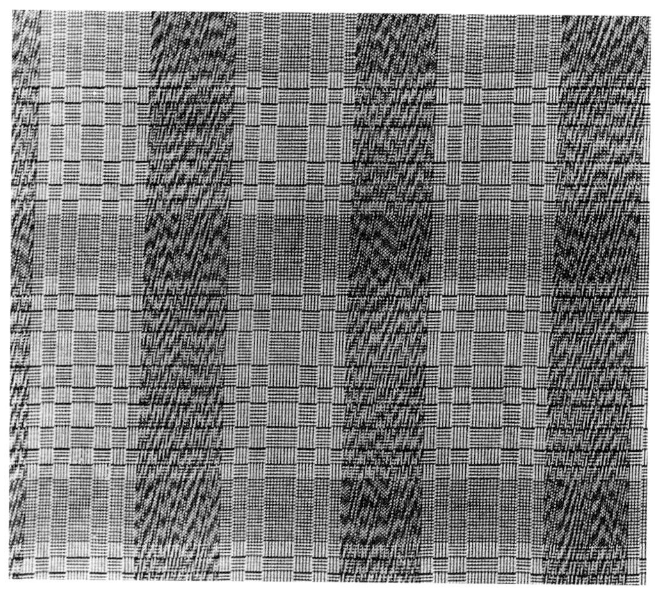

Fig. 59. A very interesting plaid pattern which has been created by setting up the loom with regular bands of two-tone twisted warp thread and alternating the colour and size of the yarn in both the warp and the weft: blue and white man's silk *lon-gyi*, Inle Lake, Burma.

When woven, these two-tone yarns create an interesting wavy pattern in the cloth. This type of pattern is very popular in the Shan States of Burma and throughout Thailand (Fig. 59). Interesting plain weave patterns are also obtained in Burma and Thailand by alternating the colours in the warp and the colour and size of the yarn in the weft.[15]

Tapestry Weave

Tapestry weaving produces mosaic designs by weaving with discontinuous wefts in various colours. In this technique, which produces a weft-faced fabric, a single weft pick does not travel completely across the warp. Instead, a number of individual weft threads are woven by hand back and forth across the warp, each in its individual pattern area (Fig. 60).[16] The weft threads may consist of 30–60 cm lengths of coloured thread, either loose or wound onto small shuttles to facilitate passage through the warp. Where a slit occurs between two discontinuous wefts, the tapestry weave is referred to as KELIM. The kelim technique is known on the islands of Timor and Sumba. The Atoni people of Timor at one time used this technique to create their *meo* warrior costumes, called *pilu saluf*, consisting of red, yellow, and blue tapestry weave ribbons woven on a band to go around the waist, shoulders, and head.[17] The Sumbanese once used this technique to decorate waist cloths, called *rohubanggi*, which were worn when going into battle.[18] The Ibans of Borneo make use of this technique to decorate the back border of men's *kalambi* jackets in brightly coloured geometric designs.

Since a slit weakens the fabric, many weavers prefer to interlock neighbouring weft threads by winding them either around a mutual warp thread, or by interlocking opposing weft threads at the end of each colour zone. The weavers of Lombok and Sumbawa use the interlocking tapestry weave technique to delineate the centre field of their distinctive *selampe* shoulder cloths. The people of Ceram in the Moluccas at one time made extensive use of tapestry weaving to create brightly coloured sarongs patterned in geometric designs.[19] This technique is also used in Sulawesi and by the Angkola Batak of Sumatra. The Muslim peoples of the southern Philippines decorate their *malong* sarongs and colourful head-dresses using a tapestry weave, while the Thai Lu of northern Thailand and Laos decorate their sarongs with bands of zig-zag tapestry weave (see Fig. 146). The Burmese also use this method to weave brightly coloured wave patterns on silk sarongs (see Plate 4).

Fig. 60. Tapestry weaves.

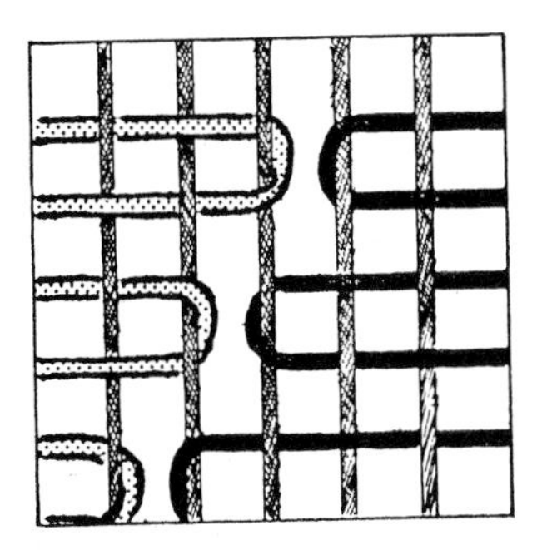

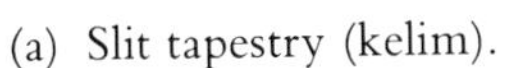

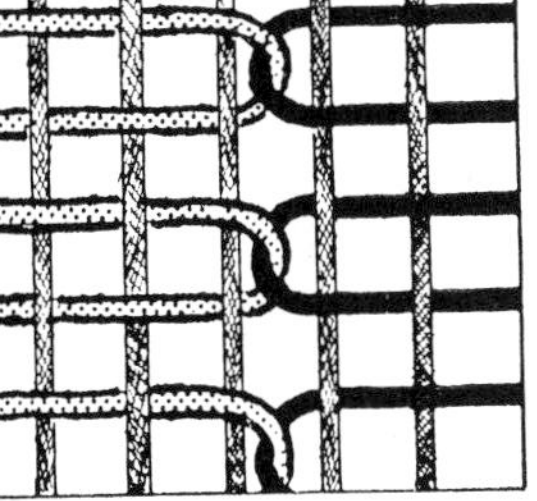

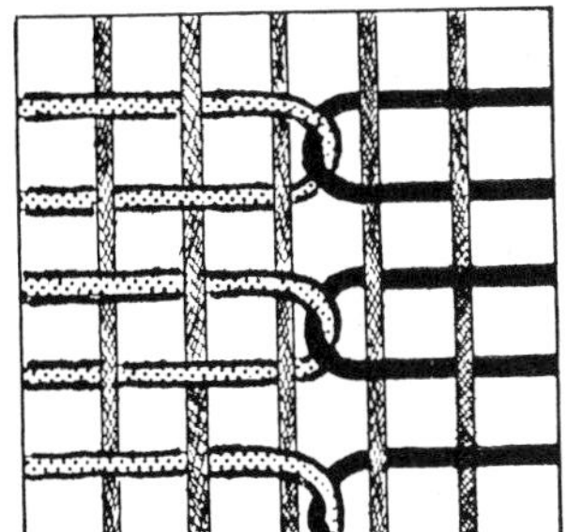

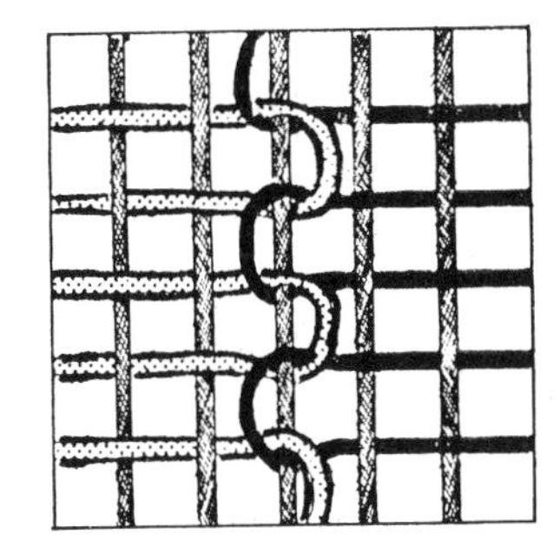

(a) Slit tapestry (kelim).

(b) Interlocking tapestry.

Supplementary Thread Techniques

Some of the most beautiful textiles in South-East Asia are decorated with brightly coloured yarns or metallic threads added in various ways to form a design. The patterns are usually supplementary or, occasionally, complementary to the plain ground weave. Every country of South-East Asia produces textiles which use a supplementary thread technique. In addition to clothing, a number of items woven in this technique are of considerable ritual importance, which suggests that this technique has been long established in South-East Asia.

Supplementary Weft

The most important supplementary thread technique is that of decorating cloth with threads which are inserted into the same shed as the ordinary wefts, but are allowed to FLOAT over selected warp threads to form distinctive patterns of contrasting colour and texture relative to the ground cloth. When supplementary threads are carried continuously back and forth over and under warp threads across the full width of the fabric, they are said to be CONTINUOUS. The supplementary weft pattern floats over the surface of the cloth and the thread is hidden on the under-side of the cloth when not required as part of the design (Fig. 61). To avoid possible snagging of long threads, the supplementary threads may be DISCONTINUOUS in that they are worked back and forth across the weft in small pattern areas only, and do not extend the full width of the fabric. The discontinuous technique is generally preferred over the continuous when a design is widely scattered or confined to a small area of the cloth.[20]

Textiles patterned in the supplementary weft technique include the famous ship cloths of South Sumatra, the ends of important Batak ceremonial cloths, and the sumptuous *kain songket* textiles of Malaysia and Indonesia, which are patterned with

Fig. 61. The front and reverse side of a design made by the continuous supplementary weft technique. This pattern depicts a well-known Thai–Lao motif in the form of two snakes (*nak*). It was woven in the Pu Wieng district of north-east Thailand.

gold and silver yarns and, occasionally, coloured threads.[21] The Yakan of the southern Philippines make extensive use of this technique to pattern ceremonial textiles. The Thai people, including the Shan and Lao, have a tradition of making exquisitely patterned clothing and ceremonial cloth in a variety of supplementary weft techniques. The Arakanese, Chin, Kachin, and Karen of Burma, and the Lawa of Thailand also use this technique to embellish their textiles.

Supplementary weft patterns may be created in a variety of ways. Once the threads for the background warp have been set up in the loom, the weaver, working from memory, graph paper, or a textile sample, carefully counts the warp yarns to work out the combinations of threads which need to be raised to form various pattern sheds. The simplest means of doing this is by selecting and raising the proper warp threads with a fine stick and passing the supplementary weft threads through the shed thus created. This method is used by some Montagnard groups in Vietnam, such as the Jarai (Fig. 62).

For more complicated patterns requiring a large number of warp thread combinations, a series of bamboo pattern rods may be inserted behind the two main heddles used to make the ground weave. For a single design, up to 200 pattern sticks might be used (Fig. 63). During the weaving process, the weaver lifts each pattern rod in order, and by inserting and turning her sword stick on its side she raises the appropriate warp threads to create the desired pattern shed. After being used, the pattern rod is pushed further back towards the warp beam to be stored for future use. The decorative weft thread, which may be loose or wound on a small

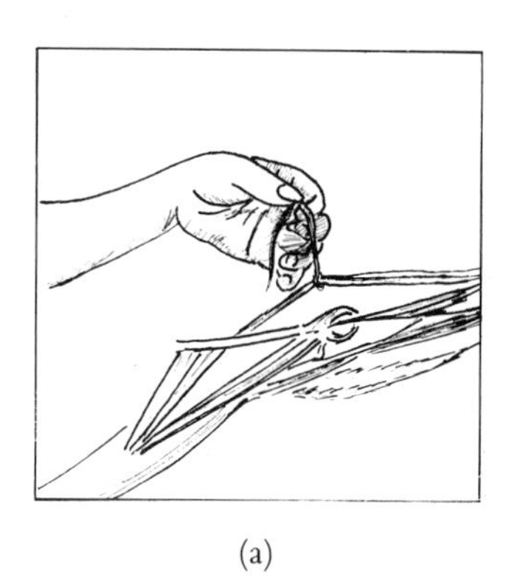

(a)

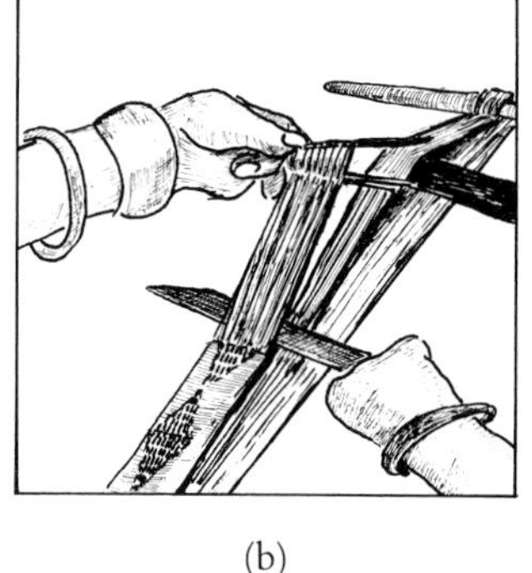

(b)

Fig. 62. Simple method of supplementary weft patterning as used by the Jarai people of South Vietnam.

(a) Carefully counting and lifting the warp threads which will form the design for the supplementary weft threads.

(b) The selected warp threads are raised with a fine stick and the supplementary weft threads pass through the shed created. (After Dournes.)

Fig. 63. Yakan weavers of Basilan Island in the southern Philippines making use of two different methods of supplementary weft patterning on their large body-tension looms. The woman in the foreground has the supplementary weft pattern in the loom set with numerous bamboo sticks, while the woman in the background is using the Malay *ikat butang* method of setting up patterns in the weft by looping the appropriate warp threads with string to mark out the pattern. Photograph courtesy of Mrs Lelita Klainatorn, Bangkok.

shuttle, is passed through the shed created by the pattern rod. A regular pick to make the ground weave is passed through the warp to anchor the supplementary thread before the next pattern rod is selected to continue the pattern. When one-half of the design is completed, the stored rods may be reused in the reverse order to create a mirror image of the pattern. This method is particularly suited to weaving Malaysian and Indonesian *kain songket*, for the patterns are traditionally repeated on both ends of the textile.[22]

In parts of north-east Thailand and Laos, pattern rods may be pre-set by means of threads set vertically rather than horizontally relative to the warp (Fig. 64). Each vertical thread is attached to a warp thread which, when selected by the pattern rod, can be pulled up and the sword stick inserted and turned on its edge to create a shed. The weaver may be helped by one or two assistants who pull down the pattern rods as required. After use, the rods are dropped below the warp.[23]

In north-east Malaysia, the warp threads are carefully counted in groups of three or five and the pattern is picked out section by section, from left to right, with a fine bamboo stick to make *kain songket* patterns. Each section is lifted with the sword stick

Fig. 64. In north-east Thailand and in parts of Laos, supplementary weft pattern rods may be pre-set by means of threads set vertically rather than horizontally relative to the warp.

Fig. 66. A young woman from Kelantan setting up *ikat butang* loops in the warp to create supplementary weft patterns for *kain songket*. She is wearing the Muslim woman's *selayu* head scarf.

to form a shed. A fine cord is passed through the shed to be looped under appropriate warp threads forming the pattern. These loops are grouped together and then tied in bundles called *ikat butang*, which may be suspended on longitudinal cords, or may rest on the warp behind the main heddles (Fig. 65). A complicated pattern may have up to 100–150 rows of *ikat butang*. Prior to weaving, the appropriate row of *ikat butang* is raised and a pattern rod is inserted to transfer the pattern to the horizontal warp. The sword stick, when inserted, creates a shed for the supplementary weft threads, which are held in a small shuttle while they pass through the warp. The supplementary threads are held in place by two shots of plain weave. *Ikat butang* patterns can be stored at the back of the warp to be reused, if necessary (Fig. 66).[24]

To give greater flexibility in supplementary weft weaving, pattern sticks may be replaced by rods with string heddles (Fig. 67). The warp threads are first selected with a stick according to the pattern. A rod is attached and the leases are threaded through the selected warp threads. These secondary heddles

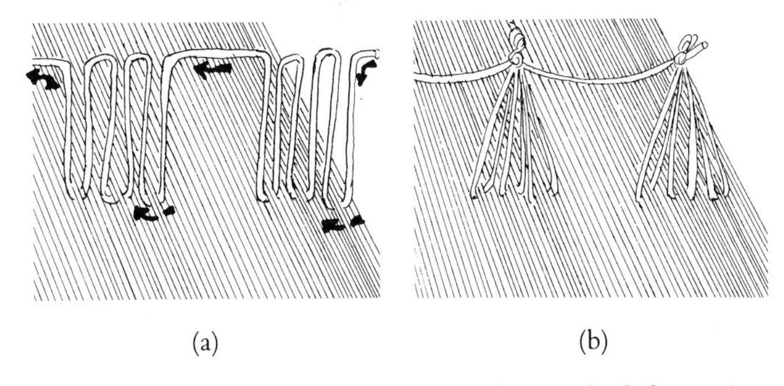

Fig. 65. The *pungut* process is a Malay method for setting supplementary weft patterns for *kain songket* in the warp prior to weaving.
(a) Loop leashes are made on the warp with thick string.
(b) The string loop leashes are tied together to form *ikat butang*. (After Kraftangan.)

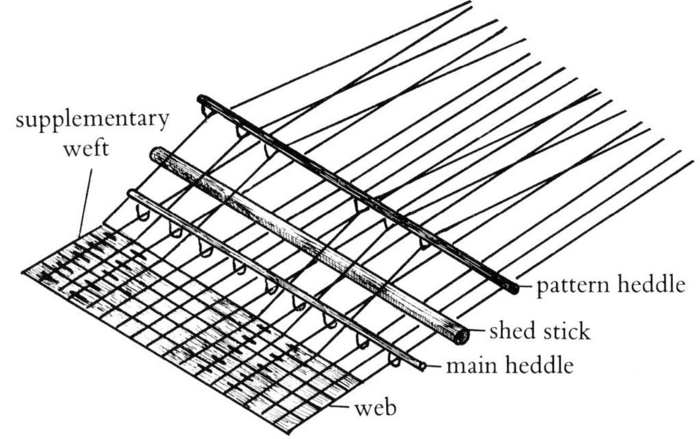

Fig. 67. Supplementary weft patterning with the use of an extra heddle. (After Hitchcock.)

Fig. 68. At Phum Rieng in southern Thailand, supplementary weft patterns are set with extra heddles placed behind the two main heddles. These are suspended over the warp by a support hung from the top of the loom frame.

lie behind the two main heddles. Although this method is very time-consuming to set up on the loom, it offers more flexibility, for it is possible to use the string heddle pattern rods in any order or combination with the main heddle rods. This method is used for patterning cloth in Palembang, Sumatra, southern Thailand, parts of Burma, and the Philippines (Fig. 68).[25]

The Ibans of Borneo make use of a unique supplementary weft technique, called *pilih*, to decorate some of their blankets. Narrow bands of supplementary weft threads in two or three colour combinations appear as a background of long continuous floats across the width of the cloth face. By contrast, the main pattern motifs appear outlined in a plain, warp-faced weave on the face of the cloth. During the weaving process, the supplementary threads are carried by small shuttles across the weft and disappear on the under-side where the main motif appears. This technique is unusual in that the supplementary weft threads are subordinate to the main design.[26]

Fig. 69. A Lao Phuan woman from Hat Sieo in Central Thailand making a *tdinjok* border for the Thai woman's *pha sin*. The pattern is laboriously picked out through the warp threads with a porcupine quill and various coloured weft threads are inserted over and under different warp thread combinations to form the design. After each row, a tabby shot is passed through the weft to hold the supplementary threads in place. At Hat Sieo, the weaver works on the right side of the cloth which is unusual. Note that the weaver has a small sampler to her left resting on the cloth bar to which she may refer for a pattern.

Another very common supplementary weft technique is that of inserting, by shuttle or by hand, different-coloured supplementary threads to create various designs. In this method the weaver, working either from memory or from a sample, slowly and laboriously takes each coloured thread lying in the path of the shed and, with the aid of a small quill or needle, coaxes the thread into place (Fig. 69). She wraps it over and under various warp thread combinations according to the dictates of the design. After each row is completed, a tabby shot is passed through to hold the supplementary threads in place. Excess threads may be left hanging on the underside or be secured with knots. Cloths patterned in this technique are not usually very wide. This weaving technique is often referred to as 'embroidery' weave, or 'brocading'.[27]

This supplementary weft technique is widely known in Thailand and Laos, where it is called *tdinjok*, and is used to make border strips for traditional sarongs. The Atoni people of Timor, who refer to this technique as *sotis*, or *pa'uf*, use this method of decoration on blankets, sarongs, and small bags.[28] The Iban use a similar technique, called *sungkit*, where the coloured threads are inserted with a bone needle during the weaving process. The discontinuous pattern threads may pass back and forth two or three times within a design area before being secured by two weft picks. Motifs produced by the *sungkit* technique appear exactly the same on both sides of the cloth.[29]

The Pwo Karen of Thailand decorate the shoulders and lower part of their shift dresses with geometric chenille-like patterns which are formed by inserting thick tufts of red thread into the weft while weaving, and later cutting them short to create a fuzzy pile effect.[30] The Batak of Sumatra and the Gaddang of the northern Philippines incorporate small beads into the supplementary weft of some of their distinctive cloths.

Supplementary Warp

Supplementary warp techniques are also known by a few groups of weavers in the more remote areas of South-East Asia. The best known and possibly most complicated method is done by the women of Sumba (Fig. 70). They weave a ceremonial sarong called *lau pahudu* (formerly worn only by noblewomen of East Sumba). This sarong is decorated with delightful neolithic-style animal and human figures in white on a dark ground. To weave a *lau pahudu*, the body-tension loom is equipped with two heddles to make the basic ground weave. The loom is set with two circular warps consisting of a basic warp of ordinary dark-coloured threads which form the ground weave. Over these is laid a supplementary warp of thicker white yarns. A stick is inserted towards the warp beam to prevent any tangling between the two sets of warp threads. Additional sticks are inserted to control the tension and keep the threads taut and regularly aligned. A temple keeps the selvages straight. Using a small stick and string model as a guide (Fig. 71), the weaver inserts numerous small pattern rods to lift up and mark out the appropriate supplementary warp threads which will appear on the surface to form the design. When not required on the surface, the supplementary warp appears as a continuous float on the reverse side. The pattern rods are used in succession. With the aid of a sword stick, the weaver raises the warp threads set by each pattern stick to create a shed for the passage of the weft thread. The design on the cloth face appears in a diagonal twill on a dark plain weave ground. After use, the pattern rod is stored in the warp if the design is to be repeated.[31]

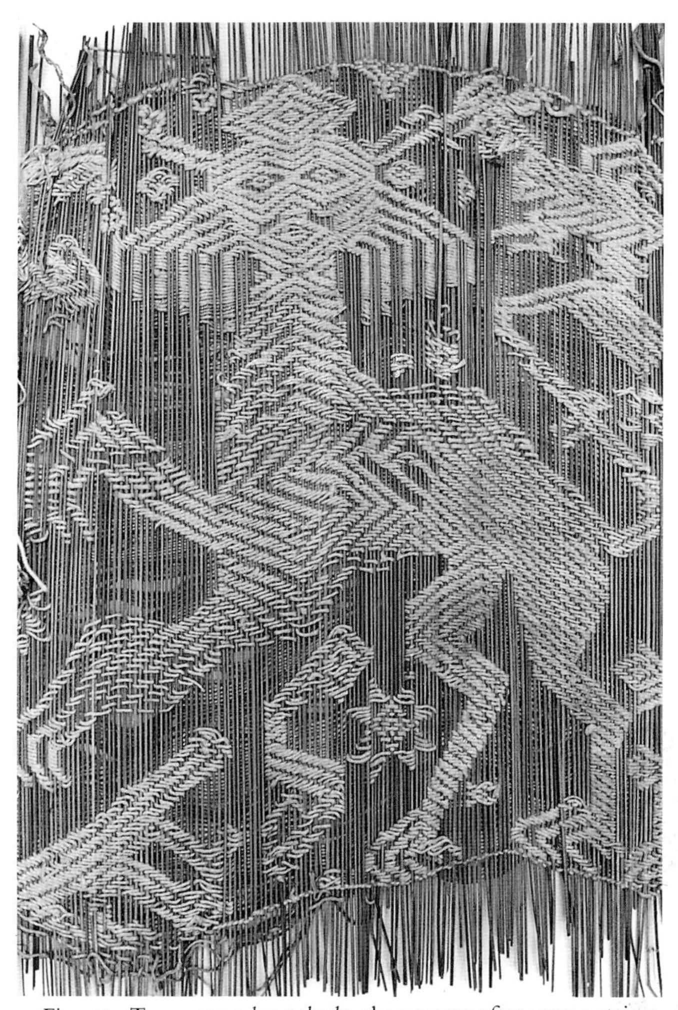

Fig. 71. To weave a *lau pahudu*, the weaver often uses a string model as a guide to placing pattern rods to set the design.

Fig. 70. A woman from Pao in East Sumba weaving the *lau pahudu*, a woman's sarong which is patterned by a very complicated supplementary warp technique. A thicker supplementary set of white warp threads are placed on top of the darker finer warp threads which will form the ground of the cloth. Numerous sticks are inserted into the warp to control tension and to prevent the tangling of the two warps.

The supplementary warp technique is also known on Bali. It was once used to weave *lamak*, the banner which is traditionally suspended from high poles or altars during important festivals, such as Galungan, the Balinese New Year (Fig. 72). Although usually made from palm leaf, a few *lamak* have been found to be woven of cotton and patterned with white supplementary warp yarns on a dark blue ground. These cloths have wide vertical borders filled with triangular and lozenge shapes. The lower centre field may have a triangular hour-glass form in combination with a disc. This motif might be a *cili*, or female figure of Dewi Sri, the Goddess of rice and cultivation.[32]

The weavers of northern Luzon embellish some of their textiles by using supplementary warp stripes. The warp is set up in the usual way for the ground

Fig. 72. A Balinese *lamak* which was at one time woven using a supplementary warp technique. National Museum, Jakarta.

weave. Quadruple black threads and double white threads are inserted into the design areas of the warp and are attached by twisting to the ground warp threads. The designs, set with pattern sticks or heddles, are formed by raising both the black and the white threads. The quadruple threads, which are raised a few times depending on the design, obscure the white threads which are raised only once. The white threads appear as specks when the dark blue threads are not raised. The design appears only on the cloth face.[33]

Float Weaves

A float weave is a method of patterning where one weaving element, such as the warp or weft, skips or 'floats' over two or more threads in its path before being interwoven. Float weaves traditionally have not been widely used in South-East Asia.

Until late last century, the women of Minahassa in North Sulawesi wove a special textile called *kain pinatikan* with longitudinal stripes in a warp-faced weave (Fig. 73). Bands of simple repetitive geometric patterns were created by 'floating' warp threads over a number of weft picks in various combinations. This was done with the aid of a main heddle rod and a shed stick to create the ground tabby weave. Two additional heddles and three or four pattern sticks 'picked out' the warp thread combinations needed for the pattern (Fig. 74). To create a shed, six laze rods were inserted one by one and removed immediately after use, to be reinserted later for a pattern repeat. Pleasing bands of zig-zags, lozenges, honeycomb, and cross patterns in white, light and dark blue, and reddish brown were painstakingly created by this unique technique.[34]

The weavers of Timor and nearby islands also use a special decorative warp patterning technique referred to as a 'warp face, alternating float weave'. Two parallel bands of different-coloured warp threads lie more or less on top of one another. As weaving progresses, the pattern is picked out from the two different-coloured warp threads with a pattern stick. The pattern occurs in the alternate colour on the other side.[35]

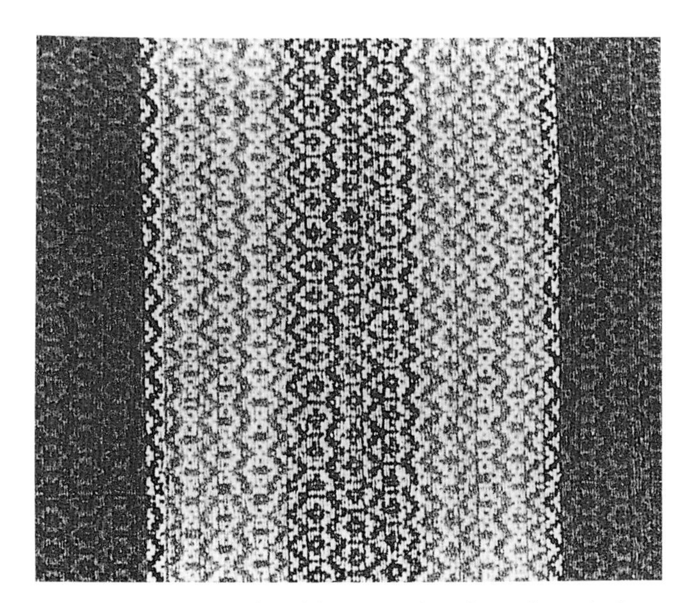

Fig. 73. An example of *kain pinatikan* from the Minahassa region of North Sulawesi. This warp-striped textile is patterned by 'floating' warp threads over various weft threads.

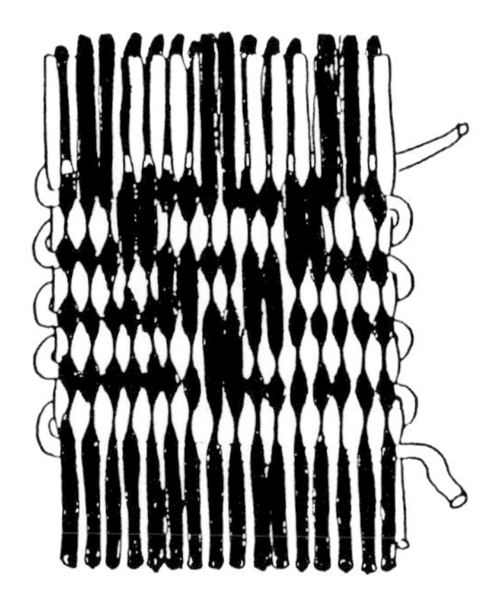

Fig. 74. A warp pick-up weave. (After Brown, Emery, Hitchcock, and Larsen.)

A diagonal float weave, called twill, is known to a few groups who use the body-tension loom. Writing in 1952, Jager Gerlings reported that this weave is known to the Kayan and Pnihing Dayaks of Borneo.[36] The Angkola Batak of Sumatra have sometimes used a twill as a ground weave on their ceremonial cloths.[37]

Over recent years, float weaves have become more widely used in parts of mainland South-East Asia where frame looms are common. Shan weft *ikat* has traditionally been woven in a twill weave. This weave is now being used in Thailand to create weft *ikat* and furnishing fabrics. In float weaves, the number of warps over which a weft passes is expressed as a ratio of the number of warps under which it passes. A twill weave is commonly woven in the ratios of 2 : 2, 1 : 3, and 1 : 4 (Fig. 75). In Thailand, a 1 : 3 weft-faced twill is a very popular weave for weft *ikat*. Unlike a plain weave, the two fabric faces are not the same. This is particularly noticeable in a twill woven weft *ikat*, where there is a distinct 'right' and 'wrong' side to the fabric. Textiles woven in a twill weave are strong, durable, and firm. They generally have better draping qualities than fabrics woven in a plain weave. However, a twill-patterned fabric requires more time and yarn to weave.

Satin weave, where there is a long float of one element, such as a weft followed by a single crossing of a warp of both sides of the fabric, is used in Thailand on occasion for weft *ikat*.[38] Float weaves are made possible by adding extra heddles to the loom. A skilled weaver in Thailand and Burma can successfully manipulate up to eight heddles on a floor loom. In areas of South-East Asia where hand-weaving has become more commercialized, patterns such as herringbone, zig-zag, and diamond weaves, are made by changing the direction of basic float weave patterns.

Twining

The process where pairs of adjacent elements of one set (such as a weft) are twisted around each other in their passage may be referred to as twining.[39] This simple, ingenious method of patterning is widely used in Indonesia as a border element to finish off a textile by securely binding the loose warp ends to prevent fraying. Batak, Sumba, and some Timor textiles are noted for their finely twined borders which feature geometric diamond and hook patterns (Fig. 76).[40] Straps of bags and the edges of jackets may be finished in a twining technique. Some Montagnard people of Vietnam, such as the Jarai, finish the fringes of some of their textiles with twining. In the Philippines and over much of mainland South-East Asia, fabrics are finished with separately woven, closely patterned bands which are stitched to the main textile at the conclusion of weaving.

Fig. 75. Float weaves.

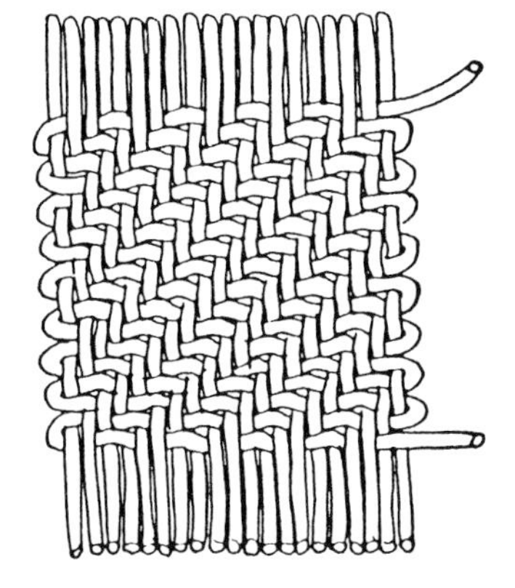

(a) A 2 : 2 twill weave.

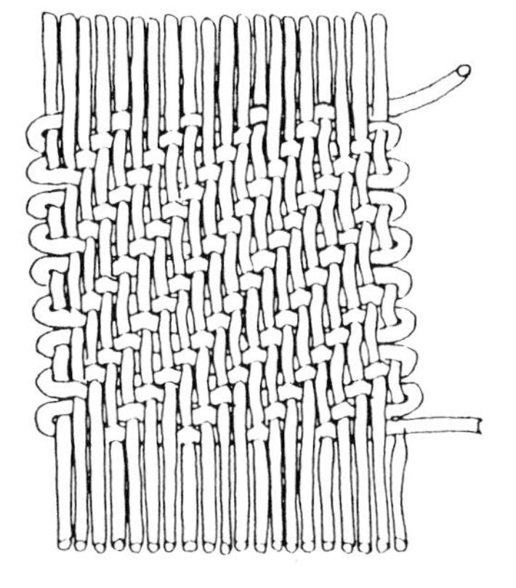

(b) A 3 : 1 twill weave.

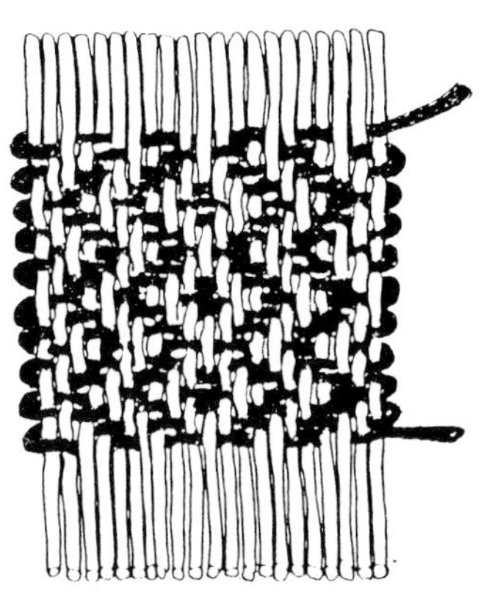

(c) A diamond twill.

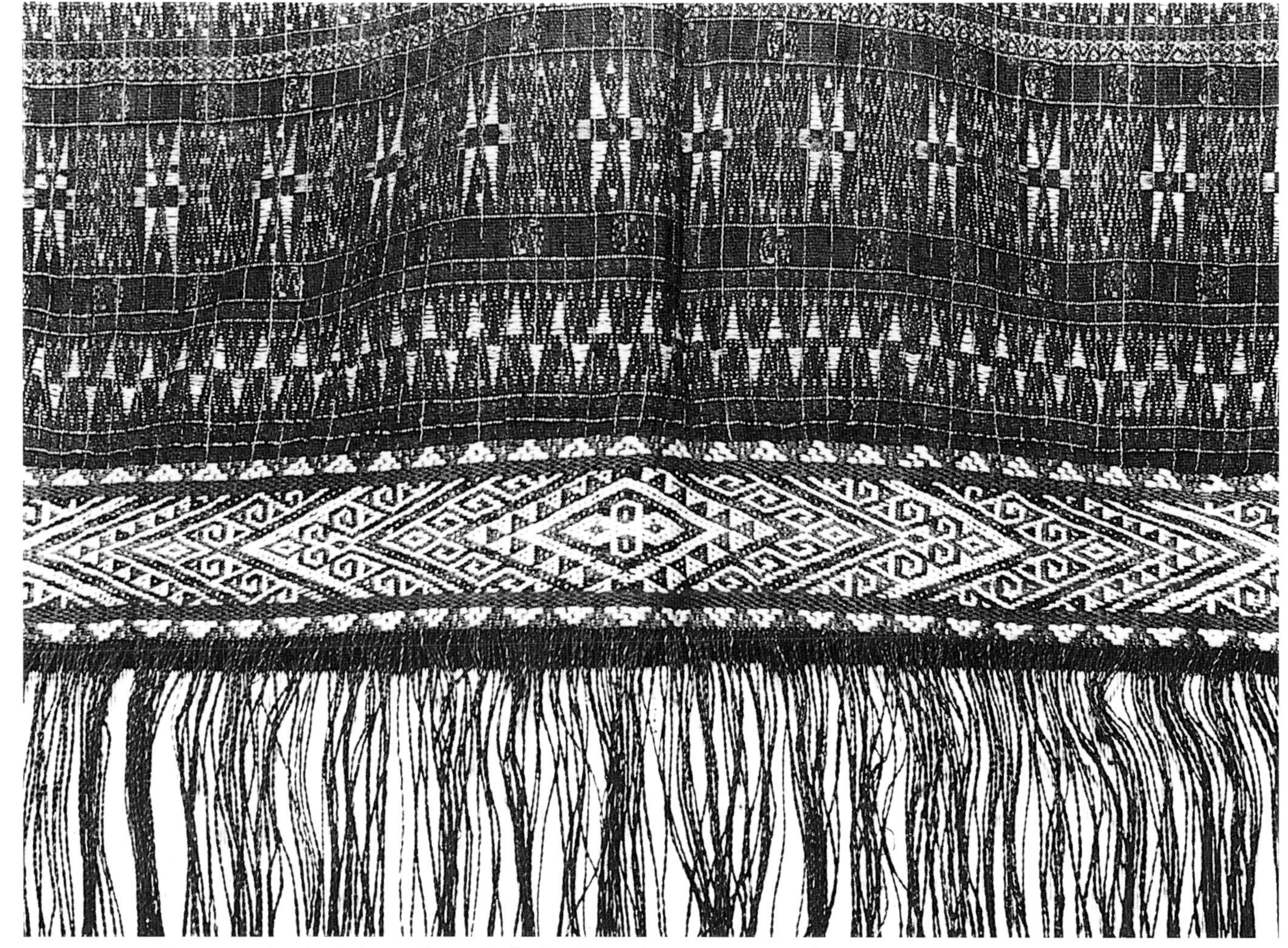

Fig. 76. A fine example of twining at the end of a Batak *ragi hotang* cotton textile.

Combination Techniques

Embroidery, Appliqué, and Beading

Although the major methods of patterning for South-East Asian textiles have been described separately, the weavers themselves often combine two or more decorative techniques to achieve stunning results. A number of people, such as the Batak, Timorese, and Sumbanese, combine warp *ikat* with supplementary thread techniques to produce notable cloths. Thai weavers like to combine a tapestry weave with weft *ikat* and different supplementary weft techniques to create remarkable formal wear. Muslim weavers of the southern Philippines and Palembang also combine weft *ikat* with a supplementary weft of coloured silk threads or metallic yarns.

Weaving techniques can be combined with embroidery, couching, appliqué, and beading techniques. The peoples of West Flores embellish their warp *ikat* with spidery beading designs, while the Iban and Maloh peoples of Borneo and the hill people of the southern Philippines (the Bukidnon, Bagobo, and Mandaya), complement warp *ikat*-patterned cloth with bright appliqué, coloured wool, metallic yarns, shell, mica chips, and elaborate beading. In South Sumatra, homely black and brown handwoven cloth may be transformed into sumptuous textiles by the addition of couched gold yarns. In the Lampung area, subdued warp *ikat* is combined with semi-flamboyant embroidery to produce some very interesting sarongs.

In Bali and parts of Java and Malaysia, woven cloth is sometimes embellished with a gold substance. A glue in the form of the desired pattern is spread on the cloth and gold dust, gold leaf, or gold paint is applied to the design area. The gold may be used to create new decorative patterns or to highlight designs already there. With the current widespread availability of machine-woven cloth, however, gold is now only rarely applied to handwoven textiles.

1. Jack Lenor Larsen *et al.*, *The Dyer's Art: Ikat, Batik, Plangi*, Van Nostrand Reinhold, New York, 1976, p. 129.

2. Mattiebelle Gittinger, *Splendid Symbols, Textiles and Tradition in Indonesia*, Textile Museum, Washington, DC, p. 13.

3. Personal communication from Dr Jesus Peralta, Curator, Anthropology Division, National Museum, Philippines, 21 October 1986.

4. Gittinger, *Splendid Symbols*, p. 17.

5. Only Aceh in North Sumatra, Banka Island in East Sumatra, and Donggala in Central Sulawesi have been known to make warp *ikat* in silk. Larsen *et al.*, op. cit., p. 130.

6. Alfred Buhler, 'The Ikat Technique', *Ciba Review*, No. 44, 1942, p. 1587.

7. Wanda Warming and Michael Gaworski, *The World of Indonesian Textiles*, Kodansha, Tokyo, 1981, p. 62.

8. Larsen *et al.*, op. cit., pp. 161–2.

9. Warming and Gaworski, op. cit., p. 117.

10. Susan McCauley, 'Thai Mudmee', *Sawaddi*, American Women's Club of Thailand, Bangkok, November-December 1982, p. 23.

11. Warming and Gaworski, op. cit., pp. 109–13; and Alfred Buhler, Urs Ramseyer, and Nicole Ramseyer-Gygi, *Patola und Geringsing*, Museum fur Volkerkunde, Basel, 1975, photographs, plate nos. 31–77.

12. Inger McCabe Elliott, *Batik: Fabled Cloth of Java*, Clarkson and Potter, Inc., New York, 1984, p. 167.

13. R. A. Innes, *Costumes of Upper Burma and the Shan States in the Collections of Bankfield Museum*, Halifax, p. 42, fig. 46; and Thelma R. Newman, *Contemporary Southeast Asian Arts and Crafts*, Crown Publishers Inc., New York, 1977, photograph, p. 95.

14. Irene Emery, *The Primary Structure of Fabrics*, Textile Museum, Washington, DC, 1966, pp. 76–7.

15. The last piece of information was provided to the author by Dr Mattiebelle Gittinger, Research Associate for Southeast Asian Textiles, Textile Museum, Washington, DC, who very kindly analysed some cloth samples from Burma and Thailand.

16. Emery, op. cit., p. 78.

17. Gittinger, *Splendid Symbols*, pp. 180–1; and Bronwen Solyom and Garrett Solyom, *Fabric Traditions of Indonesia*, exhibition catalogue, Museum of Art, Washington State University and Washington State University Press, Pullman, 1984, p. 47.

18. Warming and Gaworski, op. cit., p. 139.

19. Gittinger, *Splendid Symbols*, p. 196.

20. Emery, op. cit., pp. 140–3.

21. There seems to be some confusion regarding the term *songket* with respect to supplementary weft-patterned textiles. Western scholars, such as Dr Mattiebelle Gittinger, tend to define *kain songket* as supplementary weft patterning with gold and silver threads (see *Splendid Symbols*, p. 234), while Indonesian textile scholars use the term *kain songket* to include all textiles patterned in the supplementary weft technique (see Suwati Kartiwa, *Kain Songket Weaving in Indonesia*, Penerbit Djambatan, Jakarta, 1986, p. 15).

22. Mattiebelle Gittinger, 'An Introduction to the Body-Tension Looms and Simple Frame Looms of Southeast Asia', in Irene Emery and Patricia Fiske (eds.), *Looms and their Products: Irene Emery Roundtable on Museum Textiles, 1977 Proceedings*, Textile Museum, Washington, DC, 1979, p. 56.

23. Nancy Charles, 'Textiles of Laos and Thailand', unpublished notes from a lecture given to the Southeast Asian Textile Group of the National Museum Volunteers, Bangkok, May 1985, p. 2; and Anne Binks, 'An Unusual Pattern-Loom from Bangkok', *Bulletin of the Needle and Bobbin Club*, Vol. 44, Nos. 1–2, 1960, pp. 18–19.

24. Perbadanan Kemajuan Kraftangan, Malaysia, *Serian Songkit*, pp. 32–5. This method of supplementary weft patterning is also known in the southern Philippines; see Newman, op. cit.

25. Gittinger, 'An Introduction to the Body-Tension Looms', p. 57.

26. Laurens Langewis and Frits A. Wagner, *Decorative Art in Indonesian Textiles*, C. P. J. van der Peet, Amsterdam, 1964, p. 43.

27. These terms are disliked by textile scholars due to their confusion with unrelated textile processes. See Emery, op. cit., p. 171.

28. Kartiwa, op. cit., p. 87.

29. Langewis and Wagner, op. cit.; and Kartiwa, op. cit., p. 29.

30. Paul Lewis and Elaine Lewis, *Peoples of the Golden Triangle*, Thames and Hudson, London, 1984, p. 74.

31. Newman, op. cit., p. 104; Warming and Gaworski, op. cit., p. 136; and Rita Bolland, 'Weaving a Sumba Woman's Skirt', in Th. P. Galestin, L. Langewis, and Rita Bolland, *Lamak and Malat in Bali and a Sumba Loom*, Royal Tropical Institute, Amsterdam, 1956, pp. 49–56.

32. Galestin, Langewis, and Bolland, op. cit., pp. 31–47; and Bronwen Solyom and Garrett Solyom, 'Notes and Observations on Indonesian Textiles', in Joseph Fischer (ed.), *Threads of Tradition, Textiles of Indonesia and Sarawak*, exhibition catalogue, Lowie Museum of Anthropology and the University Art Museum, Berkeley, California, p. 28.

33. Francis Lambrecht, 'Ifugaw Weaving', *Folklore Studies*, Society of the Divine World, Tokyo, Vol. 17, 1958, p. 22.

34. Rita Bolland, 'Weaving the Pinatikan, a Warp Patterned Kain Bentenan from North Celebes', in Veronika Ververs (ed.), *Studies in Textile History*, Royal Ontario Museum, Toronto, 1977, p. 2.

35. Gittinger, *Splendid Symbols*, pp. 176–7.

36. Gittinger, 'An Introduction to the Body-Tension Looms', p. 57.

37. Gittinger, *Splendid Symbols*, p. 100.

38. Ardis Wilwerth, 'Basics of Weaving', unpublished notes for members of the Southeast Asian Textile Group of the National Museum Volunteers, Bangkok, 26 April 1985, p. 6.

39. Emery, op. cit., p. 196.

40. Gittinger, *Splendid Symbols*, pp. 46–7.

CHAPTER FOUR

Clothing and Ritual Use of Textiles

Textiles as Clothing

DUE to a warm tropical climate, traditional South-East Asian clothing is fairly simple. Many garments consist of untailored pieces of cloth which rely on folding, wrapping, and draping for a comfortable fit. Because textiles are woven on looms of limited width, larger garments are comprised of a number of panels skilfully stitched together so that design components match exactly.

Untailored Clothing

The most important garment is the sarong, which is still widely worn throughout South-East Asia. It may be referred to as a *lon-gyi* in Burma, a *pha sin* in Thailand and Laos, a *kain sarong* or *kain panjang* in Indonesia and Malaysia, and a *tapis* in South Sumatra and the northern Philippines. It usually consists of a waist-to-ankle-length rectangle of cloth measuring approximately 200 cm long by 100 cm wide. It may be worn wrapped in the style of a kilt to overlap in the front before being tucked into one side at the waist, or the ends may be sewn together into a tubular skirt which, when worn, is stepped into and wrapped around the body before being drawn tight on one side. For women, the excess cloth is folded over in a single pleat or gathered into a series of small pleats which are secured by a band of cloth, or a belt, or merely tucked in at the waist. The women of Burma, Thailand, Laos, and Kampuchea usually add a band of soft cotton material to the top of their sarong to make it easier to fold at the waist. For a man, the excess fullness of the cloth around the midriff is kept in place with a belt or a knot.

Sarongs are commonly plain or patterned with stripes and plaids. Inspired by Indian *patola* cloths, some may be divided into distinct design areas. Traditional *kain sarong* or *kain panjang* of Malaysia and Indonesia have borders along the selvages of the cloth. There may be a central panel, called the *kepala* or head, which differs in pattern (and possibly colour) from the main field, or *badan* (Fig. 77). The *kepala* often consists of a vertical border, or *papan*, with two rows of facing triangular or *tumpul* patterns. Sometimes *tumpul* patterns appear at the ends of the cloth. This *tumpul* pattern is usually worn in the front by women and at the back by men. Some traditional weft *ikat*-patterned cloth of Thailand and Burma also has intricate borders and a special design area at one end of the cloth.

Shorter knee-length sarongs, averaging 140 cm by 75 cm, are also common in South-East Asia. In the northern Philippines, they are widely worn in wrap-over style around the waist and hips. The Dayak women of Borneo, the Lawa of Thailand, and some Vietnamese tribes prefer to wear them as short tubular skirts. The Thai–Lao people of north-east Thailand put on a short cotton sarong for working in the fields but otherwise wear an ankle-length *pha sin*. On formal occasions and for festivals, Malay men wear a short sarong, called a *sampin*, over a loose-fitting shirt and trousers (Fig. 78).

On the south-eastern fringes of South-East Asia, a long, narrow, tubular sarong is worn by most women. This sarong consists of two to three lengths of cloth sewn together. In parts of Timor and West Flores, this sarong is worn knotted over the breast. It may also be worn with a blouse. On the islands

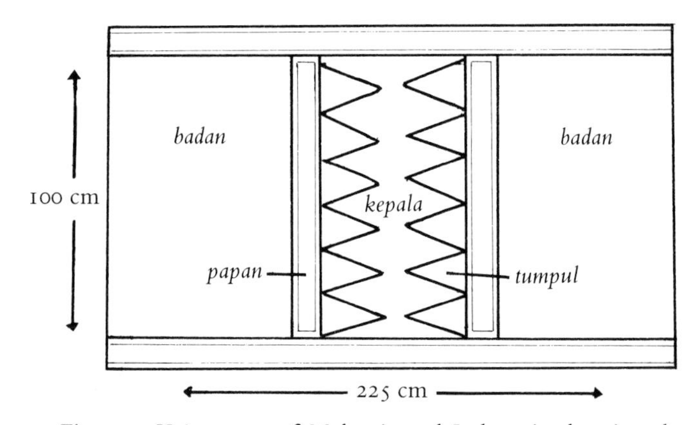

Fig. 77. *Kain sarong* of Malaysia and Indonesia showing the *kepala* and *badan*.

of Savu and Roti, this approximately 160 cm long by 100 cm wide tube of cloth is secured under the armpits, with the excess fabric falling back over the chest and hips to be cinched at the waist with a belt. In East Flores, an even longer tubular sarong is worn. Approximately 200 cm by 130 cm in size, this sarong is worn over the shoulders. It completely covers one arm, leaving the other free. In the Philippines, a tubular garment called the *malong*, measuring approximately 165 cm square, is worn (Fig. 79). It may be secured at the waist, above the breast, or at the shoulders, depending on the style of draping. There are at least six different draping styles.[1]

Exceptionally long rectangular cloths are worn for formal occasions by the Burmese, Thai, and Javanese, who at one time had royal courts which sponsored elaborate rituals for certain events. On Java, royalty and the aristocracy, court dancers, and a bride and groom on their wedding day, wore an exceptionally long length of cloth called the *dodot*, which could be up to six times the length of an ordinary sarong. It was comprised of two lengths of cloth (usually batik) joined together and draped around the body in various ways according to court etiquette. Men wore the *dodot* knee length over a pair of *patola*-patterned *ikat* trousers, while women wore it as a long strapless dress, pulled tight across the breast with flowing drapery down the back or front.[2]

Inspired by Cambodian court costumes, Thai aristocracy and officials during the Ayuthia period (AD 1350–1767) wore exceptionally long lengths of cloth, either pleated in front or gathered up between the legs to form knee breeches. This style of Thai court dress was revived by King Chulalongkorn (1868–1910) to serve as uniforms for Thai officials.[3] Some Burmese men wear an extra long sarong length, called *pah-soe*, for formal events.

The sarong is an exceptionally versatile garment. In addition to its uses for everyday wear and formal events, it is also used for sleeping. It is worn to preserve one's modesty while bathing by the river or at a public well. Labourers clad in shorts at job sites carry a sarong to work to provide extra warmth against the early morning and late evening chill, and to serve as a makeshift head-cloth as protection against the sun. To prevent chafing when carrying heavy loads, it can be worn as a hood over the head or as a pad on the shoulders. When travelling, it may be arranged as a diagonal sling over the shoulders and waist to carry small personal belongings. It may also serve as a temporary stretcher to

Fig. 78. Malay *baju kurung* with sarong *sampin* for men.

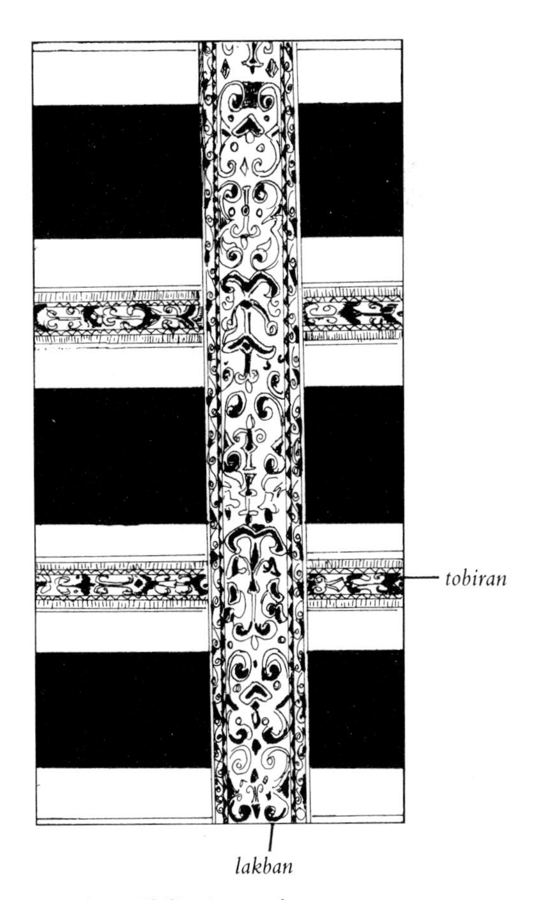

Fig. 79. Southern Philippines *malong*.

carry the injured. When wading through water or playing sports, the sarong is pulled between the legs and the ends tucked in at the waist to form a pair of shorts. The sarong is a favourite garment in which to relax. Even the most urbane, Western-educated professional man will happily change out of his suit and put on a sarong upon returning from the office.

Women, too, find a myriad of uses for the sarong. It provides the swaddling cloths for a newborn. Suspended between two posts, it cradles the sleeping infant. Stitched across one end, it functions as a sack. Filled with locally grown kapok, it makes a cushion or a small mattress. In rural areas, clean sarongs are always on hand to wipe down wet, grubby children or to refresh a perspiring bedridden relative with a change of clothes.

Blankets may also function as clothing. The men of Timor wear a rectangular 180 cm by 120 cm blanket composed of two to three panels joined along the selvages. This cloth is wrapped around the hips and secured with a belt so that the end fringes fall down the front. Another cloth may be worn for extra warmth around the shoulders early in the morning or in the evening. During the day, this extra cloth is tied around the waist over the first cloth.[4] The men of Sumba wear a pair of matching blankets called *hinggi*, approximately 250 cm long by 140 cm wide. One is wrapped tightly around the waist and hips with the ends falling between the legs, while the other is worn in a variety of ways over the shoulders. Blankets are particularly important in ceremonial events throughout South-East Asia and serve as bedding, room dividers, and baby carriers.

Heavy cotton loin cloths were formerly widely worn in the interior of South-East Asia. Today, they continue to be worn by some tribal groups in northern Luzon, central Borneo, South Vietnam, and on a few outer islands of Indonesia. Averaging 250–350 cm in length and about 25–30 cm in width, these narrow strips of cloth are wound around the hips and through the legs in a variety of ways so that at least one end of the loin cloth falls over the front of the body to preserve the modesty of the wearer. The ends of the loin cloth are often beautifully decorated with traditional patterns in supplementary weft yarns, and embroidery, and are weighted with seeds, beads, and shell work. Colour, patterning, and methods of tying readily identify a wearer as being a member of a particular tribal group. In the highlands of Vietnam, where trousers have become popular, the loin cloth may be worn around the waist as a mark of identification and pride.[5]

Breast cloths, long strips of cloth approximately 250 cm long by 50 cm wide, are wound tightly two to three times around the chest, leaving at least one, if not both, shoulders bare. They were at one time widely worn with the sarong by women throughout South-East Asia, especially in Java, Bali, Burma, Thailand, Laos, and Kampuchea. This garment may still be worn on special formal occasions and by classical dancers; for the most part, however, the breast cloth has been replaced by blouses or neat-fitting jackets worn over a padded brassiere.

Shoulder cloths have always been an integral part of dress for South-East Asian women. They range from a symbolic sliver of lace across one shoulder, to substantial cotton and silk textiles beautifully patterned with *patola*-inspired designs and metallic threads. Some come with matching sarong material. They serve as scarves or shawls draped around the head and upper torso in a variety of ways. Sturdy shoulder cloths are sometimes tied into slings and used as a carry-all for babies, food, and personal objects.

Sashes and belts are important for securing clothing around the waist. Where short skirts are worn by women, such as the Hmong and Akha in Thailand, long sashes help to preserve the modesty of the wearer when she squats down. Belts may be plain, as with the *stagen*, a long, tightly woven band which is wound many times around the waist to secure the *kain panjang* sarong worn by Javanese ladies. Sashes and belts vary in size from fine, delicately patterned cords woven on a small band loom, to long, wide strips of cloth embroidered at the ends. In Bali, men and women are expected to wear a sash around the waist when visiting a Hindu temple. Sashes may be worn diagonally over the upper torso by Iban women on ceremonial occasions. In the courts of South-East Asia, dancers sometimes wore—and still wear in public dance performances today—long sashes tied around the hips with the ends trailing to the floor. The skill with which the dancers manipulate the sash is an important element in South-East Asian dance and drama. The mere flick of a sash by an imperious king or overbearing official can spell death or banishment for a less fortunate character in a play.

The peoples of South-East Asia are readily identifiable by their head-cloths. They range from the brightly coloured, shop-purchased Turkish towels worn by the Pa O of Burma, to the elaborate *songket*-patterned horned head-dresses of the Minangkabau of West Sumatra (see Plate 37). The Karen and Shans of Burma wear a simple head-cloth

Fig. 80. Head-dress worn by the White Hmong people.

Fig. 81. Bag from Timor patterned in the *sotis* supplementary weft technique and decorated with beads, pieces of lead, and hair. Collection of Mrs Marta Ostwald, Jakarta.

consisting of a strip of plain or patterned cloth wound once or twice around the head. The ends hang loose over the shoulders or are tucked into the head-cloth. Longer lengths of dark indigo cloth are used by the Lisu and Mien people of Thailand to form large doughnut-shaped turbans. They are rivalled in splendour by the White Hmong, who take numerous strips of black and white plaid and painstakingly fashion a mitred turban (Fig. 80). Large squares of silk or cotton, beautifully patterned in a supplementary weft, tapestry weave, or resist technique, are particularly popular with the Muslim peoples of South-East Asia. These cloths are folded in innumerable ways to produce a most distinctive turban called *iket* in Indonesia, *destar* in Malaysia, and *pis* in the southern Philippines. At one time, it was possible to tell the rank and provenance of a man by the way he tied his turban. Today, turbans can be purchased already folded into a cap of stiffened material. The Burmese male wears a turban of light silk or stiffened muslin, called a *gaung-baung*. The revival of Islamic fundamentalism in recent years has led to the increased use by Muslim women of the *selayu*, a light-coloured silk cloth which covers the head and shoulders, leaving only the face exposed (see Fig. 66).

Bags

Since traditional untailored clothing has little in the way of pockets, handwoven bags have always been very important for carrying personal items, such as containers for betel-nut paraphernalia, tobacco, a few coins, a sharp knife, and possibly a protective charm. The most beautiful bags seen in South-East Asia are those made by wives and sweethearts for their menfolk. The amount of decoration that has gone into some of these bags clearly classifies them as 'labours of love'. Emblazoned in the style, colour, and motifs of the man's tribe, these bags are intensely personal items and are carried everywhere with pride. No self-respecting young Lisu male of Thailand would leave his house to meet his lady love without his colourful, appliqué-patterned courting bag, decorated with a riot of multicoloured tassels, dangling silver, and row upon row of beads. In Timor, a man receives a handwoven bag from his bride-to-be as part of the nuptial gift exchange (Fig. 81). At the time of death, a man's bag is hung on the main pole of the house and remains there in memory of him.[6]

Small bags, designed to hold only a few coins, are made from a rectangle of cloth folded in half and sewn at the sides. They may or may not have a shoulder strap. Such bags are popular amongst the Hmong and Lisu, and are usually decorated with fine embroidery and appliqué. Similar bags are made in Timor. Sometimes, a length of cloth may be folded in four and stitched along the sides of the two centre folds to create a bag with an overlapping flap on both sides. Timor bags are decorated by combining twining and tapestry weave and the *sotis* supplementary weft patterning technique.[7]

In mainland South-East Asia, the hill tribe groups

make their larger bags from two pieces of handwoven fabric (Fig. 82). One consists of a 12–18 cm wide, 150–180 cm long strip of cloth, fringed at both ends, which forms a long continuous strap. Another 18–12 cm by 60–90 cm rectangle of cloth forms the body of the bag. The bag is assembled by folding the long strip in half along the warp, sewing the two ends across the bottom, and joining the sides to the selvages of the wider piece of folded cloth. These bags are extremely strong and durable. For ease of carrying, the strap may be worn around the neck, over the shoulder, or around the waist (Fig. 83).

People of the southern Philippines, such as the Bagobo, make a very attractive bag from a strip of abaca or cotton fabric, which is folded twice to create a front flap, and which is beautifully embellished with elaborate beadwork. Small brass bells are attached to a pair of finely woven handles sewn along the front and sides of the bag. The bag may be worn knapsack fashion on the back (Plate 31). The Ifugao women of the northern Philippines make a *butong*, a small, distinctive, triangular-shaped betel bag, by taking two pieces of handwoven cloth (approximately 25 cm square) with warp fringes at

Fig. 83. Kachin *n'hpye* shoulder bag from northern Burma patterned with typical Kachin geometric designs in a brightly coloured wool supplementary weft on a cotton warp. The name of the town where the bag was made is inscribed in Burmese characters in the centre of the bag. A modern zipper has been added in the front. Length 30 cm, width 30 cm, length of shoulder strap 90 cm.

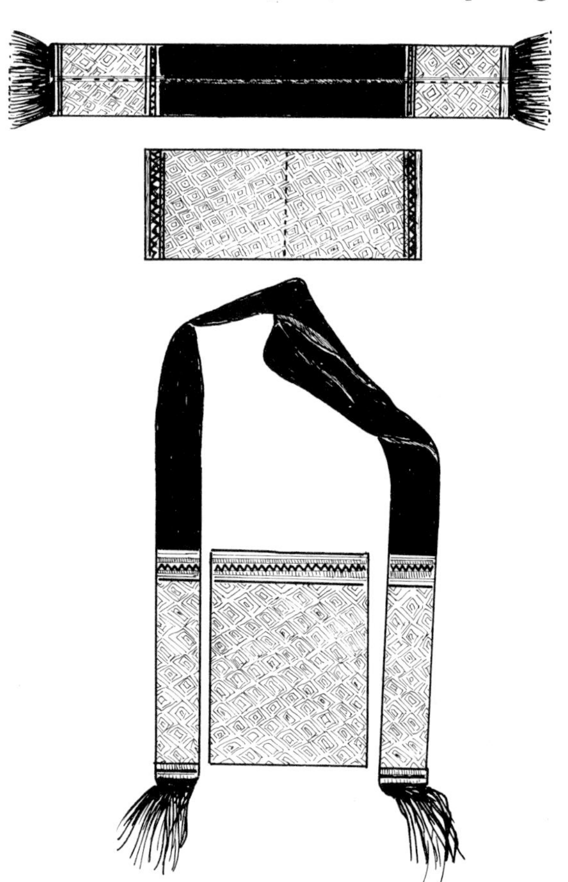

Fig. 82. Hill tribe bag of mainland South-East Asia.

Fig. 84. Ifugao *butong* bag, northern Philippines. Length 20 cm, width 27 cm, length of fringes 35 cm.

one end, and folding each piece in half lengthwise along the selvages. The two pieces of cloth are joined together by sewing the bottom half of each selvage with an embroidery stitch. The top half, which is not sewn, provides the opening for the bag. The warp ends are cleverly twined and twisted to join the bag along the bottom and to provide long decorative fringes. The material forming the top of the bag is pleated, and the topmost ends are sewn together before being covered with a coil of brass, which serves as a handle (Fig. 84).

Tailored Clothing

In addition to untailored cloth, a variety of simply tailored garments are worn in South-East Asia. Apart from a few very basic blouses and jackets, much of the tailored clothing seems to have been inspired by Muslim, Chinese, or European methods of cutting and assemblage.

The simplest blouse or jacket is one made from a folded rectangle of cloth with a hole or slit for the head (Fig. 85). Sometimes, two longer parallel rectangular lengths of cloth may be sewn together with holes left for the head, arms, and front opening, if it is to be a jacket. For greater ease of movement, small gussets may be added under the armholes. These garments may be tailored to fit snugly at the waist or hips, or they may be loose and bulky according to the dictates of fashion or tradition.

Fig. 85. Sleeveless blouses.

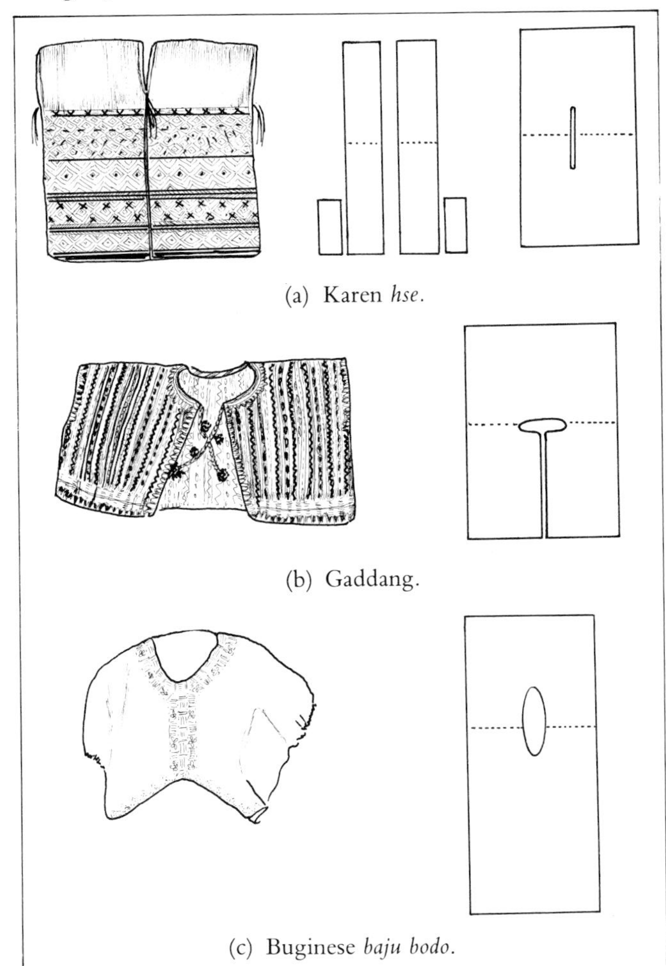

(a) Karen *hse*.

(b) Gaddang.

(c) Buginese *baju bodo*.

Sleeves may be created by sewing two tubular pieces of cloth to the basic jacket or blouse. Every country in South-East Asia has groups of people who add sleeves in this way. Sleeves may also be formed by cutting the entire jacket, both front and back, from a single piece of cloth, to include 'batwing'-style sleeve forms. The piece of cloth is then folded at the shoulder line and stitched along the sides and lower sleeve seams. This method of tailoring is thought to be inspired by Chinese and Mongolian jackets. The Burmese woman's *ein-gyi* is sometimes styled with a continuous sleeve (Fig. 86). The Toradja of Sulawesi at one time made their *lemba* jackets in this way from *fuya*, or beaten bark.[8] The Shans of Burma and the Thai Lu of Thailand and Laos use the 'batwing' style for the upper sleeve and sew on two tubes of material to complete the lower sleeve (Fig. 87). Some of these blouses and jackets fasten diagonally across the front with cloth buttons, Chinese style. The hill tribes of Thailand

Fig. 86. Burmese woman's 'batwing'-style *ein-gyi* with a continuous sleeve.

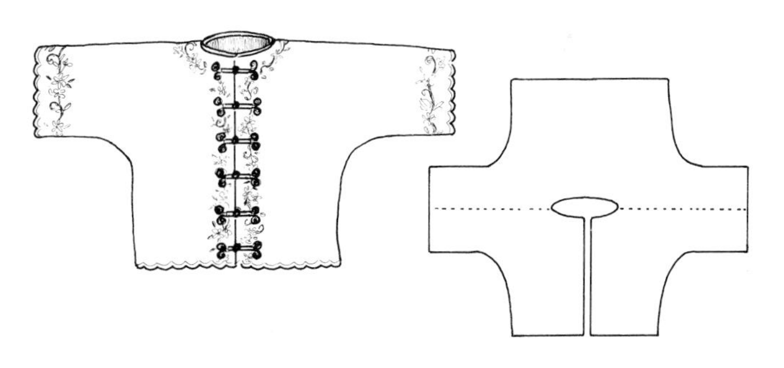

Fig. 87. Thai Lu jacket, also worn by some Lahu women.

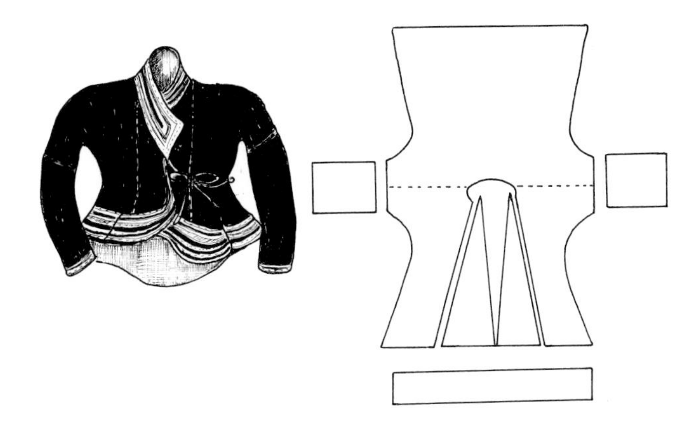

and Burma fasten their jackets with large finely chased silver buttons secured with a pin. Additional pieces of embroidery and appliqué are added as front panels, lapels, and collars (Fig. 88).

Waistcoats or vests are worn by various tribal peoples in the Yunnan border area of China. Men wear the waistcoat as an outer garment for extra warmth. For women, waistcoats are largely ceremonial. The Lisu like to wear a black velvet waistcoat covered with silver ornaments for a festive event (Figs. 89 and 90). Both sexes of the Blue Hmong wear a finely appliquéd vest strung with silver coins for New Year celebrations.[9]

Fig. 88. Jackets with sleeves.

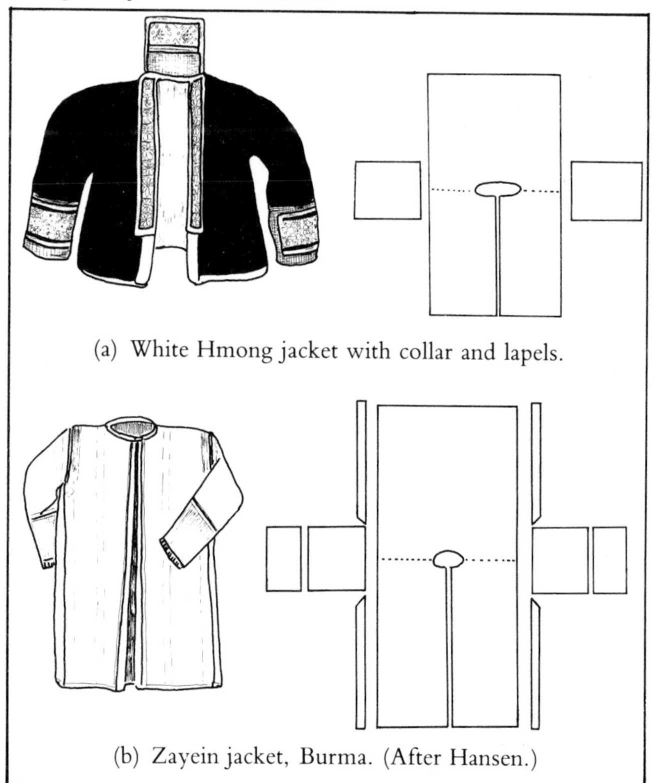

(a) White Hmong jacket with collar and lapels.

(b) Zayein jacket, Burma. (After Hansen.)

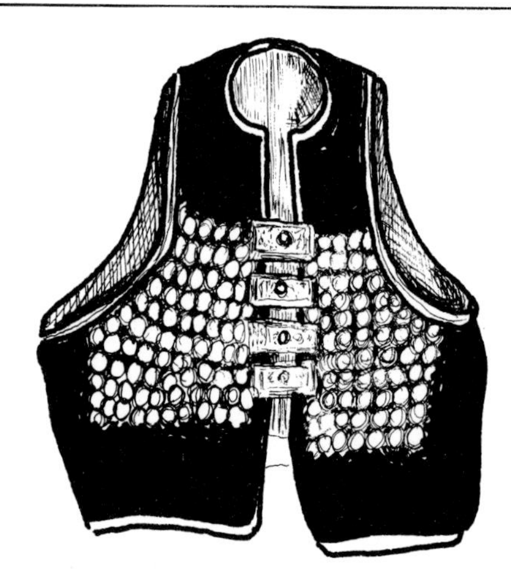

Fig. 89. Lisu waistcoat. (After Lewis.)

Fig. 90. Lisu woman from Burma in her traditional costume.

Another piece of clothing inspired by Tibeto-Chinese influences is a type of caftan worn by the Lisu and Lahu Na of Burma and Thailand. It consists of a length of cloth folded in half to form the front and the back. There is a long opening in the front, beginning at the neck and extending across the chest down the right side. It is secured from the chest to the waist with loops attached to small cloth or silver buttons. An extra width of cloth is sometimes added to create an additional, but invisible, front section on the inside. Long sleeves are attached at right angles to the shoulder sections of the main length of cloth. In the Lisu gown, side panels may be added for extra width (Fig. 91), while in the gown of the Lahu Na, the seams from the waist are left open at the side (Fig. 92). The caftan may be worn over pants or a sarong.[10]

The Vietnamese national costume, the *ao dai*, is a graceful variation of the form-fitting Chinese dress, the *cheongsam*. The *ao dai* has a mandarin collar and fastens at the side. It fits neatly at the waist, has long side splits, and is worn with long loose pants (Fig. 93). The Filipina *terno*, a long dress with rounded neck, fitting waist and large, stiff, bell-shaped sleeves is one of the few national costumes to be derived from European sources (Fig. 94).

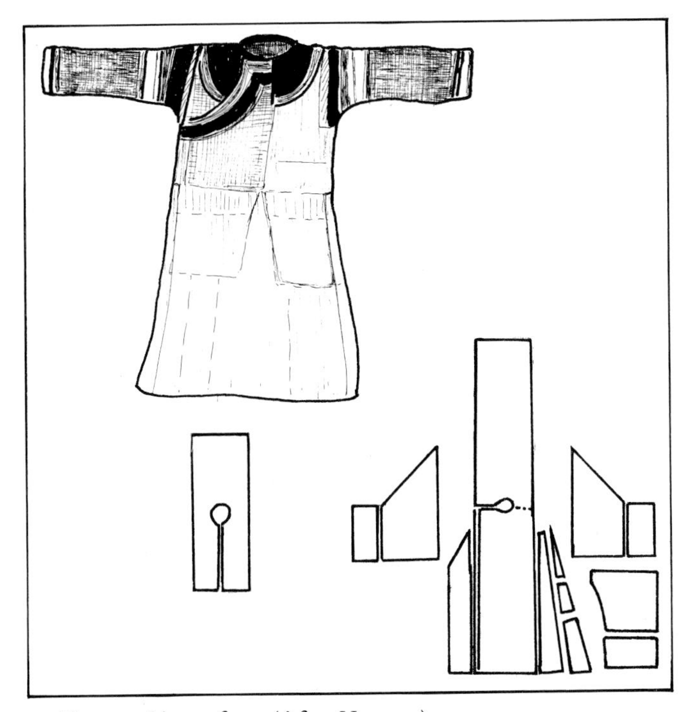

Fig. 91. Lisu caftan. (After Hansen.)

Fig. 92. Lahu Na woman from Burma wearing a caftan which has side splits. It has a front opening which is secured by a silver button, and is worn over a sarong.

Fig. 93. The Vietnamese *ao dai*.

With their sarong, Muslim women of Indonesia and Malaysia wear either a *kurung* or a *kebaya*, two garments possibly from Arab sources.[11] The *kurung* consists of a knee-length tunic with a split opening at the neck (Fig. 95). Side panels and gussets are added for comfort. Men wear a shirt length version of the *kurung* as formal wear. The *kebaya*, usually fashioned in lace or some fairly sheer fabric, is a contoured, long-sleeved jacket with a collar which continues as plain lapels down the front (Fig. 96). It has shaped front panels or darts at the bust line to give a neat, tailored fit at the waist and hips. The *kebaya* may have a panel inset across the bust (common in Indonesia), or it may be fastened down the front with a set of three ornamental pins.[12] The *kebaya* is also worn by *nyonya* or *Baba* Chinese women who have lived in Malaysia, Singapore, and Indonesia for many generations and have adopted the Malay language and many local customs. In Indonesia, Dutch and Indische women of mixed descent also at one time wore the *kebaya*.

Tailored skirts are not widely worn as traditional clothing in South-East Asia, although fashion-conscious Burmese and Thai girls are not beyond sewing a few darts into the waist of their sarongs to give them a neat, smooth fit. The Akha and Hmong wear pleated skirts made from a rectangular piece of cloth which is gathered at the waist or hips with a drawstring (or, more recently, elastic) (Fig. 97).

Fig. 94. Philippine national dress. The woman is wearing the *terno* dress, while the man is wearing the Filipino shirt, the *barong tagalog*.

Fig. 95. Malay woman's *baju kurung*.

Fig. 96. The *kebaya*.

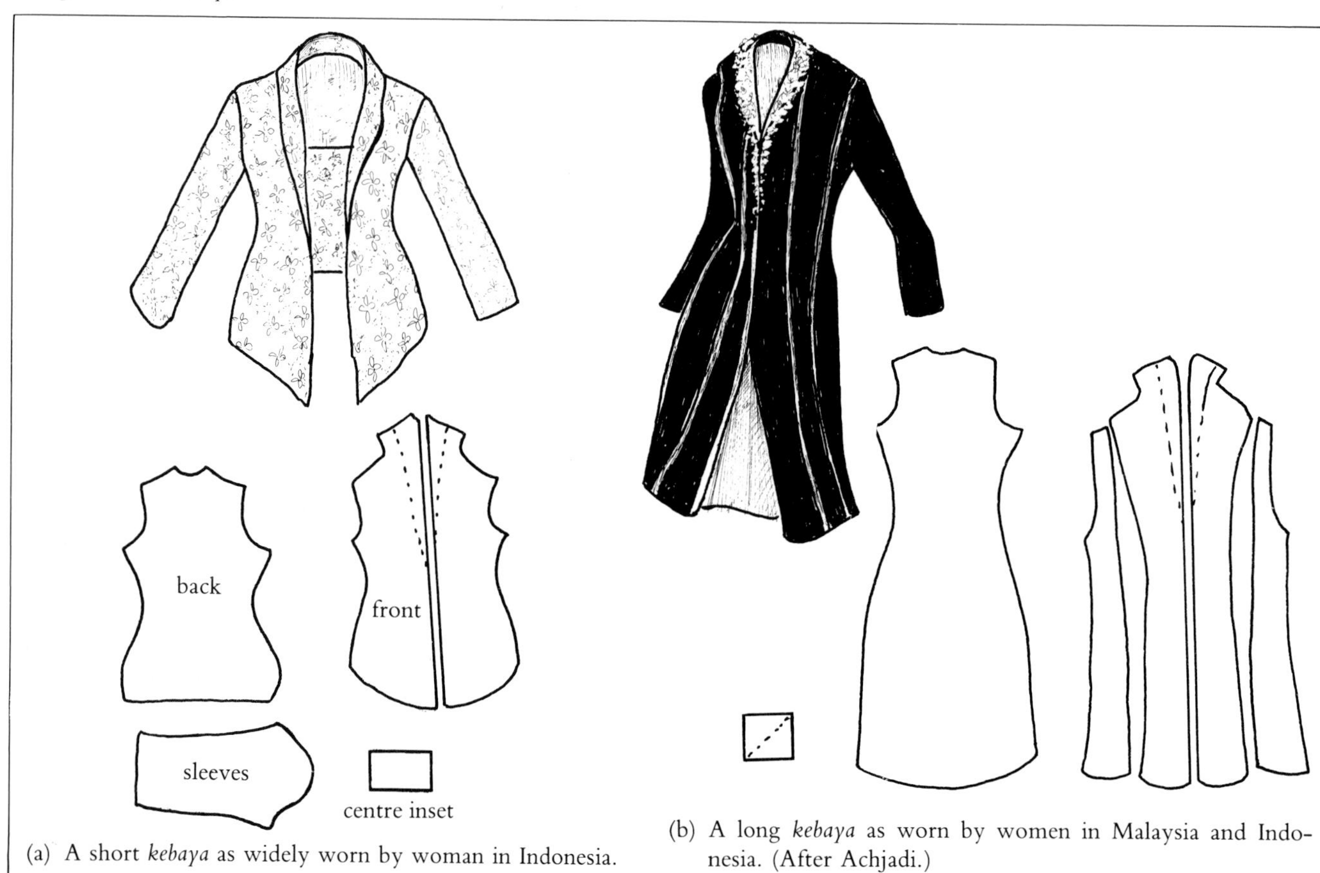

(a) A short *kebaya* as widely worn by woman in Indonesia.

(b) A long *kebaya* as worn by women in Malaysia and Indonesia. (After Achjadi.)

Fig. 97. A Hmong girl from a village near Chiangmai dressed in her best clothes for the New Year celebrations. Note the short accordion-pleated batik skirt embellished with applique and embroidery, and the long apron.

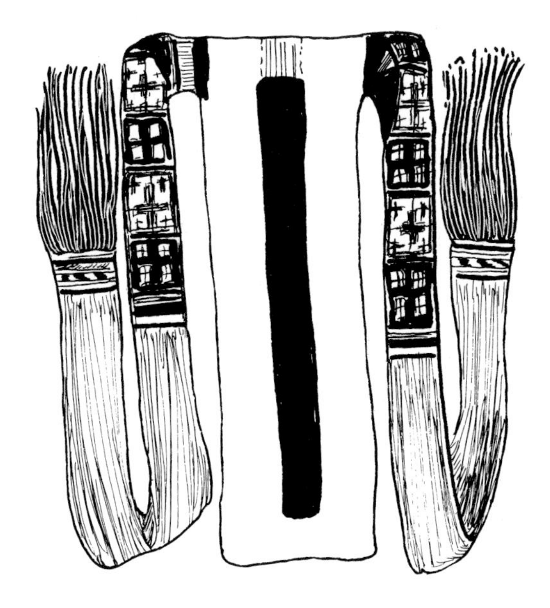

Fig. 98. White Hmong apron. (After Lewis.)

Fig. 99. Hill tribe pants.

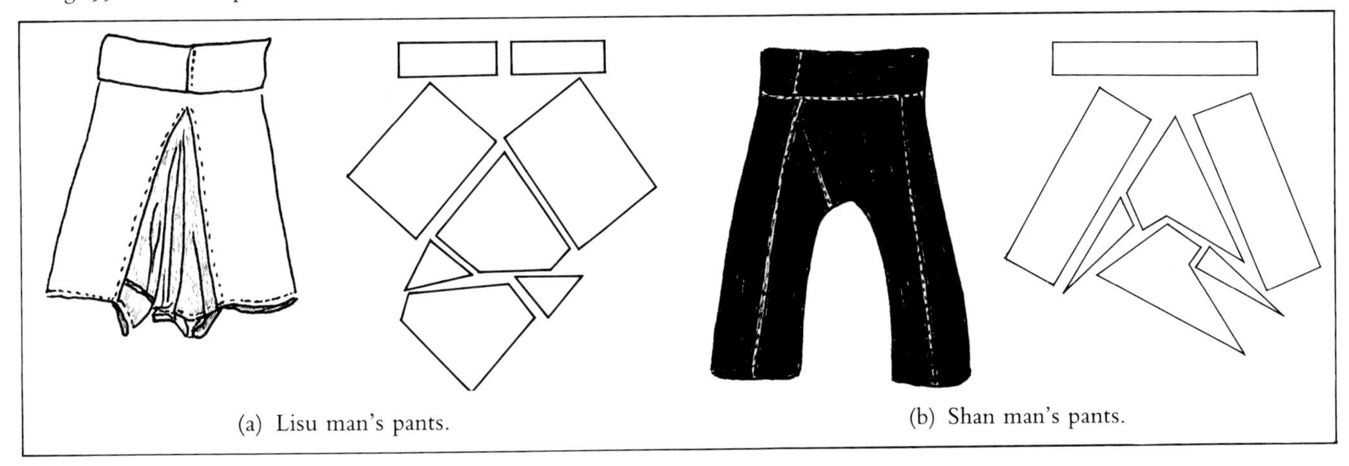

(a) Lisu man's pants.

(b) Shan man's pants.

Some of the hill tribe peoples of mainland South-East Asia like to wear finely decorated aprons at ceremonial events. These aprons also double as baby carriers or capes (Fig. 98).

Loose, comfortable, Chinese-inspired cotton pants are widely worn by men (and some women) throughout mainland South-East Asia and parts of the southern Philippines (Fig. 99). They are usually tailored from two rectangular strips of cloth for the outer leg and are inset with four additional pieces of cloth cut diagonally across one end to form the inner leg and crotch, which hangs low between the legs. The pants are attached to a wide band, the excess of which is folded over in a pleat in front and is secured by knotting or with a belt or piece of string. The crotch size and length and width of leg vary with different groups. The beautiful pants of the Mien women are tailored and folded in such a way as to ensure that the beautiful embroidery is shown to its greatest advantage. Lisu women wear trousers which have a smaller crotch and a leg portion made in two parts. Lisu men, and some other tribesmen of Burma, tend to favour a more flared style of pants with wide legs joined by a large, triangular crotch.[13] The trousers of the hill people of the southern Philippines generally fit more neatly than those of mainland South-East Asia and, in many cases, they are beautifully decorated with embroidery. Those of the Yakan are especially tight and have colourful cuffs patterned in supplementary weft decoration.

Tubular leggings are widely worn by the hill tribe people of South-East Asia as protection against prickles and thorns along mountain trails and in the swidden fields. These may be strips of cloth wound round the legs and fastened with string or cord below the knees. More common are colourful cylindrical leggings, usually about 30 cm wide by 45 cm long, made from coarse homespun cotton and patterned with supplementary weft, embroidery, and appliqué decoration.

Traditional dress in South-East Asia is rapidly being replaced by Western-style clothing, particularly amongst the men who increasingly prefer trousers, shorts, and shirts for everyday wear. Western clothing is widely worn by people in authority, such as government servants, teachers, the police, and members of the armed forces. Jeans and 'tee' shirts have found their way into virtually every corner of South-East Asia, and are the 'in gear' for young people eager to cut a figure in the modern world.

Brightly patterned, loose-fitting Western-style blouses are widely worn with sarongs by women, even in remote rural areas. In Thailand, fashion-conscious *kunying* (titled ladies) like to team their traditional silk dress sarongs with the latest *haute couture* Western blouses. Even traditional tailored clothing has been subjected to the whims of fashion. For example, hemlines of the *kurung* and *kebaya* may be longer or shorter according to current trends. Neckline styles, collars, and lapels come and go on traditional clothing. Sleeve styles are also subject to change.

Symbolism and Ritual

Apart from their great visual beauty, much of the fascination with South-East Asian textiles amongst Western scholars arises from the fact that textiles in this part of the world have always been closely associated with traditions, festivals, and religious ceremonies.

Magic and Protective Uses

In the past, South-East Asians have been firm believers in the magical and protective qualities of certain handwoven cloth. Threads, colours, and design motifs have all been credited with supernatural powers.

In Burma and Thailand, no self-respecting man would intentionally walk under a woman's sarong hanging on a clothes-line or pole. In Thailand, it is not fitting that the head, the most sacred part of the body, should brush against the nether garment of a woman.[14] Folktales abound in Burma of Samsonesque heroes who have been lured by a temptress into losing their supernatural powers by inadvertently passing under a woman's sarong.[15] In a popular Burmese folktale, a vindictive consort reduces her husband, a formerly charismatic king, to the laughing stock of the kingdom by sewing replicas of his 'magic eyes' along the hem of her skirt.[16]

At most Thai religious ceremonies invoking blessings and protection, a big ball of unspun cotton threads, folded many times and formed into a long string, is used. This *sai sin*, or sacred cord, is draped around the perimeter of the area to be blessed. Anything within the circumference of the sacred cord is considered consecrated and safe from harm. The *sai sin* may also be seen as a mystical 'electric wire' conveying the protective power and blessings from the sacred texts as intoned by the monks to everything within the sacred cord.[17]

As a mark of respect, saffron cloth may be wrapped around Buddha images and sacred banyan trees (*Ficus religiosa*) (Fig. 100). In Thailand, the

Fig. 100. An image of Buddha from the late thirteenth century, Sukhothai period, at Wat Chang Lom, Si Satchanalai, near the famous Lao Phuan weaving village of Hat Sieo. It is draped in a saffron monk's robe as a mark of respect and is regularly presented with floral garlands by the faithful.

highly revered Emerald Buddha image at the Royal Palace in Bangkok has its robes changed three times a year by no lesser person than His Majesty, the King of Thailand. In Burma, *nat* images representing spirits of the locality and former legendary or historic personages who were creatures of misfortune, are adorned with a fillet of cloth around the head. Offerings, such as coconuts, may be tied with strips of red and yellow cloth.[18] Ratu Kidu, the Goddess of the South Seas, formerly regarded as a formidable deity in Java, was regularly placated with offerings covered by special textiles.[19]

The Toradja of Sulawesi traditionally respected a variety of sacred textiles. These included imported *patola* and patterned Indian cloths, their imitations, and boldly patterned local textiles, such as batik *sarita* cloths and *pelangi*-patterned fabrics, collectively called *maa* cloths. These textiles were regarded as being imbued with special powers and were put on display at major ceremonial events. They were used when making offerings to the spirits controlling health, fertility, and rain. They could form a protective cover over the bier of the deceased, or be used as a head-cloth for a deceased noble. Shamans wore them during sacred rites.[20]

The Muslim Sasak people of North Lombok, Indonesia, have woven shawl-sized striped cotton textiles in a continuous warp. Although they appear very unassuming, these weavings could only be started on certain days. While setting up the loom, and throughout the course of weaving, various offerings had to be made and special rituals performed. On completion, these cloths were presented to family members at rites-of-passage ceremonies. They were believed to be effective against such illnesses as head- and stomach-aches.[21]

The *geringsing* double *ikat* from Tenganan Pageringsingan, Bali, due to the time and purifying rituals involved in their manufacture, are highly revered in Bali. In addition to magical and protective powers, they are credited with the ability to heal and may be used to wrap the sick. They are important in rites of passage. During the marriage ceremony, the warp may be cut. They may cover the pillow during a tooth-filing ceremony, or be one of the cloths placed on a corpse at a funeral. *Geringsing* cloths also serve as banners and flags during these ceremonies.[22] Bali at one time produced a special, loosely woven, gauze-like cloth, called *bebali*, in light shades of blue, pink, and brown on a beige ground patterned with zig-zag borders and hexagons in the centre field. The latter were worked in weft *ikat* touched with over-painting. These cloths were purely ritual in function. They served as offerings at death ceremonies and clothing for temple deities.[23] *Bebali* are no longer made.

In the central highlands of South Vietnam, the Jarai mystical leader, the King of Fire, was acknowledged for his special powers, not only amongst other Montagnard tribes, but by the royal courts of Laos, Cambodia, and Vietnam. This leader made special lustral water by praying and touching the water with an article of clothing. Washing in the water was thought to offer protection against evil spirits.[24]

In Thailand, a *pha parichat*, or protective piece of cloth about the size of a handkerchief inscribed with special protective mantras, may be worn as a neck- or armband for protection against weapons and malignant spirits.[25] Burmese soldiers at one time were tattooed, or wore handwoven jackets inscribed with mantras, to make them invincible against the foe. The Yakan of Basilan were known on occasion to wear a shirt covered with Arabic writing which was considered effective against bullets.[26] Important ritual weapons and amulets were often carefully wrapped and stored in cloth that was considered to have magical and protective powers.

In Indonesia, certain motifs and designs have been credited as being protective or fortunate patterns. A batik *geringsing* pattern, consisting of a series of nucleated semicircles placed close together like fish scales, was at one time considered effective against sickness, so it was often worn in the hope of keeping illness at bay. Another design (*sidho mukti*) composed of small motifs, such as pavilions, garuda wings, plant tendrils, butterflies, and other insects placed at regular intervals over the cloth, is associated with a prosperous and untroubled life and may be worn by a bridal couple on their wedding day.[27]

Colour is also significant. In Bali, black and white checked cloth, called *poleng*, is credited with the power to ward off evil spirits. It may be seen draped around temple deities and guardian figures and over lintels (Fig. 101). It sometimes forms the costume for temple dancers and priests and adorns the drums of the *gamelan* orchestra. It is thought to symbolize the duality of life—good and evil, life and death, etc.[28] White in many South-East Asian societies is associated with both purity and mourning. In Bali, white cloth banners and umbrellas may be suspended over sacred areas in a temple compound. Participants in certain ceremonies cover themselves in white cloth.[29] White is the predominant colour for Hindu and Buddhist cremations. Cosmic unity

Fig. 101. Balinese temple guardian wearing black and white checked *poleng* cloth credited with protective powers.

in the Toradja universe is represented by four colours: black, red, white, and yellow. Black is commonly associated with death, while yellow is the symbol of growth and fertility. Red represents the blood of sacrificial animals, and white the Toradja God Puang Matua.[30]

In Thailand and Kampuchea, it was formerly believed that the heavenly bodies favoured those who wore their colour on a particular day. It was considered auspicious for Thais on Sunday to wear red; Monday, yellow; Tuesday, pink; Wednesday, green; Thursday, orange; Friday, blue; and Saturday, black or purple. Some Thai people still consider it fortunate to be attired in the colour of the day.[31] Cambodian royalty adhered quite strictly to the colour of the day when attending certain ceremonies.[32] The colour of the day on which one was born is considered particularly auspicious. At a ceremony for raising the main post when building a Thai house, cloth the colour of the owner's birthday is wrapped around the post with pieces of gold and silver tucked inside.[33]

Wealth and Status

Textiles, their colours, and designs often indicate status and group membership. They also reflect distinctions based on sex, marital status, kinship, and ethnicity. The quality and quantity of textiles a person owns also offer tangible proof of status and wealth.

The royal courts of South-East Asia at one time had extensive sumptuary laws regarding dress. Certain materials, colours, and designs were reserved for royalty and the nobility. Throughout South-East Asia, the wearing of imported Indian textiles was generally limited to royalty and the aristocracy. The colours yellow and gold have largely been reserved for these groups. For example, in Vietnam, gold brocade with dragons was reserved for royalty. High-ranking mandarins wore purple, while those of lower rank wore blue. On the islands of Sumba and Roti, particular motifs indicated commoner and royal status. To distinguish differences in rank between various members of the Sultan's family and high officials, ordinances were periodically passed at the courts of Jogjakarta and Surakarta listing batik patterns which could be worn by persons of various relationship to the Sultan.[34] More recently in South-East Asia, political affiliation has been marked by the use of cloth. The Khmer Krom, a Cambodian nationalist group active in South Vietnam in the late 1950s, was called *Can Sen So*, or 'White Scarves', because they wore white strips of cloth inscribed with cabalistic symbols as a badge of identification.[35] The Khmer Rouge during their rule in Cambodia were distinguished by wearing checked scarves.

On the island of Savu, maternal clan relationship was indicated by the use of both particular motifs and nuances of colour on certain textiles. On the islands of Solor, north-east of Flores, the design layout of some ceremonial cloths indicated an individual's membership in a particular kinship group. A weaver would learn to make these patterns first from her mother, then from the women of her father's family. To wear designs from an unrelated lineage was forbidden and could have unpleasant repercussions.[36] Designs from Sikka, in East Flores, were placed in an ordered sequence on ceremonial cloth. Patterns from a man's clan as well as certain ancestral motifs had to be included in specific areas of the overall design. An Iban woman held the 'patent' on her own designs. She could either pass them on to her daughter or 'sell' them to other weavers.[37]

Ownership of vast numbers of textiles in the northern Philippines was a symbol of social prestige. At key ceremonial events, an important elder of the Isneg tribe in Apayao Province would sit flanked by ceremonial blankets as proof of his earthly attainments. When signing a peace pact, a host from the Kalinga tribe would put an array of his best blankets in full view of his guests. Additional textiles would serve as head cushions for the drinking of ritual wine.[38] A Kalinga man, if he wished to become a resident of another town, could display a blanket in a public place accompanied by an appropriate announcement, to make his relocation effective.[39]

Textiles might be used to pay fines and make marriage payments. In the northern Philippines, payments could include textiles in association with the transfer of certain types of property. Some cloths and blankets could be given to represent 'wrappings' in which to carry the articles purchased. A loin cloth could represent a 'rope' with which to tow away newly acquired animals.[40] Members of the Rhade tribe of the central highlands of South Vietnam might rent out land they were not using to a fellow villager for a jar, a pig, or a woman's skirt.[41] On the island of Buton and parts of Sulawesi, small pieces of handwoven striped and plaid cloth were backed with coins and used as currency by nobles and administrators. Their use was noted from the seventeenth to the twentieth century.[42]

A seventeenth-century chronicle of the Sultan of Ternate in the northern Moluccas relates how the admiral of its fleet, after fighting a successful military campaign on behalf of the Sultanate of Bone, was entrusted with collecting tribute for that state. He is reported to have collected 300 pieces of woven cloth from the Crown Prince of Koloncucu of Buton Island. He stopped at two other places on Buton and received 300 more sarongs at the first stop and 170 pieces of cloth at the second. From Buton, he went to other places in Sulawesi and then on to Flores. Altogether, he acquired 1,000 sarongs, ten shiploads of bolts of cloth as well as mats and pillows.[43]

In the Vogelkop region of Irian Jaya, imported textiles, such as the *patola* and cloths from eastern Indonesia, were purchased from traders from the seventeenth century onwards. These textiles were highly prized by the Mejpat peoples, who had a bark-making rather than a weaving tradition. Over time, an elaborate system of classification was built around these imported cloths, which became the epicentre of a complex cycle of exchanges. Fines, obligations, and general payments were discharged by offering cloth. Certain textiles could be given in exchange for food. A few months later, the recipients had to return a similar number of textiles, plus interest in additional cloths. Under such a system, a man's status came to be associated with the number of cloths he had in circulation, coupled with the number he had at home. Enterprising *popot* (cloth grabbers) operated a profitable business in cloth brokerage, complete with loans, receipts, and commodity displays. Like an overheated stock market, the system got out of hand, and the Dutch, in 1954, attempted to bring the system of cloth exchange to an end.[44]

Rites of Passage

Throughout South-East Asia, ceremonies are held to mark each important stage of the journey through life. A human being is regarded as being most vulnerable at these times. To ensure success, a number of precautions must be taken prior to these important events. Special festive food, such as glutinous rice cakes or the flesh of animals specially slaughtered for the event, are eaten. Various ceremonies are performed by ritual specialists, such as shamans, spirit mediums, priests, imams, or monks. Religious observances include ritual offerings, readings and recitations from sacred texts, and visits to shrines, temples, pagodas, or mosques. At these events, the participants are always beautifully attired in their best raiment. Special sashes, head-cloths, and armbands may also be worn. At the main event, ceremonial textiles, usually in the form of large blankets and long banners, are prominently displayed, some delineating areas of particular religious significance. Formal textiles exchanges are an important aspect of many rites of passage. In traditional South-East Asia, textiles at one time permeated all stages of the life cycle from conception of death.

Birth and Childhood

After pre-marital love-making, a young Akha woman may ask her lover for a piece of clothing or jewellery to keep as proof of paternity should she become pregnant.[45] Throughout pregnancy, some groups observe taboos concerning cloth. A Yakan woman is forbidden to cut cloth in case her child will be born with a hare lip.[46] For an easy delivery, a Thai woman avoids sewing up the end of a pillow or mattress. She also fastens a needle to the waist of her undergarment during an eclipse to prevent the unborn child from being cross-eyed or having facial deformities.[47]

The Batak woman heavy with her first child is presented, at the seventh month, with a special cloth (*ragidup*) called *ulos ni tondi*, or soul cloth. This textile is thought to have special protective powers from the accumulated life force of her lineage. The *ulos ni tondi* is worn at times of greatest vulnerability, such as childbirth or sickness. The cloth's protective power also extends to the woman's children.[48] In preparation for an approaching birth, Javanese at one time celebrated the seventh month of pregnancy with a special ritual called *Selamantan Tinkeban*. A thread was wound around the mother-to-be's waist and cut by the father for the baby to 'emerge'. The baby, represented by a shuttle, was dropped from inside the woman's sarong, to be caught by the mother-in-law, who cradled it in a shawl.[49] Amongst the northern Tinguian of Luzon, a special white blanket, called an *inalson*, with blue and red warp stripes, was worn by a medium while celebrating the *gipas* rituals immediately prior to the birth of a child (Fig. 102). At the conclusion of the ceremony, this blanket was placed on the floor and covered with offerings to herald the birth. At the same time, the blanket offered proof to the spirits that all necessary formalities had been followed.[50]

A new-born child has a highly tenuous hold on human life in the first few days. Apart from attaching protective strings or amulets and burying the placenta, special welcoming ceremonies are generally postponed for at least three days (Fig. 103). If the child survives, celebration is in order. The Bimanese of Sumbawa lay the new-born child on a white cloth, the symbol of goodness and purity.[51] The Iban welcome the little newcomer by laying it on a special textile, called the *pua kumbu belantan*, or *lebor api*, and giving it a ceremonial bath.[52] After three days, the Thai hold a *kwan* or soul-strengthening ceremony. Offerings are made to the spirit of the locality, and a woman who has successfully raised many children is invited to tie a sacred thread (purchased from a monastery and blessed by monks) around the wrist of the child to secure the *kwan*, or soul, of the child. The child is placed in a cradle which may be hung with white cloth bearing, on one side, an effigy of King Wessuwan (the Hindu God Kuvera) who, in Thailand, is regarded as chief of the evil spirits and guardian of new-born babies. On the other side may be a picture of Mae Sue, the 'purchasing mother', a surrogate godmother who supposedly 'buys' the child before the evil spirits can take it.[53] Relatives come to visit the new baby bearing gifts, which may be in the form of textiles. The Batak maternal grandparents often present a *suri suri* cloth or an *ulos mangiring*, credited with protective powers.[54] The Sasak of Lombok at this time offer some of their specially woven cloths. These will be saved for the oncoming childhood rites of passage.[55] The midwife who assisted at the birth may also be paid with food, money, and textiles.[56]

It was at one time very common in Burma and Thailand to 'purify and strengthen' the mother after delivery by placing her before a very hot fire for a number of days. At the time, she might be rubbed

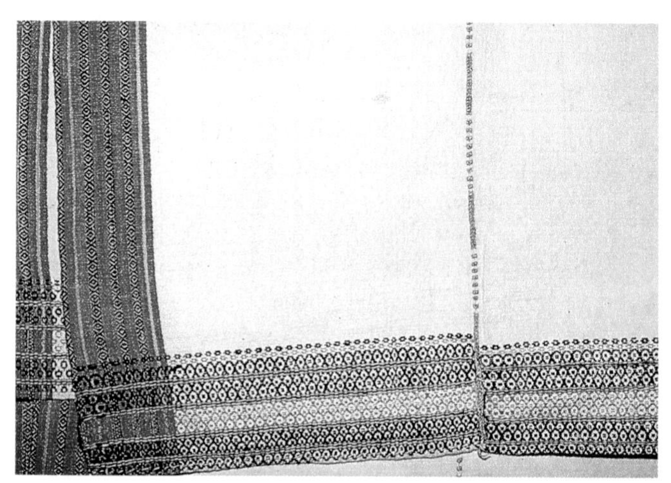

Fig. 102. This unassuming *inalson* blanket decorated with blue and red stripes and supplementary weft patterns was woven by the northern Tinguian for their *gipas* rituals held immediately prior to the birth of a child. Collection of Roland Goh, Imperial House of Antiques, Baguio.

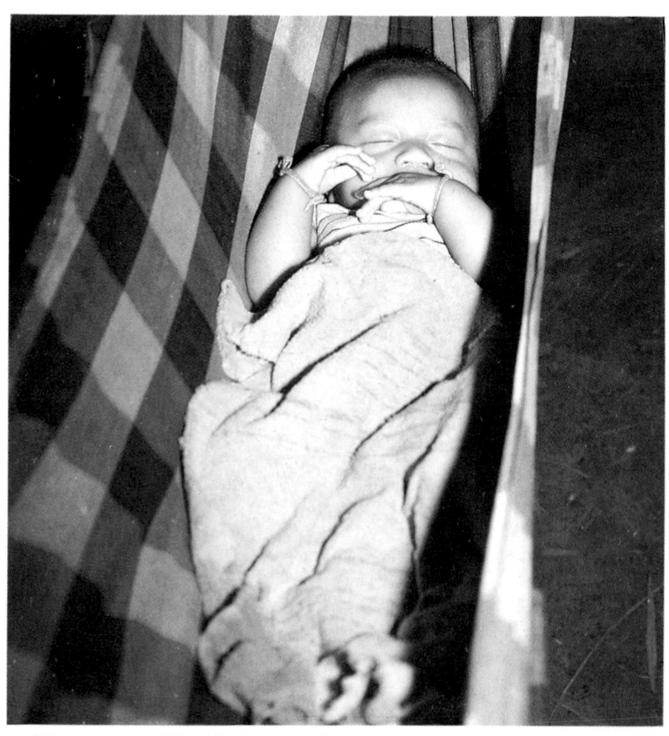

Fig. 103. A Thai baby in a hammock made from a *pha khaw ma*. Note the protective strings around the wrists.

with poultices and administered herbal potions by the midwife. During such a ceremony, the Lisu women of northern Thailand sit on an upturned pig trough inside a tent of blankets. In doing this, they bestow fertility on the livestock. Water is poured over heated stones to engulf the woman in steam. After a month of various taboos and restrictions, women are free to resume their usual load of responsibilities.[57]

In some Muslim and Buddhist societies, a child's first haircut is an important ceremonial event. Naming ceremonies are also important events. With the approach of puberty, a number of ceremonies are held to mark the transition from childhood to the threshold of manhood. When eleven to twelve years old, Muslim boys undergo a circumcision operation performed by a local doctor or circumcision specialist. The boy, dressed in his best clothes, is taken amidst much pomp and ceremony by male relatives to the house where the operation is to be performed. There is usually a bed or bench covered with special cloths and beautifully embroidered cushions. In front are containers filled with burning incense sticks, bowls for powdered leaves of various plants to be used as medicines and antiseptics, white cloth bandages, and surgical implements to carry out the operation.[58]

A Buddhist boy on approaching manhood, which can be at any time from nine to twenty, is expected to enter a monastery to become a monk, at least briefly, before he is considered ready to fulfil his obligations as a man. In keeping with the legend of the Buddha's life, young Burmese boys may arrive at the monastery on horseback (or even on an elephant), clad in rich kingly vestments. They are followed by their parents, relatives, and friends in grand procession bearing all the things that the young men will need as novices. These include items such as a begging bowl, monk's robes, blankets, sheets, pillows, an umbrella, and a fan. On arrival at the monastery, the fine clothing is discarded in favour of the monk's robe. The head is shaved and the proud parents catch the hair in a white cloth as it falls. They may keep it as a treasured momento or bury it in the monastery precincts.[59]

The attainment of womanhood is met with less pomp and ceremony. In former times, ear-piercing was a very common rite of passage for a young girl. In Burma, an ear-piercing or *nat win* ceremony is still held around age twelve or thirteen to symbolize a girl's attainment of womanhood. An auspicious day is chosen and the young participants, dressed in their best, have their ears pierced with golden needles in the presence of invited guests. Some elders utter a few words of greeting and advice to the young girls before everyone is fed at the conclusion of the ceremony.[60] In Muslim and Buddhist societies, young girls on reaching puberty are carefully chaperoned in public by elder female relatives. In hill tribe society, the opposite occurs. Young girls are encouraged to attend social events with their friends.

Tooth-filing used to be an important pre-adult ritual throughout much of South-East Asia. It was practised in many parts of Indonesia, the southern Philippines, and amongst some tribes of central South Vietnam. The fang-like appearance of certain teeth was considered unattractive, demonish, or animal-like. At the same time, some societies blackened their teeth with charcoal or lacquer. Tooth-filing is rarely practised today.

Courtship and Marriage

In traditional South-East Asia, by the time a girl was of marriageable age she was a proficient weaver. With her mother's help, she would have woven a set of clothes for her dowry which would include a full set of ceremonial textiles for herself and possibly some for her future groom. The young lady would weave herself sheets, blankets, pillows, and other textiles necessary to set up house, along with a number of extra pieces to serve as token gifts to future in-laws. Formerly, on the island of Lombok, women were expected to weave up to forty pieces of cloth for a trousseau.[61] Textiles in South-East Asia are regarded as 'female' goods and are seen as an expression of a woman's creative powers.[62] Textiles are suitable items to be used in exchange for goods, such as livestock and weapons. As far as prospective in-laws were concerned, a talented, hard-working weaver was always a welcome addition to the family.

Even today, as young men and women approach marriageable age, their clothing becomes more eye-catching and elaborate. A hill tribesman takes great care over his appearance and appears in public decked out in a colourful turban and lavishly embroidered jackets set off by brightly coloured tassels, sashes, and belts. Young women of all cultures dress in bright colours and add much embellishment in the form of silver and beads. They will spend much time applying touches of make-up and coiling their long dark tresses into the latest, most becoming style.

Marriages in South-East Asia are traditionally

arranged by the parents. Formerly, in some cases, a marriage could be contracted at birth with the bride and groom meeting on their wedding day. Today, free choice is widely practised, with young people meeting each other at work, at family and religious festivals, and at New Year celebrations. Virtually every cultural group celebrates its respective New Year in some shape or form. At this time of the year, members of the family receive new clothes and households stock up with food to offer as refreshments to guests. Everyone gets dressed up and goes visiting. Young people go around in groups and many new acquaintances are made at this time of the year.[63]

Regardless of how young people become acquainted, parental approval is necessary for a marriage to take place for young people are generally financially dependent on their family.[64] In the early stages of a courtship, astrologers may be consulted and horoscopes compared to gauge whether the proposed union is likely to be a happy one. Despite promises of connubial bliss, marriage is looked on primarily as an economic exchange and the union of two distinguished and mutually proud lineages. If the society is patrilineal, the bride's family will have to be adequately compensated for the loss of a daughter, her labour, and childbearing potential by a substantial 'bride price', which may be paid in items such as silver, livestock, and textiles. If the society is matrilineal, gift-giving weighs more heavily on the female side. Generally, both sides are expected to come up with suitable gift exchanges in keeping with the abilities and status of the respective bride and groom. Negotiations are careful and sometimes protracted: so as to extract the best deal possible, the families try to put on a good show to impress the other side with their wealth and social position.

Textiles have always played a key role in marriage negotiations. Initial contacts may be facilitated with various cloths. In former times on Sangir Island, a red cloth called *talimbuku* was carried to the prospective bride's house in procession by the elder female relatives of a potential groom. In central Sulawesi, a loose, coarsely woven textile thought to be imbued with magical powers, was used as a medium of exchange during marriage negotiations. On the island of Roti, a red cloth called *dela* was wrapped around ceremonial offerings during marriage negotiations.[65] The outcome of these negotiations sometimes involved the transfer of a large number of textiles, particularly if the families were wealthy or of noble status. One nineteenth-century account from Sumba cites that the bride's clan gave 'valuable beads, seven slaves, forty men's *ikat* mantles, and forty women's sarong' as their part of the nuptial exchange.[66]

The successful conclusion of negotiations were also marked by the circulation of cloths. For example, in Kupang, the capital of Timor, a groom once sent his engagement letter in a ritually folded cloth.[67] In upland South Sumatra, after the bride price had been paid, the groom's family presented glutinous rice cakes wrapped in a small ceremonial cloth, called *tampan*, to certain members of the bride's and groom's family who, later in the day, returned the cloths containing sweets to the groom's family.[68] In Cambodia, the period prior to the wedding was traditionally marked by periodic gift exchanges of betel-nut, *samphot* or sarong, and scarves. Scarves were a popular gift in this context. They symbolized the 'fixing of words and the tying of hearts' of an engaged couple.[69]

In modern Thailand, at the conclusion of negotiations, where in most cases the groom's family agrees to provide the house while the bride's family undertakes to donate the furnishings, a special ceremony continues to be held. It is called *sinsod tongman*, or the giving of gold and property to the fiancée. The groom and his retainers come to the bride's home carrying *kaan mak*, or ceremonial bowls covered in cloth, which contain money, gold, and sweets. At the same time, a payment is made 'for mother's milk' to the bride's family in recognition of their efforts in raising the girl.[70]

Throughout South-East Asia, after negotiations have been concluded and accounts settled, an auspicious day and time are set for the nuptials. The bride and groom, as 'king and queen' for the day, usually dress in full ceremonial costume (Fig. 104). The marriage rituals take place in either or both of the homes of the bride and groom. Amongst some peoples it is customary for the groom to 'buy' his way into the bride's house past real and imagined barriers, with tokens of money and textiles. For example, a female member of the family of a T'boli groom from the southern Philippines, on arrival at the bride's home removes a *kumo* blanket covering the bride. In this case, the relative is allowed to keep the blanket, but she must give the bride a gift of equivalent value.[71]

Most ceremonies include rituals which symbolize the union of the young couple through marriage. In Thai–Burmese Buddhist marriage ceremonies, the bride and groom are joined together with white thread. Led by a respected elderly couple with

1. A nineteenth-century double *ikat* silk *patola* cloth from Gujerat, India, featuring a *jelamprang* motif in the centre field and a *tumpul* border. Warp 450 cm, weft 100 cm. Collected on Flores. Arts of Asia Gallery, Denpasar.

2. Some Arakanese women from western Burma performing one of their national dances. They are dressed in traditional Burmese costume: silk *lon-gyi* with train, *ein-gyi* jacket, and *tabet* shawl.

3. At the conclusion of the night-long weaving competition held to re-enact a Buddhist legend, the newly woven robe is draped around a Buddha image at the Shwedagon pagoda. Photograph courtesy of Tin Maung Aye, Rangoon.

4. Burmese *acheik*-patterned silk *lon-gyi* from Amarapura featuring designs such as *nit sit lein kyo* (two twisted stripes), *nit sin ta si* (two stripes one bundle), *ta sin kyo gamoun* (one-stripe *gamoun*), and *thoun gamoun ywet set* (Burmese No. 3 joined with *gamoun* creeper). Chinese imported silk, chemical dyes. *C.*early twentieth century. Warp 170 cm, weft 113 cm.

5. Shan silk *lon-gyi* patterned with weft *ikat* and tapestry weave designs. *C.*late nineteenth or twentieth century. The Shan no longer make such intricately patterned textiles. Victoria and Albert Museum, London.

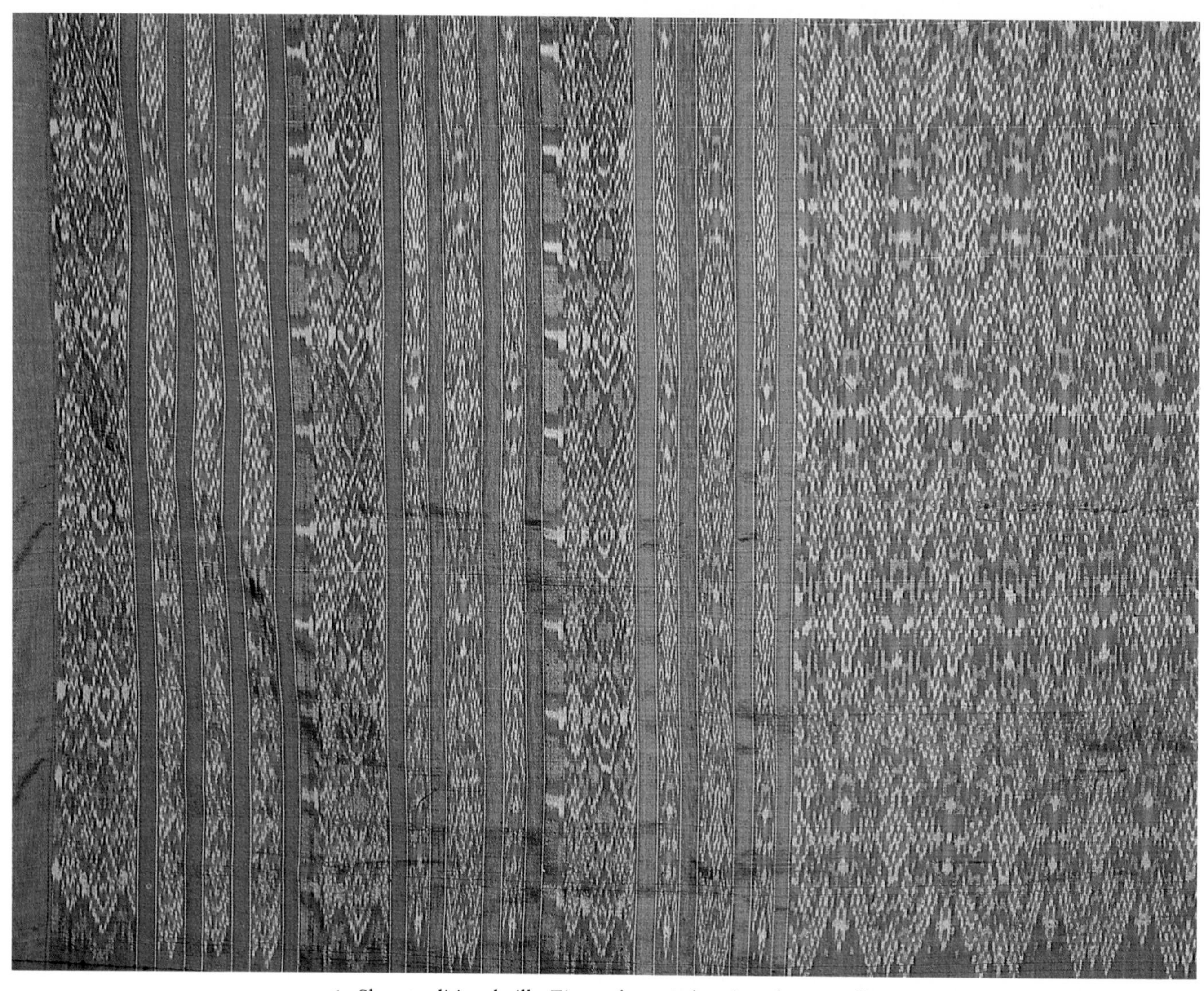

6. Shan traditional silk *Zin me lon-gyi* showing the two distinct design areas. It is from Lake Inle and is patterned with geometric weft *ikat* designs woven in a twill weave. Chemical dyes. *C.*mid-1920s. Warp 156 cm, weft 100 cm.

7. A Haka Chin gentleman wrapped in a traditional blanket enjoying an early morning pipe. Photograph courtesy of Tin Maung Aye, Rangoon.

8. Detail of a traditional Haka Chin silk blanket. It is made from two lengths of cloth decorated with regular bands of fine diamond and zig-zag motifs patterned by supplementary warps within different-coloured warp stripes. These stripes are bisected at one end by a single 9 cm wide strip of yellow woven in a twill weave. Warp 195 cm, weft 130 cm. Collection of Mr and Mrs Vum Ko Hao, Rangoon.

9. A temple fresco from Wat Phu Min temple at Nan which depicts nineteenth-century northern Thai villagers in intricately patterned sarongs. This temple was built by Chao Chetabut Phrohmin, the ruler of Nan, in 1596. It underwent renovations between 1867 and 1875. The temple frescoes date from that time.

10. A silk *mud mee pha sin* from Baan Suai, Surin. It is patterned in the *mee pai bai* or *mee hol* bamboo leaf design composed of rows of short parallel lines. Although it looks a deceptively simple pattern, up to eleven colours may be used to create this unique design. Local silk, chemical dyes, woven in a twill weave. Warp 170 cm, weft 99 cm.

11. North-eastern silk *mud mee pha sin* patterned with the *phla meuk* (squid) design. Note the intricate border pattern at the foot of the *pha sin*. Warp 170 cm, weft 100 cm. Collection of Khun Nisa Sheanakul, Bangkok.

(Opposite page)

14. Traditional Thai Lu silk *pha sin* patterned with colourful warp stripes and a band of *pha nahm laay* zig-zag tapestry weave, edged with a small band of purple *pha kit* supplementary weft geometrical designs on either side. Warp 112 cm, weft 110 cm.

12. *Pha kit* towel of unbleached cotton patterned with different-coloured supplementary weft threads depicting caparisoned elephants, *rajah singh* (Thai lions), and *hong* birds beside a plant. Warp 73 cm, weft 42 cm. Collection of the Siam Society, Bangkok.

13. *Poom som*, a weft *ikat*-patterned Thai textile from Chonnabot, north-east Thailand. In cloth layout, this former royal textile shows some similarities to Indian *patola* cloth. This particular example was woven by Khun Prappien Gaysorn in a twill weave with a Japanese silk warp and local silk for the weft. Chemical dyes have been used to create the *prajawong* weft *ikat* design. It takes two months to make a good quality *poom som* textile. Warp 364 cm, weft 100 cm.

15. Lao Phuan *tdinjok* from Hat Sieo which forms the border to a horizontally striped *pha sin*. Coloured silk threads on a red cotton ground. Warp 170 cm, weft 27 cm.

16. Lao Neua wall hanging depicting deer amidst geometric vegetal and floral forms. Mid-1930s. Collection of Khun Vanida, House of Handicrafts, Bangkok.

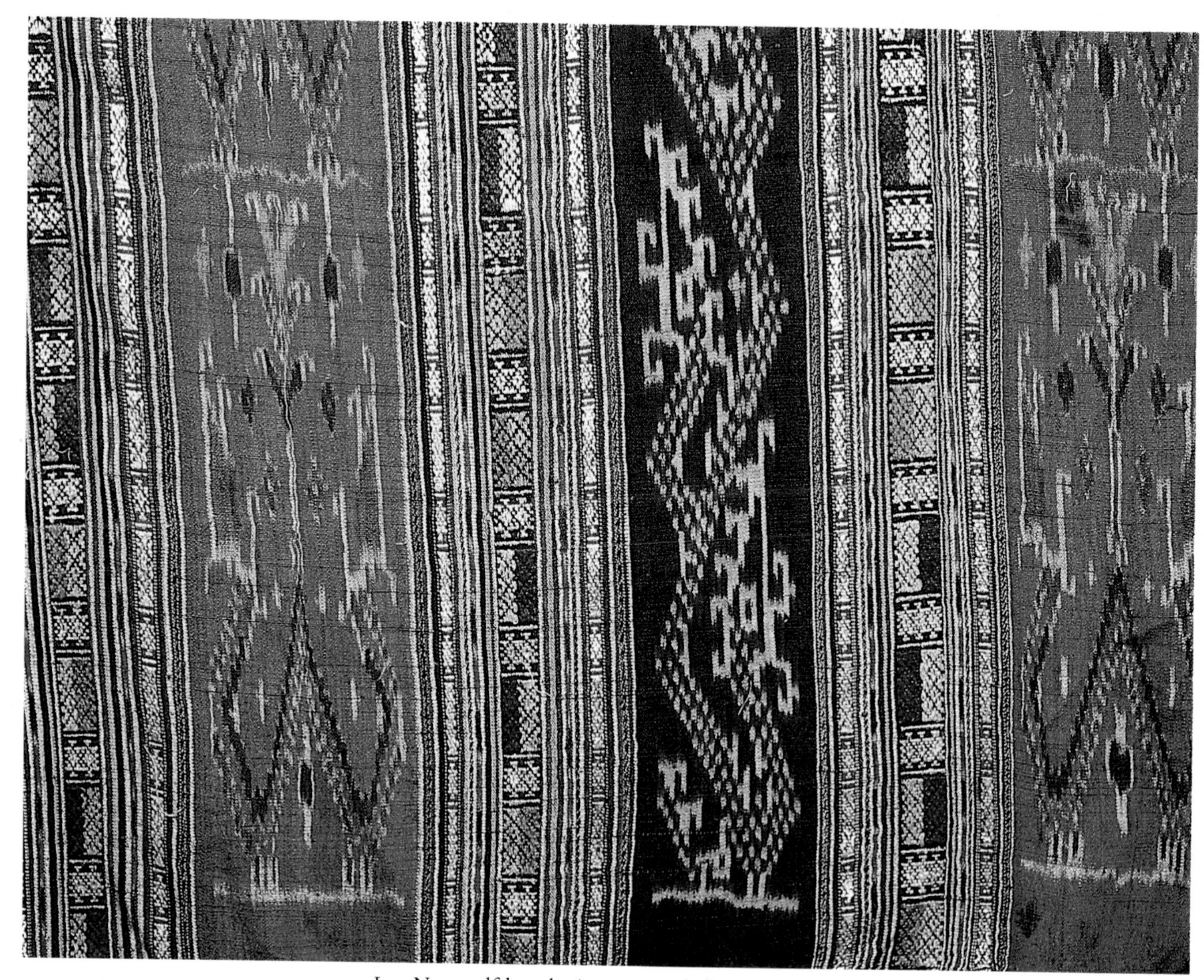

17. Lao Neua calf-length *sin* sarong which is patterned with alternating bands of different-coloured weft *ikat* and geometric supplementary weft decoration. The warp is of red silk. The weft of the red bands of *ikat* is of silk, while the weft of the indigo band is of cotton. The *ikat* bands depict the *nak* snake motif. Warp 128 cm, weft 66 cm.

(*Opposite page*)

18. Antique Lao Neua indigo *pha beang* shawl from Sam Neua. The indigo ground is of cotton patterned with supplementary weft silk yarns. It is possible to discern elephants with riders, *rajah singh* (long-nosed lions without mounts), pairs of snakes, and small *hong* birds amidst the abstract forms in the centre field. Note the sparing use of colour (from natural dyes) which has been strategically placed so as to draw attention away from the main forms. Warp 125 cm, weft 42 cm.

19. Cambodian silk weft *ikat*-patterned *samphot hol*. Collection of the National Museum of Natural History, Smithsonian Institution, Washington, DC, No. 409103.

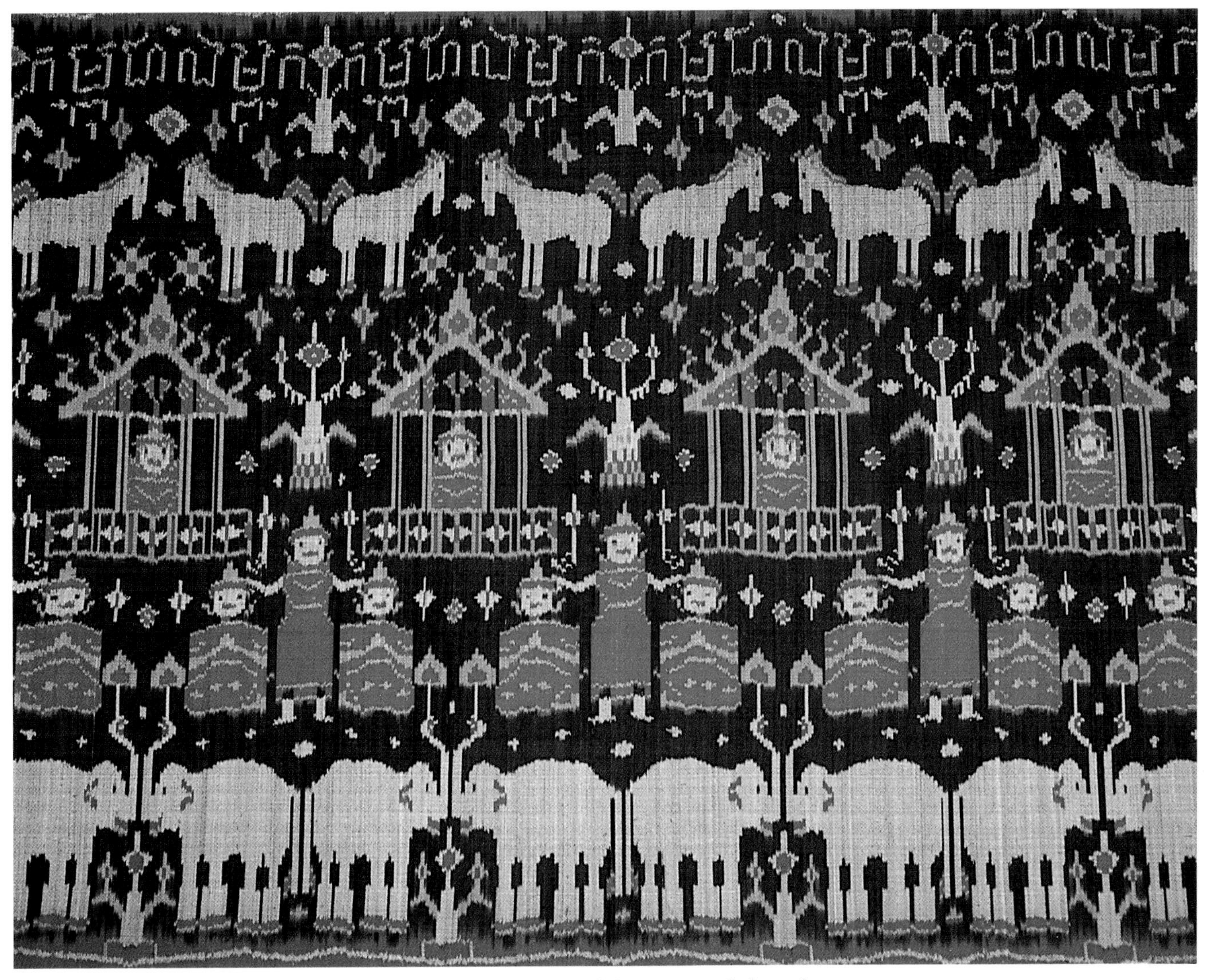

20. Modern Cambodian wall hanging made by refugees at a camp on the Thai border. Such a hanging was formerly put on display by the bereaved family during Buddhist funeral rites. Verses at the top of the hanging exhort the living to persist in good deeds. Warp 130 cm, weft 100 cm.

21. Jarai cotton loin cloth, the *toai*, which is patterned with lateral warp stripes and is finished with fringes and a band of very fine supplementary weft decoration. Darlac Province, Vietnam. Warp 480 cm, weft 30 cm. Collection of the National Museum of Natural History, Smithsonian Institution, Washington, DC, No. 415214.

Fig. 104. A Minangkabau couple from Lima Puluh Kota, West Sumatra, in traditional wedding costume, photographed in front of a sumptuous display of textiles. Photo courtesy of Rumah Adat Baandjuang Museum, Bukit Tinggi, West Sumatra.

numerous children and grandchildren, the guests at a Thai wedding take turns pouring lustral water over the hands of the newly betrothed.[72] In Burma, the guests watch as the couple eat together; they may then pelt them with grains of coloured rice.[73] An Ifugao couple are married under a special blanket.[74] At one stage in the Yakan ceremony, the groom gives the bride his mantle,[75] and the giving of the *ragi hotang* to a young couple by the bride's father is the high point in a traditional Batak wedding ceremony. An even more prestigious cloth is wrapped around the groom's mother by the bride's family to unite, metaphorically, the two lineages.[76]

Some ceremonies signify a new life and status for the young couple. At a Karen wedding in Thailand, the groom comes to the wedding in old clothes. On arrival, he changes into a new outfit that has been specially woven for him by his bride. The bride changes from the white shift dress that she wore as a single girl into the skirt and blouse of a married woman.[77]

Most Muslim and some Buddhist wedding formalities end with the preparation and blessing of the bridal bed, which is often a huge four-poster festooned with a cloth canopy and numerous costly textile hangings, rich covers, and mounds of embroidered satin pillows. In Java, many of the textiles are treasured family heirlooms, which carry with them connotations of well-being, fertility, and continuity of the generations.[78] In Thailand, various auspicious objects, such as peas and sesame seeds representing fertility, water for purity, a gourd for peace and happiness, and coins for wealth, are placed on a tray. After being blessed by respected elders with numerous progeny, they are left on the bed.[79]

Attaining a Position

Being elevated to a position which bestows new dignity and status, can also be considered a transition. Human beings derive much satisfaction and a special sense of belonging by obtaining special ranks or titles. Distinguished warriors and hunters in South-East Asia have often been accorded certain honours and privileges and, in some cases, they wear special clothing. In the southern Philippines at one time, a Mandaya warrior who had taken ten heads was allowed to wear a red shirt with brown sleeves.[80] Bagobo and Mandaya warriors also wore *pelangi*-spotted red *tangkolo* head-cloths.[81] When a warrior died, special ceremonies would be performed in recognition of his contribution to the preservation of the tribe.[82]

Throughout his life, a Mien male assiduously amasses money, livestock, and goods so that he can work his way up through the complicated rituals of the Taoist religious hierarchy revered by the Mien, to achieve status and respect for his immediate family and his ancestors in the spirit world. To attain the *tou sai* rank, a man must abstain from food and other desires of the flesh for a period, before climbing up a sword ladder covered with white cloth representing the 'pathway to heaven'. Women dressed in all their finery attend these ceremonies and proudly share in their husband's promotion.[83]

Textiles also played a role when a man in the Lampung area of South Sumatra was invested with the title of *Penyimbang*, or 'Chief', by the Sultan of Bantam. This title was originally hereditary, but during the latter years of the nineteenth century it could be purchased for a vast sum of money (usually derived from the profitable pepper trade). At his accession, a chief had to 'raise the *papadon*', an elaborate four-legged wooden chair, which was an important symbol of chieftainship. The recipient of the title was driven by chariot to the village community centre in grand procession. In front walked two women bearing a large ship cloth, a textile whose use was limited to chiefs. During the ceremony, the *papadon* was also covered with a ship cloth. The ceremony took place in front of poles festooned with shells, cloth, and coins to represent trees. These were scaled and relieved of their contents by eager children at the end of the ceremony.[84]

To reap social prestige and respect, a wealthy Ibaloy in the northern Philippines at one time held a *peshit*, a celebration which included the slaughter of large numbers of pigs, cattle, and buffalo, and the preparation of a grand feast before kinsfolk, neighbours, and certain influential invited guests, most of whom had previously held a *peshit*. Ceremonies included the blessing of the animals by a pagan priest prior to the slaughter, followed by dancing and feasting. Up to ₱ 20,000 (US$1,000) might be spent in holding this ceremony, which offered tangible proof that the sponsor was of considerable means.[85] It entitled him to wear a red head-cloth as a symbol of wealth. His wife was also permitted to wear a prestigious multi-layered *devit* wrap-around skirt instead of the plain one reserved for commoners.[86]

The End of the Cycle

The important role that textiles play in rites of passage is most evident in rituals associated with death. Held at properly auspicious times, burial and cremation ceremonies are traditionally 'great events' amongst South-East Asian peoples. The rituals reflect the importance of ancestor cults and beliefs concerning the afterlife. Ceremonies involve rites requesting forgiveness from the deceased, animal sacrifices, feasting, exchanges of gifts, and the building of megaliths. If the person is of high status, the burial or cremation may be delayed a year or so to allow time for appropriate preparations to be made.

To ensure that the last rites will be correctly performed and to ease their way into the afterlife, many individuals make preparations well before their demise. These preparations include making sure that the appropriate textiles will be on hand. In the northern Philippines, the Ifugao, Kankanay, Ibaloy, and Tinguian weave special blankets as burial shrouds. Amongst the Kalinga, death blankets can only be woven by women who are barren or past their childbearing years.[87] Among these peoples, it is the custom to wrap the body in as many blankets as the family can afford. A rich man may commission up to six specially woven blankets, while a poor man will do his best to see that he has at least one. The Ifugao make special *ikat*-patterned loin cloths for the dead, while the Kankanay make a loosely woven blue burial jacket called *losodan*, which is decorated with anthromorphic figures in warp *ikat*.[88] In Sumba, the wealthy collect burial textiles both to present as gifts to mourners and to surround the corpse for use in the afterlife.[89]

At the time of death, the body is washed, laid out, and dressed in its best garments. In the case of a Hmong tribesman, the children and grandchildren clothe the body in special hemp burial clothing, which is made by the women for themselves and their husbands. This clothing is beautifully ornamented with layers of appliqué and embroidery. Custom also decrees that special 30–35 cm square cloths made by sisters, daughters, and nieces be placed under the head of the deceased as 'pillows'. A sash may be added to the corpse of a man and embroidered collars to the body of a woman.[90] The Akha, before placing the body in a wooden, boat-shaped coffin, put in tufts of cotton 'to keep the person warm'. 'Tears' of the deceased may be also dried with cotton before the corpse is placed in the coffin. Several sets of clothing are sometimes laid on the body.[91]

Shrouds are among the most important death textiles. In Timor, warp *ikat* cloths cover the coffin prior to interment. The Batak use blue and red *ikat* cloths as a cover for those lying in state.[92] In the village of Ban Na Mun Sri in southern Thailand, a special textile called *pha pan chang*, a 1–2 m long cloth woven in two to three parts is placed on the coffin. It is patterned with Thai verses in a supplementary weft, which exhort the living to persist in good deeds.

> Keep to good deeds and thoughts.
> Death comes without warning.
> Birth and death are the common lot.
> After death only the merit of good deeds matters.[93]

The Toradja, in addition to covering the deceased with a cloth, use large *ikat*-patterned textiles as funeral decorations, which are strung around bamboo pavilions especially erected to house the mourners at funeral rites. In the northern Philippines, textiles may be prominently displayed around the corpse for a number of reasons. Some cloths serve as grave goods, while others have a protective purpose. The Tinguian believe that a spirit which wishes to harm members of the family, must first count all the threads in the textiles and holes in the nets before it can carry out any evil intent.[94] By using numerous textiles for the funeral rituals, they hope that the ceremonies will be concluded before danger arises. A fitting display of textiles also demonstrates that the deceased and his family are people of substance. In Jolo in the southern Philippines, a Muslim of noble birth lies in state in a house decorated with fabrics appliquéd with the Tree of Life design. His best clothes are piled in layers on the bed.[95]

Mourners, too, wear special textiles. In Thailand, it is *de rigueur* for women to wear black and men to wear either a dark suit or a white shirt with a black armband to Buddhist cremations. There are still a few itinerant dyers who ply their trade around the back streets of Bangkok. For a small fee, they will dye used clothing black.[96] At a traditional South-East Asian Chinese funeral, close relatives of the deceased wear coarsely woven white cloth. White cloths with messages of condolence written in black characters hang from the buses carrying mourners to the funeral. White cloth is also used for mourning head-cloths amongst the Balinese and Mien peoples. Toradja women wear a special cowl-like hood to funerals, while the men wear narrow woven head-bands with a single tassel which falls over the fore-head.[97] Priests and shamans officiating at funeral rites also wear special clothing. The Batak, in keeping with their custom of giving textiles at rites of passage, present a special ritual cloth, the *sibolang*, a pale indigo to black cloth with zones of closely spaced warp *ikat* patterns, which for mourning is worn on the head (see Fig. 219).[98]

Textiles also play an important role in some funeral ceremonies. The Mien make a ritual boat out of banana leaves, and various symbolic articles are placed inside before it is tossed into the forest (which represents the sea). A long scroll is used as a bridge, and a length of blue cloth leads out of the house to the roof to 'show' the spirit the way to the land of the ancestors.[99] Women sometimes follow the funeral cortege with a strip of white cloth attached to the coffin. This cloth symbolizes the 'bridge' by which the soul of the deceased ascends to heaven.[100]

In Thailand, a sacred cord is tied around the casket and a monk holds the end of the cord to symbolize 'pulling' the casket to the site of the cremation. Members of the family take the opportunity to present *pangsakula* robes to the monkhood by laying them on the casket, where they are taken by monks as a symbolic reminder of an injunction by Lord Buddha, who instructed his followers to obtain their clothing from the dead or from graveyards.[101] After a Hindu/Buddhist cremation, the remains are collected, laid on a white cloth shaped in the form of a human body, and sprinkled with gold and silver coins. The coins are considered lucky, and are kept by close relatives, while the ashes are placed in an urn for the family to keep or dispose of.[102]

In Timor, royal funerals used to be remembered in terms of the number of animals slaughtered and the number of textiles buried with the deceased. Some funerals were occasionally concluded with a display of competitive gift-giving between the family of the deceased and the immediate in-laws. In Sumba, over 100 textiles might be interred with a deceased noble, while an equal number would be given to guests and retainers who participated in the extensive funeral rites.[103] Sumba *hinggi* blankets often served as symbolic sails when mounted on the top of huge stone boulders as they were hauled from the quarry to the ritual centre of the village to be set up as ancestral gravestones.[104]

Ceremonies occasionally continue after burial. Twelve days after the death, the Lahu Nyi of Thailand hold a special ceremony in which they build a small hut for the deceased halfway between the village and the grave. A new set of clothes and a sacrificed chicken are left there with an order for the spirit to vacate the site, refrain from molesting the living, and make haste to the underworld.[105] Some groups, such as the Ifugao, practise secondary burial. After a few years, the remains are disinterred, wrapped in a new blanket, and placed under the eaves of the house until sufficient resources have been accumulated to provide a fitting secondary burial (Fig. 105). The Ibaloy at one time embalmed the body of the deceased by a smoking and drying process. The body was covered in expensive blankets before being put in a wooden coffin and placed in a cave. Remains of these mummified figures may be seen at Kabayan and Sagada in Benguet Province in the northern Philippines.[106]

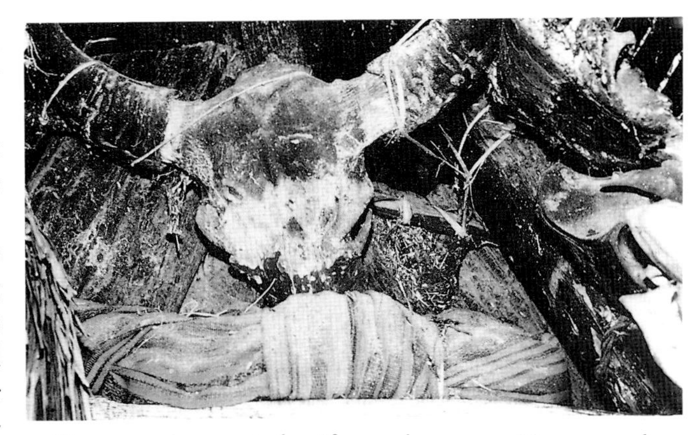

Fig. 105. Amongst the Ifugao, bones awaiting secondary burial may be wrapped in a *ga'mong* blanket and placed under the eaves of the house until sufficient resources have been gathered for a fitting reburial. Banaue.

1. Pat Justiniani McReynolds, 'Sacred Cloth of Plant and Palm', *Arts of Asia*, July–August 1982, p. 98.

2. Sylvia Fraser-Lu, *Indonesian Batik, Processes, Patterns and Places*, Oxford University Press, Singapore, 1986, p. 24. See line drawing.

3. Chira Chongkol, 'Textiles and Costumes in Thailand', *Arts of Asia*, November–December 1982, p. 125 (photograph).

4. Mattiebelle Gittinger, *Splendid Symbols, Textiles and Tradition in Indonesia*, Textile Museum, Washington, DC, 1979, p. 53.

5. Robert Mole, *The Montagnards of South Vietnam*, Tuttle, Tokyo, 1970, p. 180.

6. Gittinger, *Splendid Symbols*, p. 179.

7. Ibid.

8. Ibid., p. 63. Due to the relative remoteness and isolation of the Toradjas, Chinese influence is debatable. It is quite possible that the Toradja developed this style of blouse independent of outside sources.

9. Paul Lewis and Elaine Lewis, *Peoples of the Golden Triangle*, Thames and Hudson, London, 1984, p. 117, photographs.

10. Henny Harald Hansen, 'Some Costumes of Highland Burma at the Ethnographical Museum of Gothenburg', *Etnologiska Studier*, Vol. 24, Goteborg, 1960, pp. 28–30.

11. *National Dress of Peninsula Malaya* (in English and Bahasa Malaysia), Perbadanan Kemajuan Kraftangan, Malaysia, n.d. This publication states that the basic form of the *kurung* might have been influenced by the *galabiah* of the Middle East, and that the word *kebaya* is believed to have come from the Arabic *habaya*, meaning a long tunic open down the front. This publication also states that prior to the introduction of these two garments Malay women wore only the sarong with its top secured above the breasts.

12. Ibid.

13. Hansen, op. cit., pp. 35–6.

14. Denis Segaller, *Thai Ways*, Allied Newspapers Ltd., Bangkok, 1979, pp. 68–70.

15. Maung Htin Aung, *Folk Elements of Burmese Buddhism*, U Hla, Rangoon, 1959, p. 40.

16. Khin Myo Chit, *A Wonderland of Burmese Legends*, Tamarind Press, Bangkok, 1984, p. 64.

17. Mary Elana Ellis, 'Life Cycle Ceremonies: Courtship and Marriage', in *Sawaddi Special Edition. A Cultural Guide to Thailand*, American Women's Club of Thailand, Bangkok, 1978?, p. 76. The sacred thread is Brahmanical rather than Buddhist in origin. Since Lord Buddha did not leave clear instructions on the conduct of ceremonies pertaining to everyday life, Buddhist followers in South-East Asia, in many instances, have tended to follow pre-Buddhist Hindu practices and at times combine them with early animistic beliefs.

18. Shway Yoe (Sir George Scott), *The Burman, His Life and Notions*, MacMillan, London, 1896, reprinted Norton Simon, New York, 1963, p. 235.

19. Bronwen Solyom and Garrett Solyom, 'Notes and Observations on Indonesian Textiles', in Joseph Fischer (ed.), *Threads of Tradition, Textiles of Indonesia and Sarawak*, exhibition catalogue, Lowie Museum of Anthropology and the University Art Museum, Berkeley, California, 1979, p. 19.

20. Eric Crystal, 'Mountain Ikats and Coastal Silks: Traditional Textiles in South Sulawesi', in Fischer (ed.), *Threads of Tradition*, p. 58.

21. Rita Bolland and A. Polak, 'Manufacture and Use of Some Sacred Woven Fabrics in a North Lombok Community', *Tropical Man*, Vol. 4, 1971, pp. 149–79.

22. Laurence A. G. Moss, 'Cloths in the Cultures of the Lesser Sunda Islands', in Fischer (ed.), *Threads of Tradition*, p. 64.

23. Gittinger, *Splendid Symbols*, pp. 144–5.

24. Gerald C. Hickey, *Free in the Forest, Ethnohistory of the Central Highlands, 1954–1976*, Yale University Press, New Haven, 1982, p. 152.

25. Phya Auman Rajadhon, *Essays on Thai Folklore*, Duang Kamol, Bangkok, 1968, p. 280.

26. Andrew D. Sherfan, *The Yakans of Basilan; Another Unknown and Exotic Tribe*, Photomatic, Cebu City, 1976, pp. 160–1.

27. Niam S. Djoemena, *Batik, Its Mystery and Meaning*, Penerbit Djambatan, Jakarta, 1986, p. 12.

28. Moss, op. cit., p. 64.

29. Bronwen Solyom and Garrett Solyom, 'Bali', in Mary Hunt Kahlenburg (ed.), *Textile Traditions of Indonesia*, Los Angeles County Museum of Art, Los Angeles, 1977, p. 73.

30. Crystal, op. cit., p. 60.

31. Cherie French, 'Thai Fashions Then and Now', in *Sawaddi Special Edition. A Cultural Guide to Thailand*, American Women's Club of Thailand, Bangkok, 1978?, p. 127.

32. Thomas Fitzsimmons, *Cambodia*, Country Survey Series, Human Relations Area Files Press, New Haven, 1957, p. 285.

33. Segaller, *Thai Ways*, p. 31.

34. Tassilo Adam, 'The Art of Batik in Java', *Bulletin of the Needle and Bobbin Club*, Vol. 8, No. 2, pp. 63–4.

35. Hickey, op. cit., p. 61.

36. Gittinger, *Splendid Symbols*, pp. 41–2.

37. Ibid.

38. Marian Pastor Roces, 'The Fabrics of Life', *Habi: The Allure of Philippine Weaves*, brochure, Museum Division of the Intramuros Administration, Manila, n.d., p. 22.

39. Gabriel Casal *et al.*, *People and Art of the Philippines*, Museum of Cultural History, University of California, Los Angeles, 1981, p. 224.

40. Ibid.

41. Hickey, op. cit., p. 43.

42. Gittinger, *Splendid Symbols*, p. 201.

43. Paramita Abdurachman, 'Spinning a Tale of Yarn', *Garuda Magazine*, Vol. 3, No. 3, 1983, p. 24.

44. Bronwen Solyom and Garrett Solyom, *Textiles of the Indonesian Archipelago*, exhibition catalogue, University Press of Hawaii, 1973, pp. 5–6; and Gittinger, *Splendid Symbols*, p. 25. According to the Solyoms, it is not clear whether the situation has been resolved.

45. Margaret Campbell, *From the Hands of the Hills*, Media Transasia, Bangkok, 1978, 2nd edn., 1981, p. 60.

46. Sherfan, op. cit., p. 29.

47. Marelyn Tank, 'Life Cycle Ceremonies, Minus Nine to Plus Twelve', in *Sawaddi Special Edition. A Cultural Guide to Thailand*, American Women's Club of Thailand, Bangkok, 1978?, p. 85.

48. Gittinger, *Splendid Symbols*, p. 19.

49. Solyom and Solyom, *Textiles of the Indonesian Archipelago*, p. 11.

50. Casal *et al.*, op. cit., p. 226.

51. Michael Hitchcock, *Indonesian Textile Techniques*, Shire Publications, Aylesbury, England, 1985, p. 21.

52. Michael Palmieri and Fatima Ferentinos, 'The Iban Textiles of Sarawak', in Fischer (ed.), *Threads of Tradition*, p. 73.
53. Rajadhon, op. cit., pp. 253 and 282.
54. Mattiebelle Gittinger, 'Sumatra', in Kahlenburg (ed.), *Textile Traditions of Indonesia*, p. 25.
55. Gittinger, *Splendid Symbols*, p. 28.
56. Sherfan, op. cit., p. 41.
57. Lewis and Lewis, op. cit., p. 288.
58. Ganesha Volunteers, *Aspects of Indonesian Culture, Java and Sumatra*, Ganesha Society, Jakarta, 1979, pp. 65–7.
59. Khin Myo Chit, *Burmese Scenes and Sketches*, Nilar Publications, Rangoon, 1977, pp. 170–1.
60. Helen Trager (ed.), *We the Burmese, Voices from Burma*, Praeger, New York, 1969, pp. 140–1.
61. Director of Museum Development, *A Short Guide to Museum Negeri Nusa Tengara Barat* (in Bahasa Indonesia and English), Mataram, Lombok, 1985/6, p. 49.
62. Gittinger, *Splendid Symbols*, p. 39.
63. For Muslims, the New Year (Lebaran or Hari Raya) falls after the fasting month; for Buddhists it falls in March–April. The hill tribes of mainland South-East Asia and the Vietnamese celebrate the Chinese New Year (or Tet) in January–February. The Ibans of Sarawak celebrate Gawai Dayak in June.
64. Elopements do sometimes occur. The families are usually reconciled after a period of time.
65. Gittinger, *Splendid Symbols*, p. 24.
66. Ibid., p. 21.
67. Ibid., p. 24.
68. Mattiebelle Gittinger, 'The Ship Textiles of Southern Sumatra: Functions and Design System', *Bijdragen, tot de Taal-Land -en Volkenkunde*, Vol. 132, 1976, p. 211.
69. Fitzsimmons, op. cit., p. 267.
70. Ellis, op. cit., p. 75.
71. Gabriel Casal, *T'boli Art*, Ayala Museum, Makati, Manila, 1978, p. 75.
72. Ellis, op. cit., p. 75.
73. Trager, op. cit., p. 157.
74. Roces, op. cit., p. 22.
75. Sherfan, op. cit., p. 80.
76. Gittinger, *Splendid Symbols*, p. 20.
77. Lewis and Lewis, op. cit., p. 88. If a young girl dies before she is married, she will be buried in the clothes of a married woman to avoid being molested *en route* to the underworld.
78. Gittinger, *Splendid Symbols*, p. 29.
79. Ellis, op. cit., pp. 77–8.
80. Roberto de los Reyes, *Traditional Handicraft Art of the Philippines*, Casalinda Books, Manila, 1975, p. 65.
81. Casal *et al.*, op. cit., p. 137.
82. Gittinger, *Splendid Symbols*, pp. 179 and 226.
83. Jacques Lemoine, *Yao Ceremonial Painting*, White Lotus, Bangkok, 1982, pp. 25–8; and Campbell, op. cit., p. 46.
84. Toos van de Djik and Nicole de Jonge, *Ship Cloths of the Lampung South Sumatera*, Gallerie Mabuhay, Amsterdam, 1980, pp. 20–1.
85. Gabriel Pawd Keith and Emma Baban Keith, *A Glimpse of Benguet Culture and Artifacts*, Hilltop Printing Press, Baguio, 1981, pp. 30 and 37.
86. Information from displays at the Benguet Provincial Museum, Baguio, visited by the author, 28 April 1986.
87. Roces, op. cit., p. 1.
88. Casal *et al.*, op. cit., p. 226.
89. Gittinger, *Splendid Symbols*, p. 21.
90. Lewis and Lewis, op. cit., p. 128.
91. Ibid., p. 232.
92. Joseph Fischer, 'The Value of Tradition: An Essay on Indonesian Textiles', in Fischer (ed.), *Threads of Tradition*, p. 13.
93. Vilmophan Peetathawatchai, *Folkcrafts of the South*, Housewives' Voluntary Foundation Committee, Bangkok, 1976, p. 124.
94. Casal *et al.*, op. cit., p. 226.
95. Roces, op. cit., p. 22.
96. Denis Segaller, *Traditional Thailand: Glimpses of a Nation's Culture*, Hong Kong Publishing Co., Hong Kong, 1982, p. 53.
97. Crystal, op. cit., p. 60.
98. Mattiebelle Gittinger, 'Selected Batak Textiles, Technique and Function', *Textile Museum Journal*, Vol. 4, No. 2, 1975, pp. 26–7.
99. Lewis and Lewis, op. cit., p. 167.
100. Takuji Takemura, 'Funeral Rites', in Yoshiro Shiratori, *Ethnographic Survey of the Hill Tribes of Northern Thailand with Special Reference to the Yao*, Kodansha, Japan, 1978, p. 329.
101. Segaller, *Thai Ways*, pp. 42–3; and Violet Turner, 'The End of the Cycle', in *Sawaddi Special Edition. A Cultural Guide to Thailand*, American Women's Club of Thailand, Bangkok, 1978?, p. 83.
102. Fischer, op. cit., p. 11; and Turner, op. cit., p. 83.
103. Gittinger, *Splendid Symbols*, pp. 21–2.
104. Marie Jeanne (Monni) Adams, *Leven en Dood op Sumba* (Life and Death on Sumba), exhibition catalogue, Museum voor Land- en Volkenkunde te Amsterdam, 1965–6, p. 23 (photograph).
105. Lewis and Lewis, op. cit., p. 192.
106. Gabriel Pawd Keith and Emma Baban Keith, *A Glimpse of Benguet, Kebayan Mummies*, Hilltop Printing Press, Baguio, 1983, pp. 23–4.

PART II

A Survey of Textiles by Country

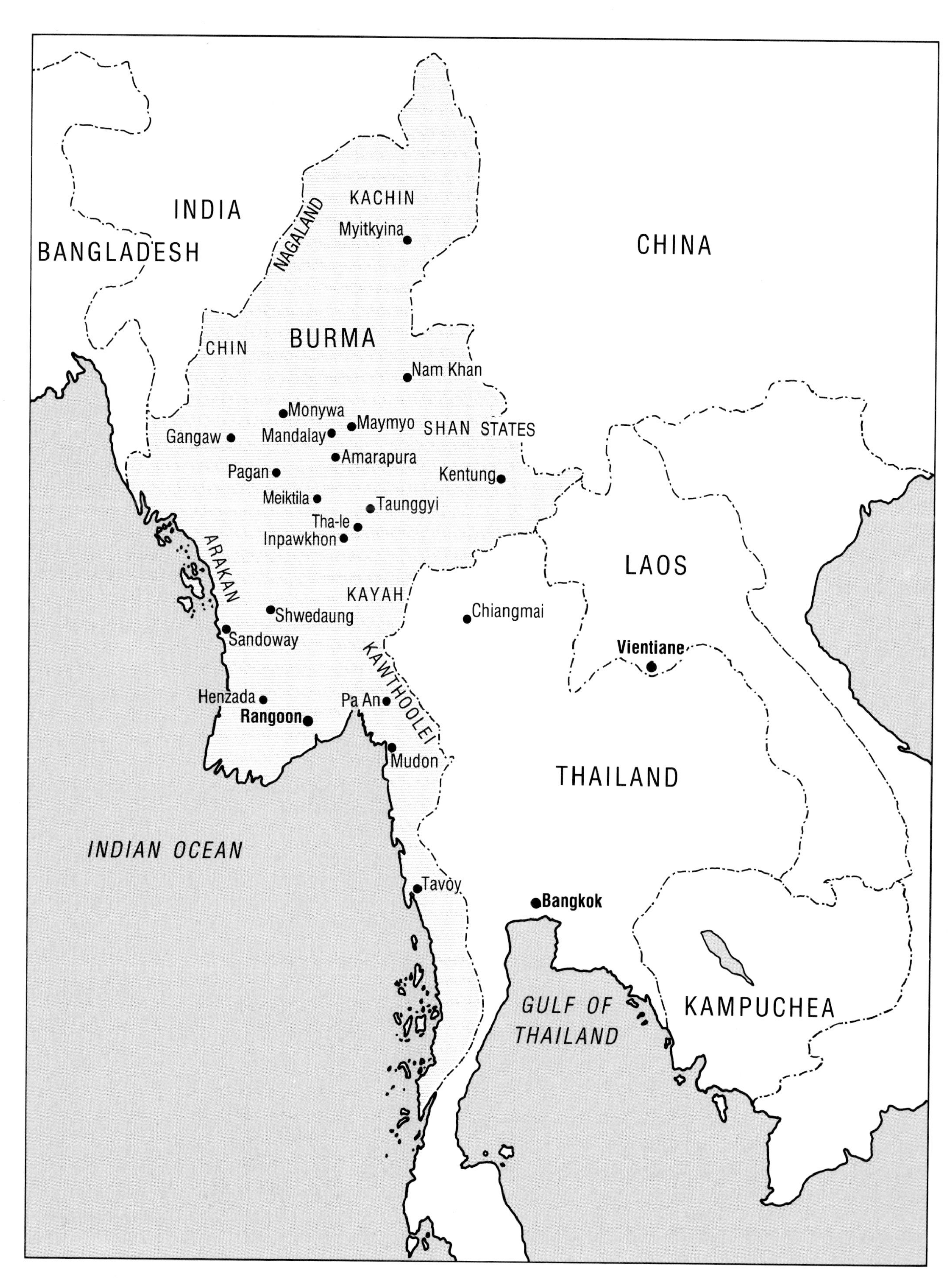

Burma: Main Textile Centres

CHAPTER FIVE

Burma

BURMA, which consists of a central plain surrounded by mountains, is home to over sixty ethnic groups, many of whom have a tradition of weaving their own distinctive textiles. Chinese sources have left us descriptions of clothing worn by the earliest Burmans, the Pyu (*c.* AD 500–900).[1] Lithic inscriptions of the Pagan Dynasty (AD 1044–1287) contain references to cotton, silk, and clothing.[2] Fragments of painted cloth banners dating to Pagan times may be seen in the Pagan Museum. The Archaeological Department at Pagan has a cloth wrapper for a Burmese religious manuscript which dates to the late Ava period (1635–1752) (Fig. 106). Eighteenth- and nineteenth-century European travellers to Burma, such as Fytche, Yule, and Symes have all provided us with colourful accounts of Burmese weaving. They comment on how widespread weaving was in earlier times, with virtually every family having a loom under the house.[3] The importance of weaving in Burmese culture is reflected in the fact that the Burmese five-*kyat* note, which was legal tender in the 1950s and 1960s, depicts a spinner and weaver on one side of the note (Fig. 107).

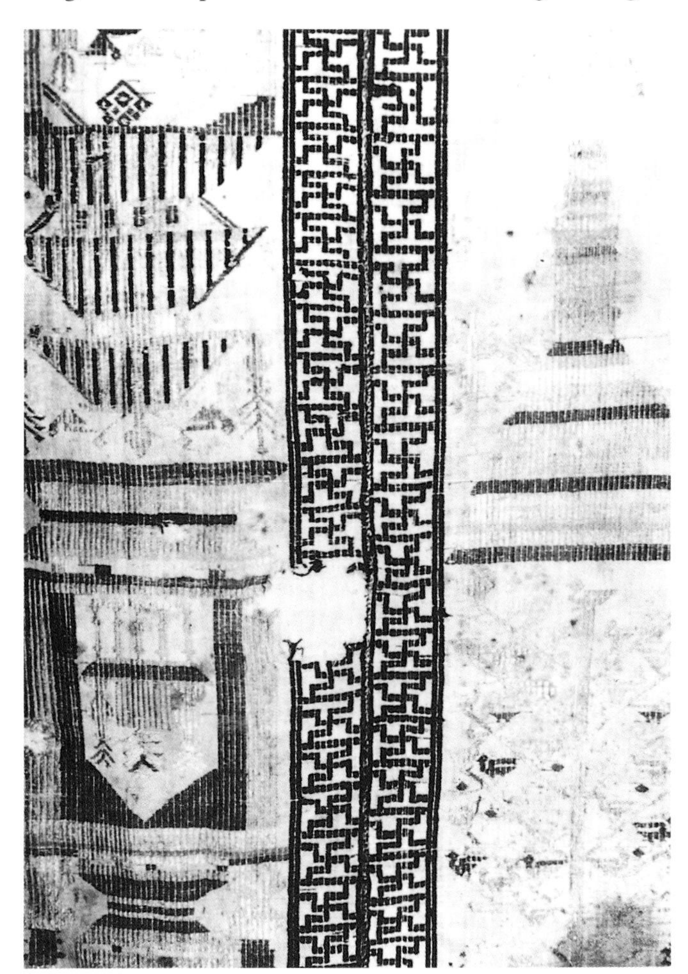

Fig. 106. A *ka-ba-lwe* cotton cloth wrapper for religious manuscripts which is patterned with pagodas in a supplementary weft. This example is thought to date back to the late Ava period (1635–1752). Department of Archaeology, Pagan.

While weaving is not as widespread today, the official Burmese policy of economic self-sufficiency and minimal foreign contacts, combined with the custom of wearing traditional dress, has created a continuing demand for local textiles. However, shortages of essential materials, such as dyes and yarn, have hampered production in many areas. In addition, local textiles also face stiff competition from cheap imported fabrics which are smuggled into Burma from neighbouring countries.

Fig. 107. Burmese five-*kyat* note depicting a girl spinning with a weaver in the background.

Cotton, silk, hemp, and wool are the traditional fibres used to produce cloth in Burma. Cotton continues to be an important commercial crop in the Mandalay/Sagaing area and small quantities are still grown by some hill tribe people on slash-and-burn plots. Silk is largely imported, while the cultivation of hemp is confined to remote hill tribe groups. Some of the northernmost hill tribe people, such as the Kachin, use wool or acrylic yarns for the weft of their traditional garments.[4] Apart from the use of indigo and lac by a few hill tribe groups, dyes currently used are mainly synthetic. Although a few hill tribes continue to use a body-tension loom, most weaving in Burma today is done on a frame loom. *Ikat*, tapestry weave, and supplementary weft patterning techniques are known to the peoples of Burma. In the nineteenth century, a batik process using a wax resist was also used in Burma to create cheap imitations of expensive locally woven products.[5] Burma is famous for embroidered wall hangings, called *kalaga*, decorated with appliqué, couched threads, sequins, and glass beads on a velvet ground.[6]

Burmese traditional dress for women consisted of a long rectangle of cotton or silk cloth approximately 60 cm by 130 cm, called the *hta-mein* (Fig. 108). It was worn tucked in at the waist and folded in front so as to slightly overlap and display a portion of the leg when walking. A length of light-coloured cloth, with horizontal stripes at the top, was attached to the base of the *hta-mein* to form a 25 cm train behind the wearer. The graceful and modest management of this train while walking and dancing was considered an important accomplishment of a Burmese woman. The *hta-mein* was worn with a long breast cloth, called *tabet*, over which was worn a semi-transparent tight fitting jacket, or *ein-gyi*, of lace or muslin (Plate 2).[7]

Men traditionally wore the *pah-soe*, an unusually long piece of cloth (approximately 700 cm by 45 cm) which was made from two lengths of cloth joined together (Fig. 109). It was worn according to the activity. For formal wear, it was wrapped around the waist to make a sarong with a large piece of spare cloth draped in folds down the front. The excess cloth could be draped over the shoulder when working. For sports, it would be passed through the legs to form bulky breeches.[8]

Today, the *hta-mein* is only worn by Burmese female dancers. Men occasionally wear a less voluminous ankle-length version of the *pah-soe* on formal occasions. However, today, both men and women wear the *lon-gyi*, a tubular sarong approximately 100 cm wide by 180 cm long. Women attach a soft, dark-coloured piece of cotton to the top of the *lon-gyi* to serve as a waistband. The *lon-gyi* is worn folded in a large pleat to one side. The excess cloth is tucked in at the waist. It is worn with the *ein-gyi*,

Fig. 108. A young Burmese woman in the early 1900s photographed in *acheik*-patterned *lon-gyi, tabet* breast cloth, and *ein-gyi* jacket. Photograph: P. Klier, Rangoon, by courtesy of the Rangoon University Library.

Fig. 109. A dapper young Burmese gentleman at the turn of the century wearing the *pah-soe*, a fur-lined *ein-gyi* jacket, and *gaung-baung* turban. Photograph: P. Klier, Rangoon, by courtesy of the Rangoon University Library.

Fig. 110. A young Burmese monk outside the Bagya Kaung monastery at Ava wearing the *sanghati*, or outer garment of monk's clothing.

which today refers to a fitting blouse, with or without sleeves. It may be fastened Chinese-style diagonally across the front with jewellery studs or with tiny cloth buttons. A *tabet* (the former breast cloth) is worn as a stole over or around the shoulders for special occasions. Men wear a plaid-patterned *lon-gyi* fastened in front with a tuck or knot. It is worn with a collarless Chinese-style jacket, which is also referred to as *ein-gyi*. A turban, or *gaung-baung*, of silk or muslin is worn by men on formal occasions.

Blankets, sheets, towels, and all-purpose lengths of cloth continue to be woven in simple cotton plaids and stripes by village women in different parts of Burma. One occasionally sees these country textiles for sale at pagoda festivals, which are held regularly throughout the country. A number of ethnic groups weave colourful shoulder bags, which are popular with both locals and tourists.

Living in a Theravada Buddhist country, Burmese laity have the responsibility of supplying the monkhood their basic necessities. At one time, women made robes for the monks which consist of three basic garments: an undergarment, the *uttarasanga*, which is wound around the upper torso leaving the right shoulder bare, the *antaravasaka*, a length of cloth wound around the loins and reaching to the ankles and, over these two garments, a large rectangle of cloth called the *sanghati*, worn in a cloak-like fashion around the chest and shoulders. In keeping with the Buddhist injunction that monks should wear robes made from scraps of cloth, these rectangular garments were made from numerous strips of cloth back-stitched together (Fig. 110).[9] Today, robes are generally machine-woven and purchased from *pongyi* stores which specialize in retailing supplies for the monkhood. Blankets, cushions, towels, and quilts for monks are now purchased rather than woven at home.

There is still one small village, Shwe Yin Ma, in the Myinmu district near Monwya in Upper Burma, which continues to crochet bags to hold the monk's bowl. *Ka-ban-gyo*, or waistbands for monks' robes, are also woven here. The village at one time also wove on a card loom *sa-si-gyo*, which are long, narrow, red and white ribbons (Fig. 111).[10] The *sa-si-gyo*, 2 cm wide and 400–500 cm long, were formerly used to bind the pages of sacred manuscripts. They were usually patterned with monastic symbols and interwoven with the donor's name, status, and pious aspirations. Colourful *ka-ba-lwe* cloth covers for religious books were also woven, with a weft interlaced with small bamboo sticks for greater strength (Fig. 112).[11] Today, silk brocade and velvet cloths are generally preferred as wrappers for religious books.

One custom associated with weaving which has survived in Burma, is that of holding an all-night

Fig. 111. A *sa-si-gyo* ribbon, formerly used to bind pages of sacred manuscripts together. This *sa-si-gyo*, woven on a card loom, is patterned with a monastic symbol and is interwoven with the donor's name, status, and religious aspirations. *Sa-si-gyo* were at one time popular gifts by pious elderly ladies to their spiritual mentors. Warp 403 cm, weft 2 cm.

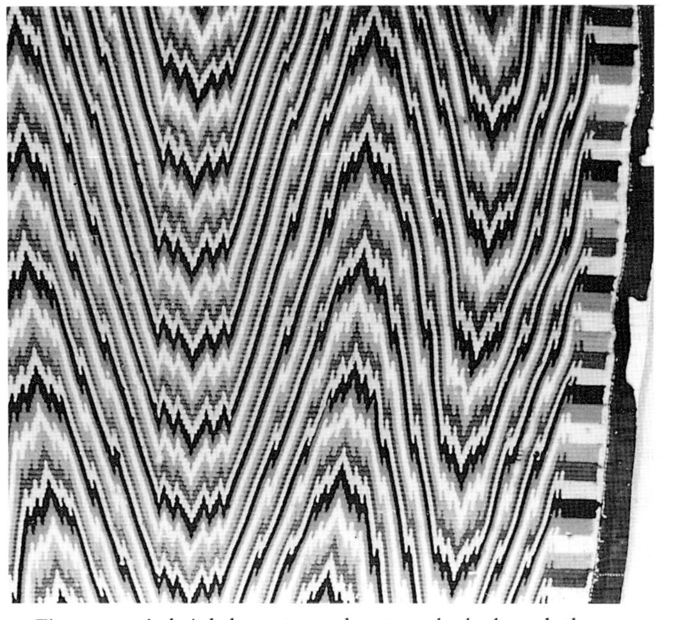

Fig. 112. A brightly patterned cotton *ka-ba-lwe* cloth cover for a religious manuscript. This textile is particularly interesting for small bamboo sticks have been interlaced into the weft to given extra strength and stiffness to the fabric. Warp 100 cm, weft 60 cm.

weaving competition amongst teams of village weavers to re-enact the Buddhist legend in which Maya, the mother of Lord Buddha, on hearing of her son's resolve to enter the monkhood, stayed up all night to weave a suitable robe. At the festival, which is held during the full moon of November, four teams of weavers under the age of twenty-six are selected from different parts of Burma to assemble their looms on the platform of the Shwedagon, Burma's most sacred pagoda. Beginning at sundown and watched by an appreciative audience, they weave in relays all night long to produce a monk's robe. At the conclusion of weaving, the robe is assembled, decorated with gold paper foil cutouts, and then borne aloft in great procession to be draped over one of the four colossal Buddha images at the cardinal directions of the pagoda (Plate 3). The winning entry is judged not only on speed, but on quality of workmanship.[12]

Central and Upper Burma

The Burmans, who constitute the largest ethnic group in Burma, inhabit the central plains drained by the Irrawaddy River. They have a tradition of weaving fine silks. Henzada, north of Bassein, Shwedaung, south of Prome, Amarapura, south of the former capital of Mandalay, and Tavoy, in southern Burma, at one time produced fine silk fabrics from imported yarns which were dyed with a wide range of natural dyes made from flowers, leaf buds, ground seeds, smoke, prawn oil, cock's blood, mosses, slime, and amber.[13]

The most important weaving centre in Burma today is in Amarapura where there are some 40,000 looms in operation.[14] Within the township itself, weavers produce their famous *acheik*, or horizontal wave-patterned silk cloth, in an interlocking tapestry weave (Plate 4). Another name for *acheik* is *luntaya*, or 'one hundred shuttles', so called because the patterns of many colours are woven into the warp with 100–200 small metal shuttles. The *luntaya* tapestry weave technique is thought to have been introduced to Burma from Manipur in north-west India. The founder of the last Burmese Dynasty, Alaungpaya (1752–60) and his descendants, invaded Manipur in 1758, 1764, and 1810, and brought back prisoners who were skilled in the arts of silversmithing, astrology, music, dancing, massage, and weaving.[15]

The weavers were highly valued by the Burmese and were settled in the Amarapura–Sagaing area to produce *acheik*-patterned silks for the royal household. At first, *acheik* was limited to royalty; with the coming of the British to Burma, however, sumptuary laws declined and *acheik* became available to all who could afford it.[16] *Acheik* has always been expensive because it is extremely laborious and time-consuming to make. A pair of girls working together at a two-harness loom, manipulating 180 shuttles containing different coloured yarns across a warp of 1,500 threads, can only weave a few centimetres a day. It takes at least fifteen days of full-time weaving to finish a simple *acheik*-patterned *lon-gyi*. Working on the wrong side of the cloth, the weavers create the designs from memory, for paper patterns are rarely used. Each small metal shuttle containing a particular colour wound around a bamboo spindle, passes over and under only five to ten warp threads at a time, according to the design. Girls begin learning the art of weaving *acheik* patterns around the age of twelve. Within two years, they are proficient weavers. Since keen eyesight is essential, it is rare to find older women weaving *acheik*.[17]

Although the weaving technique came from Manipuri, *acheik* designs are native to Burma. In all probability, they owe their inspiration to natural phenomena, such as waves, thunder, and cloud patterns. Such designs may be seen on neolithic pottery sherds and Pyu beads found in Burma. Temple frescoes from the Pagan and Ava periods are also embellished with such patterns.[18] Basic *acheik* patterns consist of weft bands of parallel 'S'-shaped, wavy, and zig-zag lines with small foliate appendages. These elements may be expanded and combined in various ways to create a series of design motifs (Fig. 113).

A series of zig-zag lines in different colours is called *hpe-lein*, or satin-wave pattern. A variation of this pattern is called *yei lein sin kyo*, or water-wave design, which is in the form of an undulating zig-zag broken by vertical lines extending from the base. Wave patterns, called *sin kyo*, are named according to the number of parallel lines in different colours which make up the design. A complicated wave pattern may have up to five or seven stripes. Some patterns have small protruding tabs or lappets at the apex and are a little reminiscent of the Chinese cloud-collar design. One such pattern is the *maha kyo shwei taik* (great line golden building) design. Other wave patterns have the tendrils of a creeper, called *gamoun*, and flowers springing from the outer undulating lines. Examples include the *thoun sin gamoun ngwe kyo* (three-stripe *gamoun*). Single flower heads are sometimes found within the wave or

Fig. 113. Burmese *acheik* patterns. (After Daw Htew Khin, U Aye Myint, and U Win Maung.)

hpe-lein
(satin-wave)

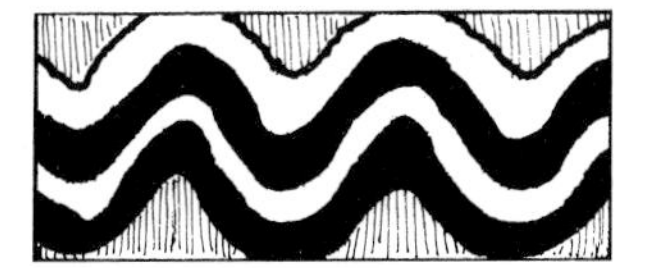
lei sin kyo
(four stripes)

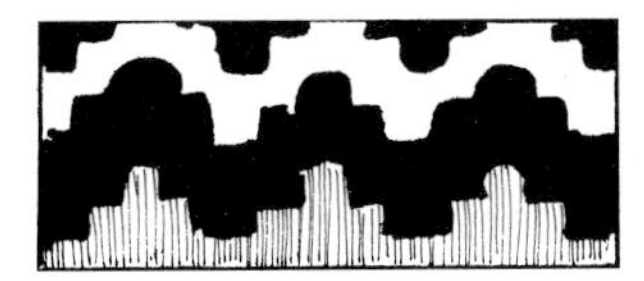
maha kyo shwei taik
(great line golden building)

gamoun
(creeper)

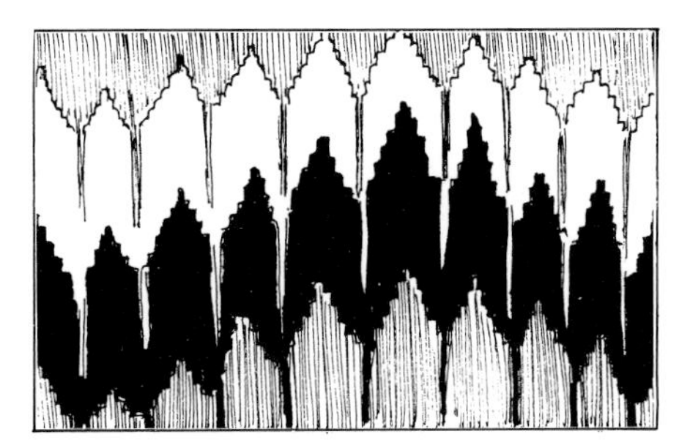
yei lein sin kyo
(water-wave)

thoun sin gamoun ngwe kyo
(three-stripe *gamoun*)

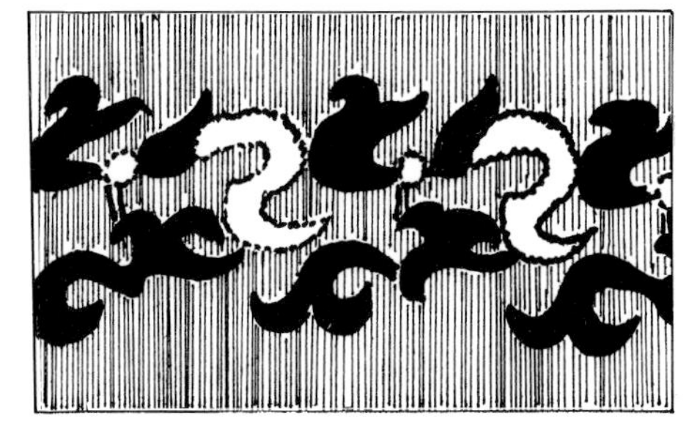
thoun gamoun ywet set
(Burmese number three joined with *gamoun* creeper)

yei lein cheik
(water-wave *acheik*)

Myin Myo gyi kyo
(five-stripe Mt. Meru)

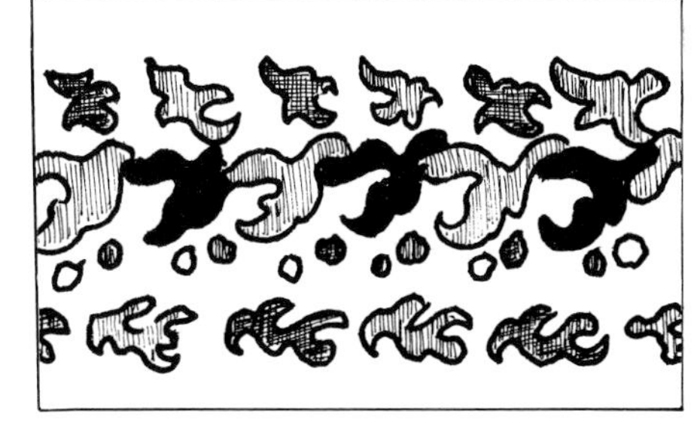
shwei ta lein, ngwe ta lein kyo
(single gold and silver twist)

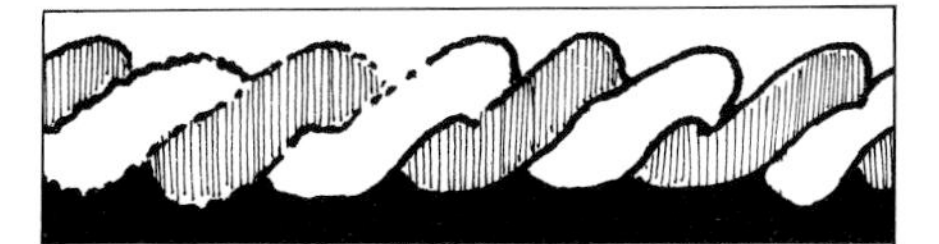
nit sit lein kyo
(two twisted stripes)

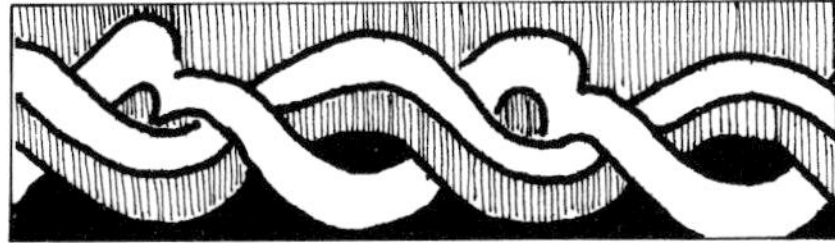
nit sin ta si
(two stripes, one bundle)

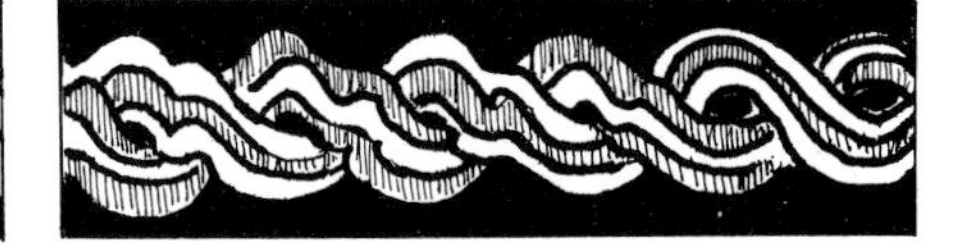
chauk sin lein kyo
(six twisted stripes)

taik kaung tin pan khet
(chief queen's floral garland)

kyo gyi cheik or *sit tit zin gamoun*
(large rope *acheik* or eleven-stripe orchid and *gamoun*)

attached to the creeper, such as in the *yei lein cheik* (water-wave *acheik*) and the *Myin Myo gyi kyo* (five-stripe Mt. Meru) designs. Occasionally, creepers and flowers may be arranged as an *acheik* pattern in their own right as on the *shwei ta lein, ngwe ta lein kyo* (single gold and silver twist) and the *thoun gamoun ywet set* (Burmese number three joined with *gamoun*) designs. Many wave patterns are arranged in a twisted cable design called *sin lein*. These designs can be very simple, as in the worm-like *nit sit lein kyo* (two-stripe twist) and the *nit sin ta si* (two stripes, one bundle) patterns, or they may be models of complexity with a narrower striped band snaking in the opposite direction over and under a wide undulating wave component. Such motifs include the *taik kaung tin pan khet* (chief queen's floral garland pattern) and the *kyo gyi cheik* (large rope *acheik*), also known as *sit tit zin gamoun* (eleven-stripe orchid pattern) (Fig. 114).[19]

Acheik patterns are enhanced by the skilled choice of colour on the part of the weaver. Since British times, silk for *acheik* has been coloured with

Fig. 114. Detail of a piece of Burmese *acheik* from Amarapura depicting the *nit sit lein kyo*, the *thoun gamoun ywet set*, the *chauk sin lein kyo*, and a variation of the *taik kaung tin pan khet* designs.

chemical dyes. Bright reds, purples, pinks, greens, and yellows have been generally preferred over subdued delicate shades. The Burmese delight in combining two to three lighter and darker shades within the same colour range, such as red, rose pink, and light pink, pink and purple, apple green, and grey to give some very striking colour combinations. Silver thread is also interwoven to add interest to an *acheik* pattern.

Acheik-patterned silk *lon-gyi* are highly prized by Burmese women, most of whom will own at least one to be worn to formal events, such as a wedding or a son's ordination as a monk. Today, there are about ten to fifteen workshops in the Amarapura–Mandalay area which are producing *acheik*-patterned silk *lon-gyi*. The standard of workmanship remains high, but due to shortages of good-quality dyes, the colour combinations are not as pleasing as formerly. Silk comes mainly from China. Silk from Maymyo, north of Mandalay, is sometimes used for the weft. Today, a larger number of silk strands (eight to ten per thread) are used as opposed to two to four strands, which was formerly the norm. This makes for a thicker fabric with less intricate patterns.

On the outskirts of Amarapura are many small workshops employing ten to twelve workers who imitate traditional designs from other parts of Burma. *Acheik*, Arakanese, Shan, Gangaw, Chin, and Kachin-patterned *lon-gyi* are woven on a four-to-eight-harness loom. Silk, cotton, nylon, and rayon yarns are used. Over recent years, plain weave geometric patterns, formed by alternating colours in the warp and changing the colour and size of yarn in the weft, have become very popular for men's *lon-gyi*. Some are further embellished with supplementary weft patterning. Most workshops come under the jurisdiction of the Ministry of Cooperatives, which is responsible for procuring supplies and marketing the finished products.[20]

In Amarapura, there is the Saunder's Weaving Institute, which was established by the British Government in 1915 to keep the art of weaving alive. Because of the influx of cheap manufactured cloth from abroad during the early years of the twentieth century, the art of weaving in Burma, particularly that of *acheik*, was in danger of dying out. It was hoped that the revival and promotion of weaving would provide extra income for a farmer's family during the slack times between harvesting and sowing. Courses were given to selected students from the rural areas in all branches of weaving, including *acheik*. Students were introduced to looms with flying shuttles and the latest dyeing techniques.

Batik and screen-printing were also introduced. Travelling demonstrations and exhibitions were organized by the Institute to keep weavers abreast of current developments. In the 1930s, there were seventeen vocational schools throughout Burma with weaving on the curriculum. At that time, some 750 students were undergoing training in weaving.[21]

After Independence, the Saunder's Weaving Institute was placed under the Ministry of Cooperatives. Lack of funds has hindered activities. At the time of this writing, some forty students from all over Burma were enrolled at the Institute (twenty-nine men and eleven women) for four-month courses in weaving. Students study basic weaving on hand-operated looms with flying shuttles, as well as Dobby, Jacquard, and Japanese power looms. Students also learn the techniques of weft *ikat* and screen-printing. Due to a shortage of funds, *acheik* weaving has recently been discontinued as a subject of study.[22]

San Khan, on the banks of the Dudawadi River some 16 km out of Mandalay, is a small village where, since 1978, some 500 people have been making cotton *si lon-gyi* patterned with weft *ikat* designs in a twill weave. The cotton yarn, purchased at Zegyo market in Mandalay, comes from India. Chemical dyes are used. Patterns consist of floral and widely spaced geometric shapes on a deep maroon, purple, green, navy blue, or green ground. Dyeing for the various colours is by the *cetak* method. During the course of dyeing, boundaries for each colour are demarcated by rubber bands. Usually no more than three colours are applied by this method. Only the ground colour is obtained by completely submerging the fabric in a dye bath. Hand-operated looms with box shuttles quickly weave the weft-patterned threads into *lon-gyi*. Two *lon-gyi* a day may be woven. A distinctive man's *lon-gyi*, using the warp *ikat* technique, is also made. Lightning-style zig-zag designs are tied into the warp prior to dyeing the ground black. The white streaks present a pleasing contrast to a weft of brown, blue, and beige stripes.

A small adjoining village (Thapye Auk) also makes weft *ikat* using identical methods and a similar colour range. The designs on the *lon-gyi* from Thapye Auk are more flamboyant than those of San Khan and owe their inspiration to Burmese *acheik* patterns (Fig. 115).[23]

A number of small villages in the Meiktila area, 80 km south of Mandalay, weave striped and plaid all-purpose material from local hand-spun cotton. Some is woven into colourful blankets (*saung*) which are widely used throughout Burma. Plain weaves predominate. Some pleasing twill weave, red plaid blankets are made in the Monywa area to the west of Mandalay.

Further south of Monywa, in the Pakokku area, lies the village of Gangaw, which is noted for producing silk and cotton textiles in a very finely patterned supplementary weft which resembles a running embroidery stitch. The most striking patterns are in a light supplementary weft on a dark ground. Designs for men are usually small repetitive patterns within a diamond frame, while designs for women may include small floral patterns, peacocks, and swastika (Fig. 116). Some *lon-gyi* have two parallel stripes with the words 'Gangaw Special', in English and, occasionally, the initials of the weaver woven into the cloth. Weaving today is mainly in cotton.

The township of Shwedaung, south of Prome, was formerly a great silk weaving centre. Now it produces plaid cotton *lon-gyi* for men on a large scale. There are some 8,000 hand looms currently registered with the Ministry of Cooperatives.

The village of Kyi Thei, a few kilometres out of Shwedaung, produces cotton, warp *ikat*-patterned *lon-gyi*. Modern methods are used to spin the thread and set up the looms. Dyeing is by the *cetak* method, and three to four *ikat*-patterned *lon-gyi* can be woven a day on modern floor looms. Simple floral patterns against a plain ground predominate for women. In recent years, the random binding and dyeing of odd weft threads to create a streaky background to the

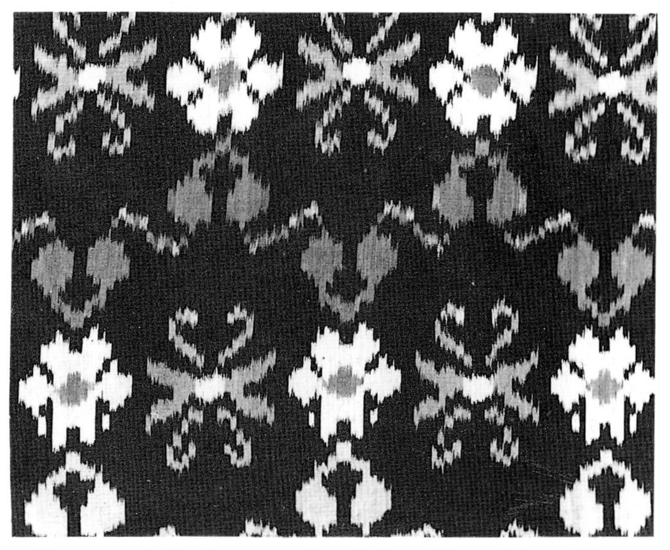

Fig. 115. A floral-patterned weft *ikat si lon-gyi* from Thapye Auk village. It is woven from Indian cotton in a twill weave. Brightly coloured floral patterns against a black ground are in four colours (including white). Resist patterns have been dyed by the *cetak* method. Warp 180 cm, weft 100 cm.

Fig. 116. Gangaw *lon-gyi* designs.

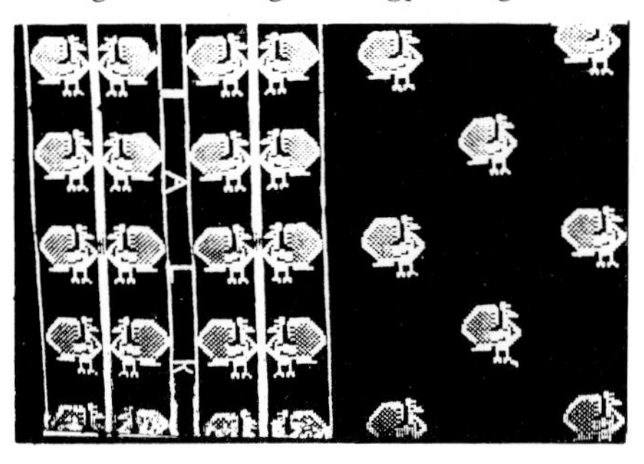

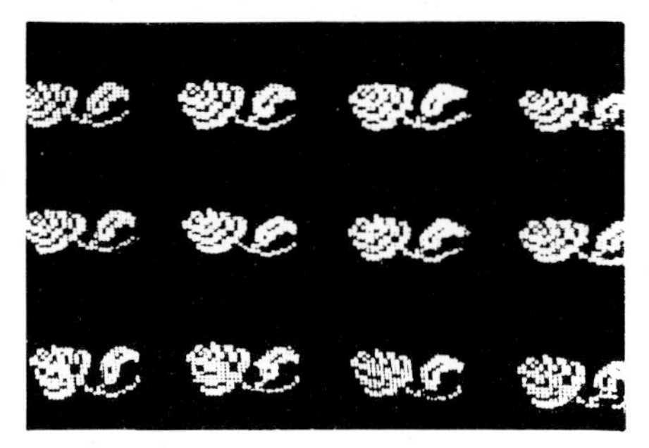

(a) Designs for women's *lon-gyi*.

(b) Diamond pattern on a man's *lon-gyi*.

main warp design has become popular. Like San Khan, Kyi Thei has begun producing striking warp *ikat*-patterned designs in bold streaks and zig-zags in white against a reserved striped weft ground for men. Currently referred to as 'breakdance *lon-gyi*', these new patterns are very popular with young men in Rangoon (Fig. 117).

South of Shwedaung, the small town of Inme specializes in making a distinctive two-colour checked *lon-gyi* for men in a plain weave, by alternating the colours in the warp and the colour and size of yarn in the weft.[24] Brown, green, blue, and grey colours predominate.

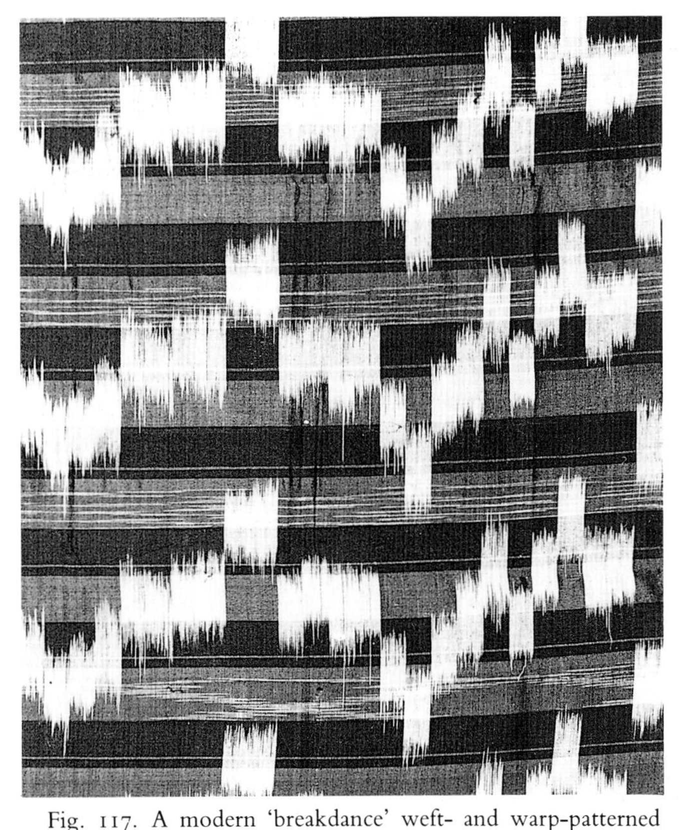

Fig. 117. A modern 'breakdance' weft- and warp-patterned *lon-gyi* from Kyi Thei village near Prome. This type of *lon-gyi* is very popular with young men in Rangoon. Warp 190 cm, weft 110 cm.

Arakan

The Arakanese, who inhabit the coastal littoral of the Bay of Bengal, had a distinct and glorious history until they were conquered by the Burmese in 1785. Their civilization was described by Maurice Collis in *Land of the Great Image*, which is based on the diaries of the Portuguese Friar Manrique, who recorded his stay in Arakan from 1630 to 1638.[25] According to contemporary descriptions, the Arakanese kings were very interested in the art of weaving. Royal weavers were permitted to reside in a particular locality. The coveted *Ayathamingyi* award was bestowed on master weavers who could speedily produce *lon-gyi* of the finest texture and design. Weaving is also portrayed in traditional Arakanese song and dance. Cotton was traditionally produced to supply local needs. Although a little silk yarn was grown in the Sandoway district in the south, most silk was imported from Bengal.[26]

Today, Arakanese weaving is done on a frame loom. Cotton *lon-gyi* lengths are patterned in a loose, continuous supplementary weft on a plain, striped, or checked ground. Black is the most popular colour for the supplementary weft. White and other light colours are used on a dark ground. Occasionally, on striped material, more than one supplementary weft colour is used.

Traditional patterns for men are strongly geometric, being based on small repetitive patterns formed in combination with lines, squares, and circles (Fig. 118). Designs focusing on lines include the *min myat chit* (king's apple of the eye) pattern, and the *ah paung let kho na* (plus sign) pattern. Circular designs within a square form the *myit ta paung ku* (love path) pattern, while squares predominate in the *sein let sut* (diamond ring) design. Rhomb-dominated designs include the *kai yu ka ma* (mother-of-pearl) pattern, the *maung chit ma chit* (lover's design), the *ka ti pa kwet* (velvet check), and the *pan bayin* (king of the flowers) motifs. Traditional patterns for women are

Fig. 118. An Arakanese women's *lon-gyi* in the *hnin si cheik* (rose) pattern, a design which owes its inspiration to European influence. Because this pattern is difficult to execute, it is rarely woven today. Warp 100 cm, weft 180 cm. Collection of Cherie Aung Khin, Elephant House, Bangkok.

Fig. 119. An Arakanese man's *lon-gyi* in the *pan bayin* (king of the flowers) motif. This design usually appears in black on a brightly coloured ground. Collection of U Mya San.

based on repetitive floral designs arranged in parallel rows across the warp of the cloth (Fig. 119). Well-known floral designs include the *yatha pan yein* (slanting flower) pattern, the *hnin si cheik* (rose pattern), and the *tha pyu cheik* (herbal flower motif). Some patterns owe their inspiration to birds and insects, such as the *daung ma cheik* (peacock tail) design and the *pin ku cheik* (spider) motif (Fig. 120). Burmese *acheik* and European flower patterns are also seen on Arakanese cloth. Arakanese weaving is centred in the townships of Kyaukpyu, Thandwei, Siddwei, and Sandoway.[27] Because of their attractive designs and durability, Arakanese *lon-gyi* are popular throughout Burma for everyday wear. To meet this demand, *lon-gyi* with Arakanese designs are now being made at Amarapura and Mudon.

Fig. 120. Some traditional Arakanese textile patterns.

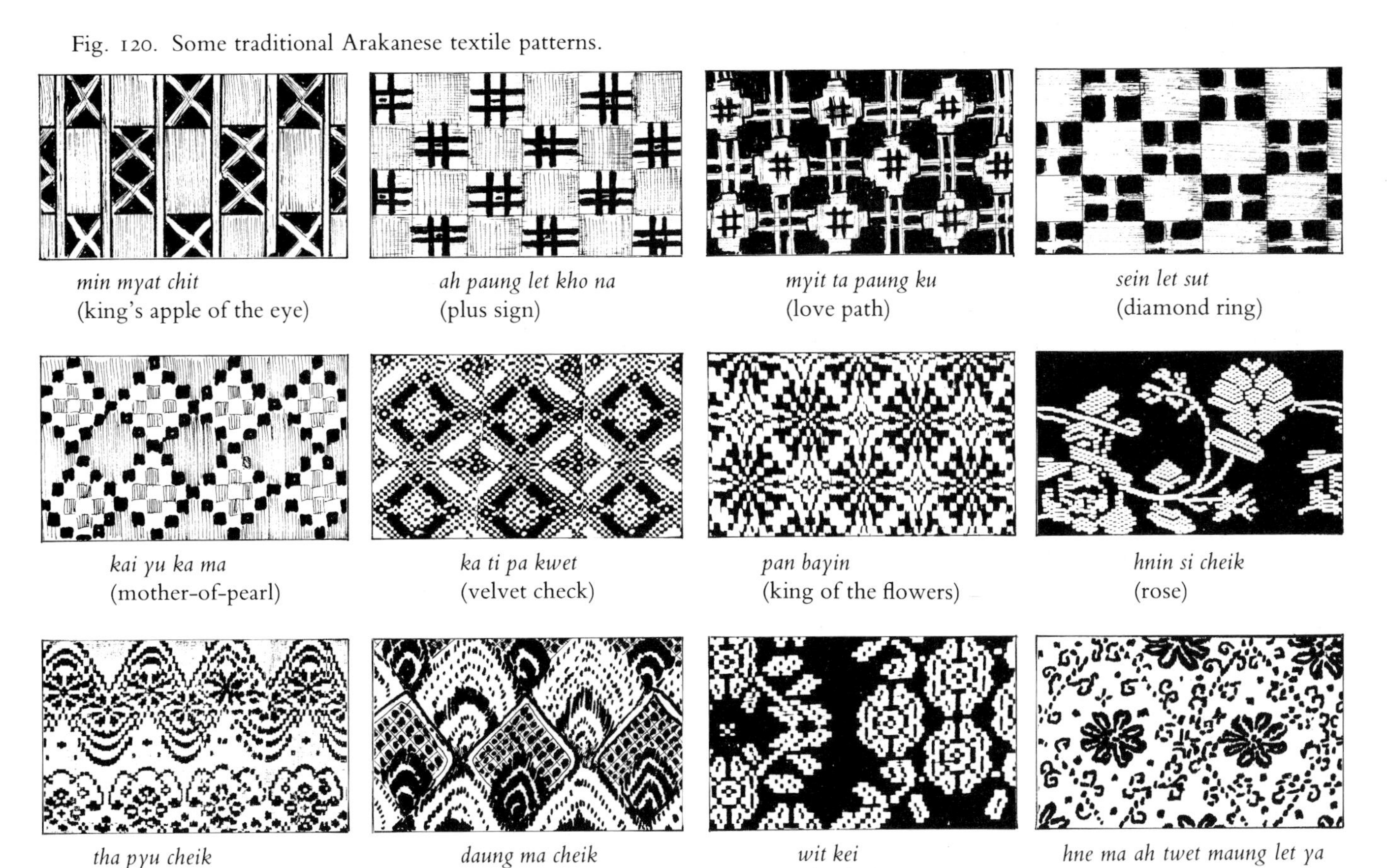

min myat chit (king's apple of the eye)

ah paung let kho na (plus sign)

myit ta paung ku (love path)

sein let sut (diamond ring)

kai yu ka ma (mother-of-pearl)

ka ti pa kwet (velvet check)

pan bayin (king of the flowers)

hnin si cheik (rose)

tha pyu cheik (herbal flower *acheik*)

daung ma cheik (peacock tail)

wit kei (floral cluster)

hne ma ah twet maung let ya

Lower Burma, Kawthoolei, and Kayah States

Lower Burma, like Shwedaung, has a tradition of weaving fine silks. Tavoy was at one time famous for producing fine silk *lon-gyi* for men. This area was once widely populated by the Mon, a proud, independent people of great cultural and artistic achievements in both Burma and Thailand. As captives of the Burmese, they provided Pagan with the Theravada scriptures and the technical expertise which enabled that Kingdom to achieve unprecedented greatness. With the fall of Pagan, the Mon once again became independent, and European travellers to the Mon Kingdom of Pegu from the sixteenth to the eighteenth centuries have left us glowing accounts of the splendours of the court, and its trade in luxury items, such as imported Indian and Chinese silks, along with snippets pertaining to everyday life (including information on clothing and weaving).[28]

While the Mon today have been largely absorbed into the Burmese population, some weaving concerns in Mudon, south of Moulmein, continue to weave Mon-patterned *lon-gyi*, such as those with a white check on a red ground, and geometric, broken-line, and square patterns in a supplementary weft for women (Fig. 121). Mudon also weaves Arakanese-style *lon-gyi* on a large scale for local use.

Kawthoolei State in eastern Burma is the home state of Burma's Karen people, who are also widely scattered throughout the delta areas of the Irrawaddy, Sittang, and Salween Rivers. The Karen may be divided into a number of groups, such as the Pwo, Sgaw, and Bwe, according to dialect differences. At one time, all Karen groups were active weavers who produced textiles for clothing, blankets, and shoulder bags from home-grown cotton on a body-tension loom with a continuous warp. Natural dyes were made from indigo, lac, and morinda. Karen handspun woven material is very firm and practically indestructible.

An unmarried Karen girl is distinguished by her clothing, which consists of an ankle-length white shift dress. Among the Sgaw Karen who inhabit the upland areas, this garment is quite plain, being relieved only by narrow bands of red at the waist and hemline. The lowland Pwo Karen girl's dress is much gayer, with bands of geometric decoration across the shoulders and below the waist, formed by fuzzy tufts of red chenille patterns inserted into the weft as the cloth is woven. The upper bodice also has strands of loose red threads inserted into it so as to create a fringe of yarn to the knees (Fig. 122).[29]

When she comes of marriageable age, the young

Fig. 121. Mon woman's *lon-gyi* patterned with traditional supplementary weft designs. Mudon, Lower Burma. Warp 100 cm, weft 170 cm.

Fig. 122. A Karen woman with her two daughters. Note the difference in clothing between married and unmarried females. This photograph was taken in Thailand. The Karen of this particular village had recently moved from Burma into Thailand. In Thailand, Karen clothing remains more traditional than in Burma.

Fig. 123. A Bwe Karen woman in traditional clothing consisting of a loose-fitting poncho-style blouse and short sarong. Photograph courtesy of Mrs Gladys Tin Aung, Rangoon.

Karen woman begins weaving herself a set of clothes of a married woman. These consist of a loose-fitting blouse called a *hse*, or *thingdaing*, and a rust-coloured skirt. She also needs to weave a *hse* for her husband-to-be. The *hse* is made by sewing together two lengths of indigo-coloured cloth to form a central seam and then folding the fabric in half. Holes are left in the upper corners for the arms and another opening is left in the centre seam for the neck. The married woman's blouse may be quite plain or elaborately decorated in diamond-shaped supplementary weft patterns, embroidered cotton yarns, and rick-rack arranged in chevron patterns, zig-zags, checks, rosettes, and stars, and highlighted by shiny white seeds called 'Job's tears'. Strips of cloth may be appliquéd around the waist area. For further embellishment, colourful tufts of yarn may be inserted at the blouse openings and along the seams (see Fig. 85). The Yang Lam (sometimes called Black Karen) have a square of bright red tassels anchored by beads of Job's tears decorating the front of their distinctive dark indigo-coloured *hse*.[30]

The Karen woman's traditional skirt, or *ni*, is made by sewing two lengths of cloth together into a tubular sarong. It is worn tight across the back, folded in front and held in place with a belt. The skirt is patterned with warp stripes. The Bwe Karen woman's sarong is patterned with alternating red, black, or white stripes (Fig. 123). The Sgaw and Pwo women wear a red, black, and white striped sarong which has a different pattern layout. A distinctive plain black stripe or a large blue area at the top and bottom of the sarong forms a boundary for a series of rust-red stripes interrupted by unobtrusive single stripes in white. Some red bands contain simple warp *ikat* patterns consisting of white streaks and dotted rhomb patterns, often referred to as 'python skin' patterns in deference to a well-known Karen legend in which a woman was captured by a fabulous white python (Fig. 124). On being released, either out of gratitude or contempt for her captor, the woman began weaving 'python' patterns. The tying and dyeing of warp *ikat* patterns amongst Karen women is accompanied by much ritual and secrecy.[31] Some Sgaw Karen in Burma used to pattern their sarongs with bands of zig-zag geometric patterns in green and yellow on a red ground. These might be further embellished with coins. On modern sarongs, it is quite common to combine narrow bands of supplementary weft with bands of warp *ikat*.

The Karen man's traditional *hse* shirt is similar to that of the woman's blouse except that it is looser and longer. It is usually woven in dark blue or bright red, broken by a few widely spaced stripes of different colours. It is traditionally worn with a red or white *lon-gyi* decorated with two narrow bands of parallel blue stripes. The Karen also wear a 180 cm long, 30 cm wide white head-cloth, called a *hko peu ki*. This head-cloth is patterned with simple supplementary weft patterns in red and may be finished with fringes. Shoulder bags and red and white striped blankets are also woven.[32]

Many of the Pwo Karen who live in the delta area of Burma have assimilated into Burmese society. With the exception of the Karen New Year Day in January, they rarely wear their national dress (Fig. 125). Apart from those who live in remote areas, few Karen weave their own clothes today. There is a thriving weaving industry in Pa An, the main town of Kawthoolei, which makes Karen *lon-gyi* for sale in other parts of Burma.

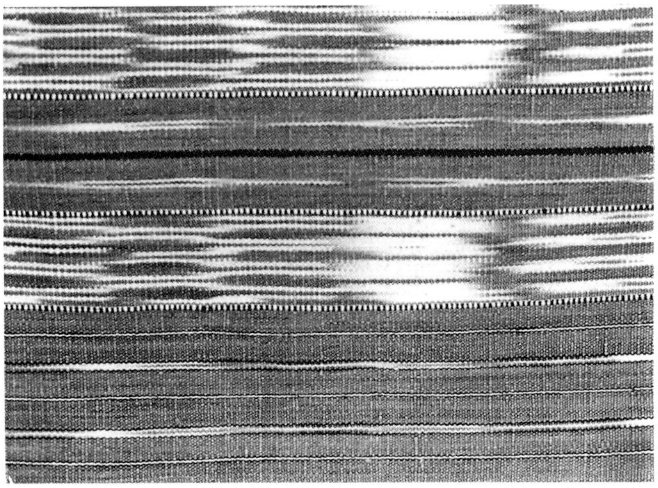

Fig. 124. Detail of warp *ikat* 'python skin' patterns on a Karen woman's skirt.

Fig. 125. Some Karen of Burma performing a traditional *doan* dance. Note how the women have remodelled their traditional blouses according to the dictates of current Western fashion.

The Kayah, sometimes called Karenni, or Red Karen, inhabit Kayah State which lies between Kawthoolei and the Shan States. They also weave their traditional clothing on a body-tension loom. The woman's costume consists of a red or black cloth tied across one shoulder. It is worn with a matching red- or black-striped sarong similar to that of the Karen. The breast cloth and sarong are held together by a fringed white sash wound several times around the waist and tied below the buttocks. The women also wear numerous lacquered cane rings around the top of the calves. The men wear baggy hill tribe pants, Western-style trousers, or Burmese *lon-gyi*. Blankets are used as cloaks for extra warmth.[33] The Kayah at one time cultivated the lac insect for the dye and sold it to neighbouring peoples, such as the Karen.[34]

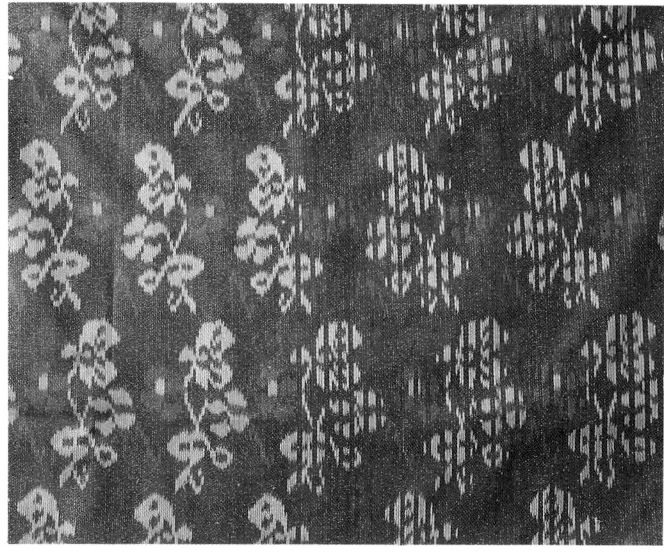

Fig. 126. Modern Shan floral-patterned *Zin me lon-gyi* from Inpawkhon, dyed by the *cetak* method and woven in a twill weave. The striped section which forms the inside pleat of the *lon-gyi*, is not seen when worn. Warp 180 cm, weft 105 cm.

Fig. 127. Shan *Zin me lon-gyi* patterns.

traditional *Zin me* designs

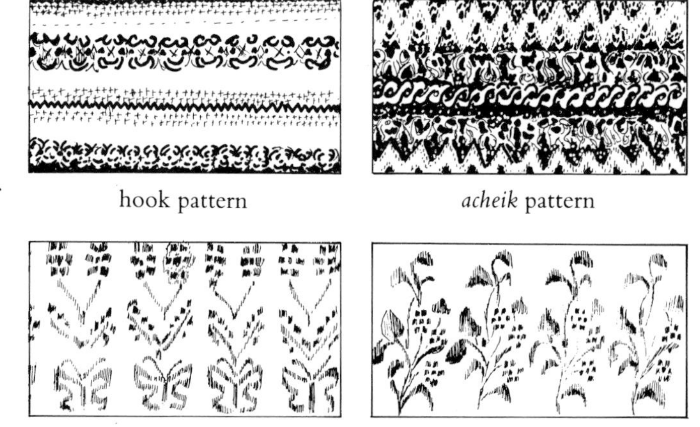

hook pattern

acheik pattern

modern floral designs

Shan States

The Shan States region of eastern Burma, because of its numerous ethnic groups, is an anthropologist's delight. Apart from the Shans, who are the largest group, a number of tribes have made their home in these hills. These include the Kachin, who form a large proportion of the population of North Hsenwi State and the Kodaung hill tracts, various sub-tribes of Karen, such as the Pa O, Padaung, Yang Lai, and La Hta, the Inthas of Inle Lake, the Lisu (also called Yawyins and Lishaw), the Akha (Ekaw), Lahu (Muhso) and Hmong in Kengtung, and the Wa and Palaung.[35] Many of these groups continue to weave their traditional textiles from locally grown cotton on a body-tension loom. Increasingly, these peoples are purchasing cloth woven in the lowlands of Burma or imported from Thailand.[36]

The most sophisticated and varied weaving is done by the Shan people who are a branch of the Thai race. They live in the river valleys and lower mountain slopes of the Shan States. Up until 1959, the Shan States were divided into thirty-four principalities ruled by Sawbwas, or hereditary princes, many of whom held sway over courts which were former cultural centres where the arts of lacquer, wood carving, silversmithing, and weaving were encouraged. Late nineteenth- and early twentieth-century examples of textiles in the Victoria and Albert Museum, London, attest to the beauty and sophistication of Shan weaving (Plate 5). Weft *ikat*, supplementary weft, and tapestry weave techniques are known to Shan weavers. All weaving is done on a frame loom.

Traditional Shan dress for women is similar to that of the Burmese. For everyday wear, a green or red cotton *lon-gyi* decorated with darker warp stripes separated by narrow yellow and blue stripes is worn. Sometimes only the top part of the *lon-gyi* may be striped from the waist to the thighs. The remainder consists of a broad piece of material which may match or contrast with a neat-fitting, long-sleeved collarless jacket which is worn with the *lon-gyi*. Silver and gold threads, embroidery, and appliqué may add further embellishment to a Shan *lon-gyi*. Women may wear the *ken hu*, a 200 cm by 30 cm head-cloth, which may be plain or striped. The Shan male wears *baung bi*, or beige-coloured baggy pants with a low crotch. These are teamed with a loose-fitting matching jacket. A large turban may also be worn.

The Intha people, originally from Tavoy in southern Burma, who inhabit the shores of Inle Lake in the southern Shan States, are also notable weavers. The village of Inpawkhon is famous for its weft *ikat*-patterned silk *lon-gyi*, which are called *Zin me*, the Burmese word for Chiangmai (Plate 6).[37] *Zin me lon-gyi* were originally woven from imported thread in subdued reds, yellows, and greens derived from natural dyes prepared from bark and seeds. The designs woven on a red warp were based on intricate hook, rhomb, and diagonal cross patterns, which owed their inspiration to the famed royal Cambodian weft *ikat* cloths. Some cloths also show Thai and Indonesian influences in their designs. With the advent of chemical dyes, the three-colour palette remained the same, but the resulting hues were much brighter. Over the years, bird and flower patterns, as well as the use of metallic weft yarns, were added to the *Zin me* repertoire. These *lon-gyi* were woven in a one- or two-ply silk thread in a 1 : 2, slightly weft-faced twill weave on a four-pedal floor loom (Fig. 126).

The body of traditional *Zin me lon-gyi* is covered with a closely patterned repetitive design. On some cloths, there is a border pattern along the base of the *lon-gyi*. There is usually a 35–45 cm *kepala*, or vertical area, consisting of closely patterned linear designs between plain weft stripes. This part is folded inside the front pleat and is not visible when worn. Popular designs on traditional *Zin me lon-gyi* include U Po Nyein, Kyaung Ama Hnyin, and Daw Gyi Hsin, which are named after the people who developed the particular motif.

During the Depression of the 1930s, inferior dyeing techniques and competition from cheap manufactured cloth virtually destroyed the art of weaving at Inle. In 1936, two weavers, both with the name of U Po Han, went to Korat in north-east Thailand and to Bangkok to study the Thai silk weaving industry. Upon their return, they revived the art of making weft *ikat* by using lightweight silk imported from China, a wide range of chemical dyes, and modern designs in a 1 : 3, balanced twill weave. While geometric motifs continue to be featured, floral and hook designs have become more popular. Modern designs include a rose pattern, sprigs of one to three flowers, floral creepers, gold and silver blossoms, a beehive motif, and imitations of *acheik* patterns (Fig. 127).[38]

The *Zin me* silk weaving industry has been modernized. Flying shuttles have been added to looms making it possible to weave two *lon-gyi* a day. To speed up the process, the *cetak* method is widely used for applying the secondary colours of modern designs. Many qualities of *Zin me* cloth are produced in both plain and twill weaves. Today, many designs have a pin-striped appearance due to the practice of breaking up the *ikat* pattern by placing a few shots of the ground colour in the weft at regular intervals. *Zin me* cloth continues to have a *kepala* of wider stripes.

To the north of Inpawkhon, the village of Tha-le specializes in producing sturdy cotton *lon-gyi* for everyday wear. Plaids and stripes predominate. Two-tone, two-ply twisted threads are added for interest to various designs. Sarongs patterned in this technique are often referred to as *Bangkauk* (Bangkok) *lon-gyi* because the technique was introduced from Thailand. *Dwei Inle lon-gyi*, with small checks in two colours enlivened with small supplementary weft designs, are popular. With four-harness floor looms, several different weaves are possible.

A few minutes' boat ride from Tha-le, lies the village of Ywa-ma. At the back of the floating market, there are a number of households which specialize in weaving Shan bags for the tourist trade. Two different types of loom are used. A traditional Burmese frame loom with disc-shaped pedals is used for weaving all-wool bags. The weft yarn is wound onto a stick and is passed by hand through the narrow (30 cm) wide warp. It is possible to weave three bags a day on this loom. Side by side with traditional looms are modern frame looms with pulley-operated box shuttles which are used to make bags with simple supplementary weft patterns in wool on a cotton warp. Five bags a day can be woven on this loom.[39]

In the Nam Khan district in the northern Shan States, there are a number of Shan weavers who weave brightly coloured, striped, 30 cm wide lengths of cloth which are used to make the upper part of a woman's *lon-gyi* (Fig. 128). Woven in four-ply cotton yarns, this cloth is often patterned with designs woven in a thick, lighter-coloured supplementary weft. Stripes may be separated by small bands of silver thread woven into the weft. Designs are strongly geometric and are no longer traditional. Bags and *lon-gyi* are also woven by the Nam Khan weavers.[40]

The Karen subgroups who have made their homes

Fig. 128. Striped *lon-gyi* material patterned with bands of simple supplementary weft decoration from Nam Khan in the northern Shan States. Warp 190 cm, weft 50 cm.

Fig. 129. Pa O woman in her traditional simply tailored long shift blouse which is worn with a short sarong. Note the towel turban and Shan hill tribe bag.

Fig. 130. Padaung couple in traditional dress. The woman is wearing a short sarong and loose-fitting blouse. Over this she has a Burmese-style *ein-gyi* jacket. Her husband is wearing Shan clothing which is typical for hill tribe men of the Shan States. Photograph courtesy of Mrs Gladys Tin Aung, Rangoon.

in the Shan States may be readily identified by the distinctive costumes of their women. The men, on the whole, have abandoned traditional garb (a loin cloth and blanket) in favour of Shan male attire. A few women continue to grow their own cotton and weave simple textiles on a body-tension loom. The Pa O people (sometimes called Taungthu) may be distinguished by their long, sombre, black shift blouses made in the style of a Karen *hse*, their matching sarongs, leggings, and colourful Turkish towel head-dresses (Fig. 129). They no longer weave but purchase navy blue or black serge cloth from the Taunggyi market.[41] They proudly display the 'Made in England' trade mark along the selvages as a border decoration around armholes and neck openings. The Padaung women, who are famous for their long brass-clad necks, weave a loose-fitting Karen-style blouse in white edged with a small yellow and brown stripe (Fig. 130). It is worn with a short white or black skirt patterned with red stripes. The Yang Lai women weave an extended version of the *hse* blouse which resembles a shift dress. It is usually patterned with white and maroon stripes. The La Hta are rarely seen because of their shyness towards other groups and their tendency to marry amongst relatives. They, too, weave a red and white shift which they decorate with running-stitch embroidery.[42]

Tibeto-Burman peoples, such as the Lisu, Akha, and Lahu, are found in Kengtung State, the easternmost state which borders Thailand. The Lisu women in the more remote areas close to the Chinese border weave hemp to produce their colourful tribal clothing, which consists of a long-sleeved jacket or caftan which fastens on the side and is decorated with pieces of colourful Chinese brocade trim. This caftan is worn with a blue-black sarong patterned with border trim and a colourful appliqué apron. The Burmese Lisu costume differs a little from that of the Lisu of Thailand.[43] Lahu dress in Burma consists of a long, black, indigo-dyed cotton coat with colourful appliqué around the edges. Woven on a rudimentary frame loom, it is fastened as far as the waist with large silver buttons and opens to reveal a matching sarong, also patterned with appliqué (see Fig. 92). A large black turban is also worn.[44] The Akha, many of whom over recent years have moved to Thailand, also weave on a simple frame loom vast yardages of cotton fabric which they later dye with indigo. This is later fashioned into clothing which is decorated with bands of colourful appliqué.[45]

The Wa, an elusive Mon–Khmer peoples once

feared for their practice of head-hunting, eke out a livelihood in the inhospitable uplands along the Burmese–Yunnan border. The women continue to weave loin cloths, jackets, and sarongs for clothing.[46] They also weave distinctive black cotton blankets patterned with two or three white pin-stripes along the selvages. The body of the textile may be decorated with small, widely scattered squares worked in a red, discontinuous supplementary weft. This textile shows strong affinities with those produced by their Riang neighbours in Burma, and with some woven by the Naga and Mizo hill tribe people in Assam.[47]

The Palaungs, who live in the northern Shan States and are the tea cultivators of Burma, unfortunately no longer weave their most picturesque costume, which consists of a velvet jacket decorated with brightly coloured appliqué, and a cotton pin-striped skirt over which is worn a colourful apron. Leggings and a long hood complete the outfit.[48] Today, the Palaung prefer to fashion their clothing from imported Chinese brocade cloth.

Fig. 131. Two Kachin women in traditional dress.

Kachin State

The various Kachin peoples who inhabit the mountain areas in the far north of Burma are also notable weavers. The women weave their distinctive costume, which consists of a dark 180 cm long, calf-length sarong, called a *pukhang* (Fig. 131). The *pukhang* may be worn either as a wrap-around or tubular skirt. Woven in a red wool weft on a black cotton warp, it is decorated with zig-zag, diamond, and Chinese-inspired key patterns in a yellow and white supplementary weft. Other small, scattered motifs also appear across the cloth. The edges of the sarong are decorated with embroidery and pom-poms. A number of lacquered rings secure the sarong around the waist. It is worn with a long-sleeved jacket of cotton or velvet which, on formal occasions, is elaborately embellished with silver bosses and coins. A large black turban and leggings complete the outfit. Men wear either hill tribe pants or a Burmese-style *lon-gyi* in an attractive black, dark green, and purple plaid, with a Chinese- or Burmese-style cotton jacket and a large turban.[49]

The Kachin take great pride in weaving their *n'hpye* shoulder bags in black or red wool patterned with geometric designs in a supplementary weft of wool or thick cotton (see Fig. 83). They may be woven in a plain or twill weave. Silver ornaments, coins, and tassels are added for further embellishment. Traditionally, the bag is a gift from a woman to a man. In Kachin society, textiles are important in gift exchanges prior to a marriage.[50]

The act of weaving was very much a social event in traditional Kachin society and took place from October to March, during the slack season in the agricultural cycle. Women neighbours set up their body-tension looms alongside each other so that they could discuss household affairs and local gossip while weaving. The women feared that evil spirits might eavesdrop on their conversations and cause trouble. To ward off possible repercussions, the women suspended a *kapang*, a star-shaped object woven from bamboo, on a pole in the belief that evil spirits fear this talisman.[51]

A young Kachin girl begins learning to weave by weaving a long, plain, 8 cm strip which will serve as the inner lining for a bachelor's belt (Fig. 132). As she becomes more proficient, she weaves the front side of the belt, which is beautifully decorated with geometric designs in a supplementary weft or tapestry weave technique. These belts are presented to young men by their relatives and sweethearts. Weaving on a body-tension loom is very time-consuming. To weave a traditional Kachin skirt may

Fig. 132. A Kachin bachelor's belt, the first item that a young Kachin girl learns to weave. She begins first with the lining in a plain weave and then progresses to simple traditional geometric patterns on the front side. This particular belt is patterned with the star-shaped *tingsan* design. Warp 150 cm, weft 8 cm.

take a full month working four hours a day. Motifs in Kachin weaving are strongly geometric, consisting of the key, lozenge, swastika, and fret patterns which show both Dongson and Chinese influences (Fig. 133). The names given to the motifs include the names of local flora, the horns of the buffalo (an important sacrificial animal), and maze patterns.

Over recent years, Kachin weaving has become quite commercialized. There are a number of weavers in the Myitkyina area who now weave full-time on frame looms to produce brightly patterned *lon-gyi* decorated with scattered traditional Kachin geometric motifs in a supplementary weft set by pattern heddles. Many motifs may be further touched up with lurex threads in keeping with the current Burmese craze for shiny metallic textiles. Modern Kachin fabrics are very popular in lowland markets.

Fig. 133. Some traditional Kachin textile patterns. (After Taung Pauline.)

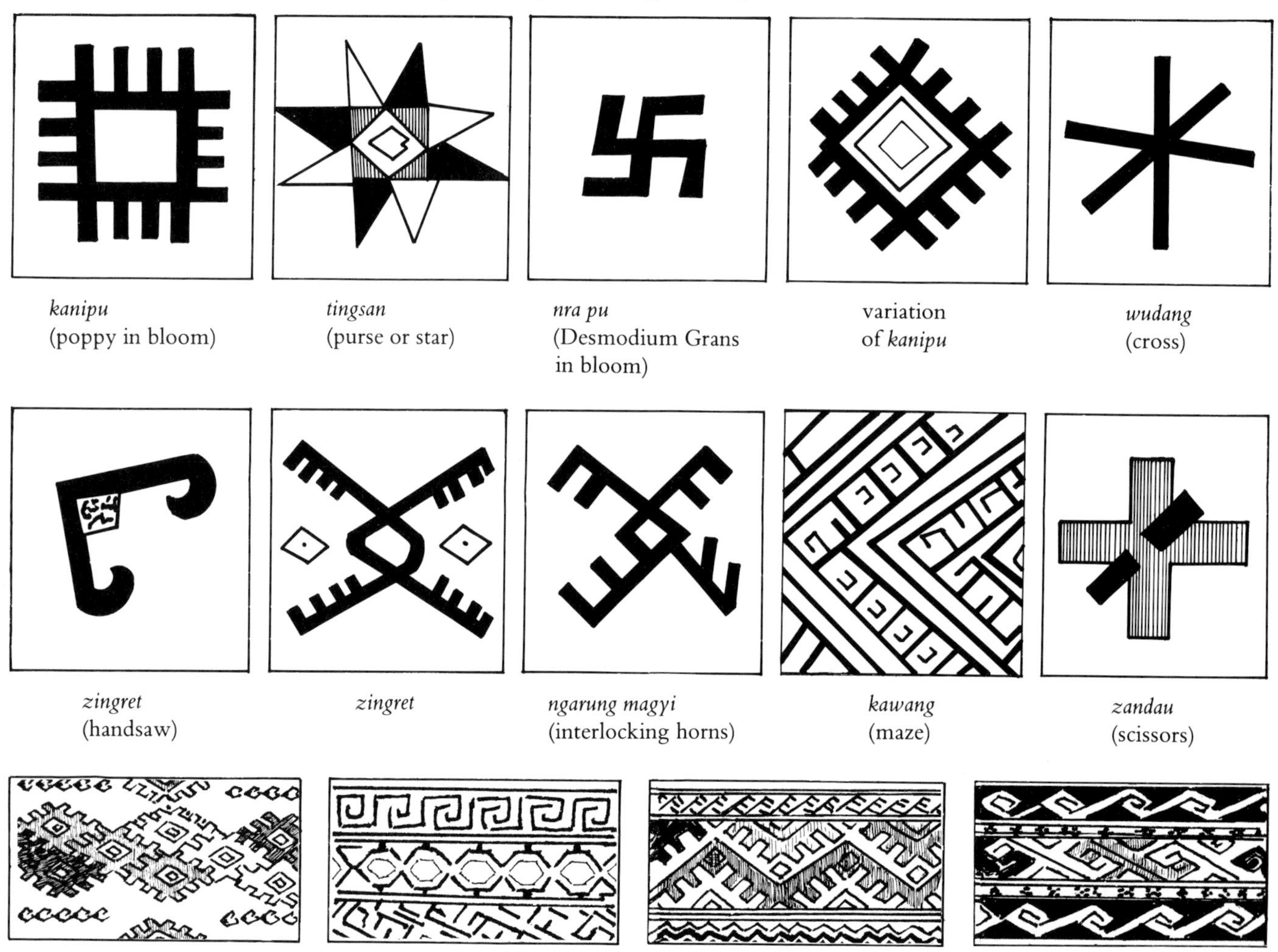

Combinations of some of the above designs to form more complex patterns.

Chin State

North of Arakan in western Burma is Chin State, where many women are very competent weavers. Using locally grown cotton, imported dyes, and embroidery threads, the women of subgroups such as the Lushei Kuki and Haka Chin weave some quite remarkable blankets on a simple body-tension loom (Plate 7).[52] Elaborately woven blankets have always been important in marriage payments and, at one time, clothing and the wearing of certain types of feathers or head-dresses distinguished chiefs, warriors, and personages of rank.

A 200 cm by 150 cm blanket worn around the body and under the right arm, with the excess tossed over the left shoulder, is the traditional costume for the Chin male. In cooler weather, more cloths may be worn, along with a white cotton jacket ornamented with bands of red geometric patterns in a supplementary weft. Originally, the blanket was worn with a loin cloth. Today, trousers or a *lon-gyi* are generally preferred. The women used to wear short wrap-around cotton skirts. The skirt was traditionally supported by a girdle of brass wire or string.[53] Today, an ankle-length sarong is worn with a short, sleeveless 'V'-necked blouse. A blanket may be worn around the shoulders as an elegant stole. Like most minority peoples of Burma, the Chin weave colourful shoulder bags.

Fig. 134. Some Chin supplementary weft textile patterns.

Geometric patterns seen on Chin shoulder bags.

Traditional geometric motifs seen on Haka Chin textiles.

A geometric border pattern sometimes seen on traditional plain weave clothes and blankets.

A popular pattern seen on modern Chin shawls.

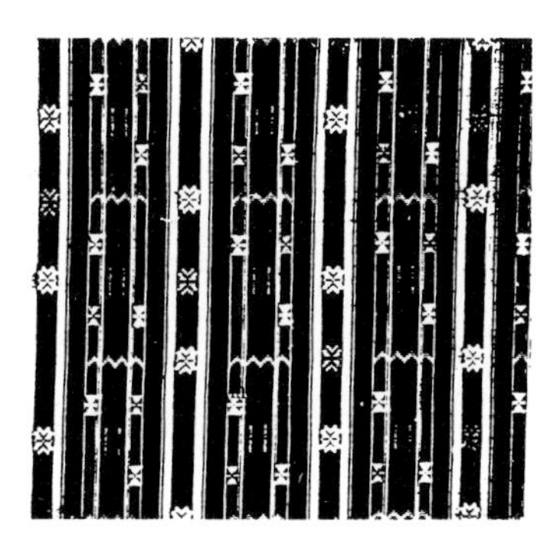

A modern Chin pattern for a woman's *lon-gyi*.

Blankets vary from coarse, white, unbleached all-purpose cotton textiles for general household use, to those gaily patterned with colourful warp stripes in black, white, green, yellow, and red. Women of the Lushei Kuki Chins at one time wove a very serviceable blanket in the form of a quilt, called *puanpui*. The padding for the quilt was made by inserting small rolls of cotton around every fourth or fifth warp thread while weaving. These tufts were held in place by the weft threads.[54]

The finest Haka Chin textile is woven from silk yarns (Plate 8). The ground colour is a rich reddish brown interrupted by a series of yellow warp bands which form a frame for extremely fine supplementary warp patterning in chains of diamonds and zig-zags. A striking weft band of four parallel lines of yellow in a twill weave completes the patterning. This particular blanket is widely used by both sexes for ceremonial wear.

Chin supplementary warp patterning on antique textiles is very delicate. As in Kachin weaving, the patterns are strongly geometric and are based on the rhomb, zig-zag, cross, and short parallel lines (Fig. 134). The Chin, too, have begun weaving for a wider clientele. Their brightly coloured all-purpose striped cloth is a familiar sight at lowland markets, as are their modern *lon-gyi* patterned with small Chin motifs worked in a supplementary warp and arranged in bands between warp stripes.

North of Chin State lies Nagaland, a remote mountainous region which separates north-west Burma from Assam. Various fiercely independent tribal groups of Nagas inhabit the area. Both men and women traditionally wear a striped *chaddar*, a rectangular length of cloth which is worn as a cape and also as a sarong by women. The *chaddar* of each tribe is distinctive in colour and in the variation of stripes. Symbolic and magical designs unique to an

Fig. 135. Naga tribesman in full regalia, including a striped handwoven blanket worn around the waist and held in place with a sash.

individual or a clan may be woven into the *chaddar*. These symbols were thought to bestow protective powers on the wearer.[55] Men may also wear a small loin cloth. On ceremonial occasions, the men also wear an elaborate head-dress made from woven cane, which is decorated with boar tusks, plumes of dyed goat's hair, and a large polished brass plaque (Fig. 135).[56]

1. Gordon H. Luce, 'The Ancient Pyu' *Journal of the Burma Research Society, Fiftieth Anniversary Publications*, No. 2, 1960, pp. 319–20.

2. Gordon H. Luce, 'The Economic Life of the Early Burman', *Journal of the Burma Research Society, Fiftieth Anniversary Publications*, No. 2, 1960, pp. 368–9.

3. U Myo Min, *Old Burma as Described by Early Foreign Travellers*, Hanthawaddy Press, Rangoon, 1947, p. 62 as quoted by Symes.

4. Ardis Wilwerth, 'Burmese, Cambodian and Vietnamese Textiles', unpublished notes for members of the Southeast Asian Textile Group of the National Museum Volunteers, Bangkok, 3 May 1985, p. 1.

5. Capt. Henry Yule, *A Narrative of the Mission Sent by the Governor General of India to the Court of the Ava in 1855*, Smith Elder & Co., London, 1858, p. 156.

6. Sylvia Fraser-Lu, 'Kalagas, Burmese Wall Hangings and Related Embroideries', *Arts of Asia*, July–August 1982, pp. 73–82.

7. Yule, op. cit., p. 155; and lithographs in Michael Symes, *An Account of an Embassy to the Kingdom of Ava, Sent by the Governor-General of India, 1795*, Nicol and Wright, London, 1800, pp. 310, 312, and 318; and Lt. Gen. Albert Fytche, *Burma, Past and Present*, Kegan Paul, London, 1878, Vol. 2, p. 63.

8. Yule, op. cit., p. 154; and Shway Yoe (Sir George Scott), *The Burman, His Life and Notions*, Macmillan, London, 1896, reprinted Norton Simon, New York, 1963, pp. 72–3.

9. Yoshiko Hokari, 'The Buddhist Robes in Japan, Korea, China and Burma', in *Proceedings of the Fifth Asian Costume Congress*, International Association of Costume, Tokyo, 1986, pp. 172–3.

10. Information from U Myint Sein of Mandalay, artist, collector, and scholar of Burmese art.

11. Shway Yoe, op. cit., p. 131.

12. San Win, 'Robes for the Buddha, Burma: Golden Country', *Horizons Magazine*, n.d., p. 15.

13. Mi Mi Khaing, *Burmese Family*, Indiana University Press, Bloomington, 1962.

14. Interview with U Than Aung, Chairman, Mandalay Division Cooperatives Syndicate, 18 July 1986.

15. G.H. Harvey, *History of Burma*, Longmans Green, London, 1925, p. 239; and H.F. Searle, *Burma Gazetter: The Mandalay District*, Vol. A, Superintendent of Government Printing, Rangoon, 1928, pp. 77 and 141–3.

16. Jane Terry Bailey, 'Burmese Textiles', *Burma Art Newsletter*, Denison University, Granville, Ohio, Vol. 1, No. 4, 1969, p. 5.

17. Visit to Amarapura *acheik* weaving workshops, 19 July 1986; and Maung Pe Kywe, 'Woven with 100 Shuttles', *Forward*, Vol. 3, No. 6, 1964, pp. 11–16.

18. Mandalay Division of Cooperatives, 'Luntaya Acheik', pamphlet (in Burmese), illustrated by U Win Maung, for an exhibition of traditional *acheik* held in Mandalay, 1985; and Amarapura Cooperatives, 'Things We Should Know About Weaving', pamphlet (in Burmese), Mandalay, 1985, pp. 2–3.

19. Daw Htwe Khin, 'Textbook for Students at Saunder's Weaving Institute', cyclostyled (in Burmese), Amarapura, 1975 (diagrams); and Aye Aye Myint (ed.), 'Burmese Acheik Patterns', from an old parabaik manuscript, cyclostyled (in Burmese), Rangoon, 1980.

20. Visit and tour of Ministry of Cooperatives weaving concerns in Amarapura, 18 July 1986.

21. Superintendent of Cottage Industries, *Report of the Superintendent of Cottage Industries*, Superintendent of Government Printing, Rangoon, 1930 and 1936.

22. Visit to Saunder's Weaving Institute, 18 July 1986.

23. Visit to San Khan and Thapye Auk, 18 July 1986.

24. Visit to the Shwedaung area, 22 July 1986.

25. Maurice Collis, *Land of the Great Image*, Alfred A. Knopf, New York, 1943.

26. W.B. Tydd, *Burma Gazetter: Sandoway District*, Super-

intendent of Government Printing, Rangoon, 1912, reprinted 1962, p. 40; and R. B. Smart, *Burma Gazetter: Akyab District*, Superintendent of Government Printing, Rangoon, 1917, p. 136.

27. This information is from an exhibition on Burmese crafts held in February 1979, Kyaikthalan Fair Ground, Rangoon.

28. See U Myo Min, op. cit., p. 16 for Varthema's sixteenth-century account, pp. 27–8 for Ralph Fitch's sixteenth-century account, and pp. 59–60 and 62 for Syme's eighteenth-century account.

29. Paul Lewis and Elaine Lewis, *Peoples of the Golden Triangle*, Thames and Hudson, London, 1984, p. 74.

30. R. A. Innes, *Costumes of Upper Burma and the Shan States in the Collections of Bankfield Museum*, Halifax, 1957, pp. 35–6.

31. Harry Ignatius Marshall, *The Karen People of Burma: A Study in Anthropology and Ethnology*, Ohio State University Bulletin, Vol. 26, No. 13, 1922, reprinted AMS Press, New York, 1980, p. 38.

32. Ibid., pp. 40–1.

33. Maj. C. M. Enriquez, *Beautiful Burma*, London, 1924, p. 101.

34. Innes, op. cit., p. 43.

35. H. N. C. Stevenson, *The Hill Peoples of Burma*, Burma Pamphlets, No. 6, Longmans Green, 1944, p. 13.

36. Unfortunately, insurgency problems in the Shan States make it impossible to travel widely and observe the various cultural groups.

37. The use of the term *Zin me* for this particular type of *longyi* is unusual, for although Burma ruled over Chiangmai from 1556 to 1775, weaving techniques of northern Thailand have not greatly influenced Shan weaving. In the case of weft *ikat*, Cambodia and north-eastern Thailand have been more influential.

38. Much of this information has been gleaned from Aye Aye Myint *et al.*, *The Structure and Designs of Zin-me Silk of the Shan States of Burma* (in Burmese), unpublished BE thesis, Rangoon Institute of Technology, 1971; and Maung Theikpa, 'Textiles from Inle Lake', *Forward*, Vol. 6, No. 24, 1968, pp. 16–20.

39. Visits to Tha-le and Ywa-ma weavers, 16 July 1986.

40. Information from Khun Pang, Curator of the Taunggyi Museum, Southern Shan States, Burma.

41. See Enrique, op. cit., p. 88; and Sir George Scott, *Burma: A Handbook of Practical Information*, Moring, London, 1906, p. 126, for further details of traditional Pa O dress as it was worn at the beginning of the twentieth century.

42. This information is based on the display of ethnographic material in the Taunggyi Museum in the Southern Shan States.

43. Lewis and Lewis, op. cit. See photograph, p. 242.

44. Innes, op. cit., pp. 31–4; and see also photograph, plate IV.

45. See Hill Tribe section in chapter on Thailand for a description of Akha clothing.

46. Scott, op. cit., p. 137.

47. Elizabeth Bayley Willis, 'The Textile Arts of India's North-East Borderlands', *Arts of Asia*, January–February 1978, photograph, p. 98 of Naga and Mizo peoples, and p. 104 of the Riang.

48. C. C. Lowis, *A Note on the Palaungs of Hsipaw and Tawpeng*, Ethnographical Survey of India, Burma, No. 1, Superintendent of Government Printing, Rangoon, 1906, pp. 16 and 24–5.

49. Revd C. Gilhodes, *The Kachins*, Catholic Orphan Press, Calcutta, 1922, p. 147, for a description of Kachin dress at the turn of the century.

50. W. J. S. Carrapiett, *The Kachin Tribes of Burma: For the Information of Officers of the Burma Frontier*, Superintendant of Government Printing, Rangoon, 1929, pp. 15–18.

51. Taung Pauline, 'The Kachin Loom', *Forward*, Vol. 17, No. 6, 1979, pp. 20–2.

52. F. K. Lehman, *The Structure of Chin Society*, Illinois Studies in Anthropology, No. 3, University of Illinois Press, Urbana, 1963, p. 165.

53. Lt. Col. J. Shakespear, *The Lushei Kuki Clans*, Macmillan, London, 1912. See chapter on Dress and Ornamentation.

54. Ibid.

55. Bailey, op. cit., p. 98.

56. Stevenson, op. cit., p. 18. The north-east borderlands of India also contain many distinct minority people who are also found in Burma, such as the Chin (called Mizo), Naga, Shan (called Meithei), and Riang. See Bailey, op. cit., for an account of the clothing and weaving of these peoples.

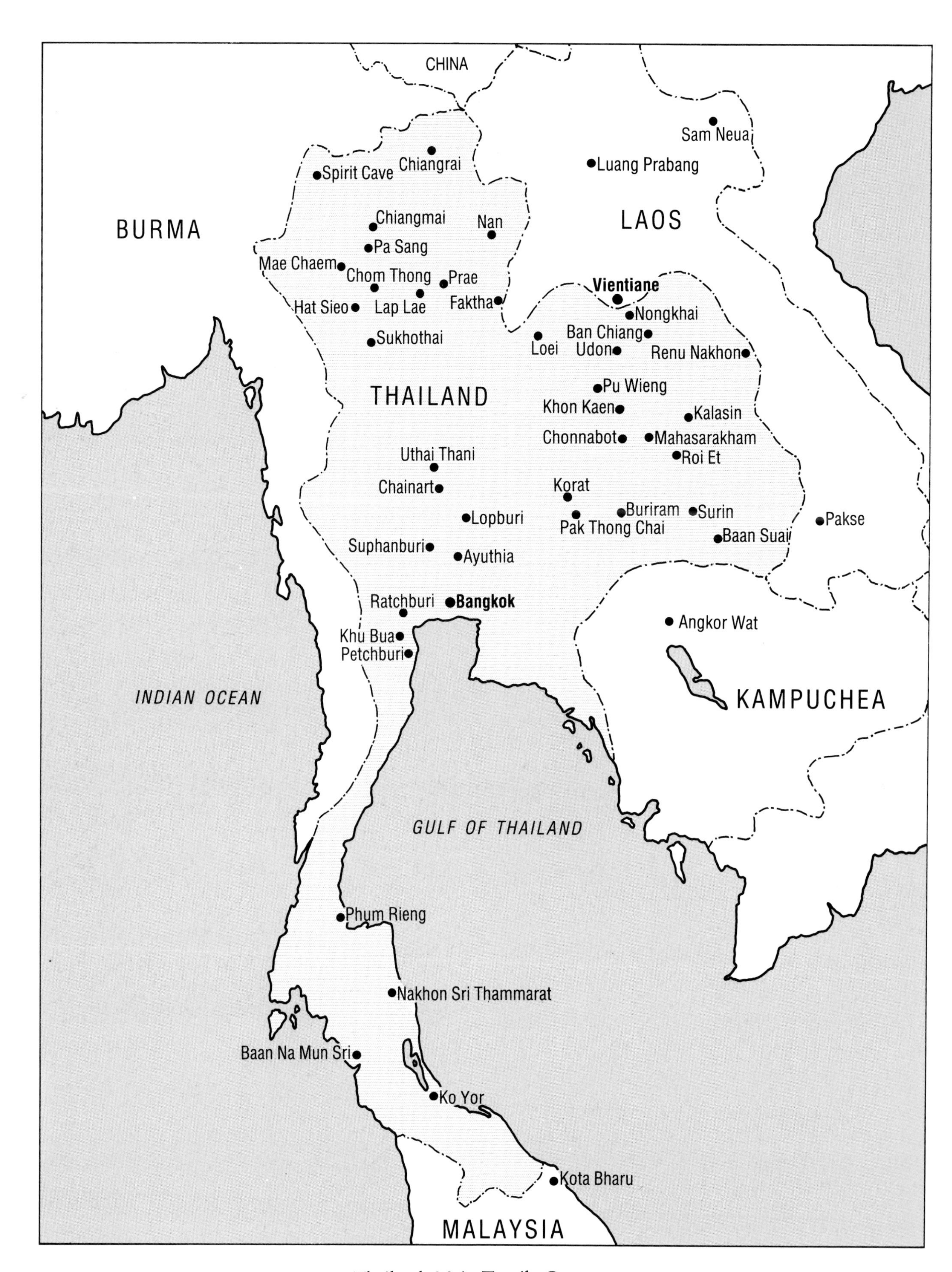

Thailand: Main Textile Centres

CHAPTER SIX

Thailand

EARLY fabric finds at Ban Chiang, Lopburi, and Uthai Thani, and more recent evidence from temple frescoes (Plate 9), strongly indicate that Thailand has had a long history of textile production.[1] Nevertheless, a complex history of migrations by various races and subgroups, interspersed with invasions and warfare with neighbouring states, make it impossible to trace the evolution of textile traditions in Thailand with any great degree of accuracy. These same historical movements, however, have resulted in the production of some magnificent and complex handwoven textiles.

Today, thanks to royal patronage and government rural development schemes, along with assistance from religious, charitable, and artistic organizations and private companies, such as the Jim Thompson Thai Silk Company, the art of weaving is being actively promoted as a source of income for rural people. For example, SUPPORT (the Foundation for Supplementary Occupations and Related Techniques), founded in 1976 by HM Queen Sirikit, has set up numerous handicraft co-operatives throughout Thailand. Equipment and materials are provided by the Foundation and goods are marketed through the Foundation's Chitralada shops. Weaving is the single most important craft to be sponsored by SUPPORT.[2] Various charitable organizations and religious groups are also active throughout Thailand in promoting the development of weaving projects. With such encouragement, the peoples of Thailand are continuing to weave their distinctive textiles.

Fig. 136. A young woman from Chiangmai wearing the *pha sin* ankle-length sarong, which is patterned with horizontal stripes and finished with a *tdinjok* border from the Mae Chaem district. Photo courtesy of Khun Duangjitt Thavisri, Chiangmai.

Hemp, silk, and cotton are the traditional fibres used to weave cloth in Thailand. Today, the use of hemp is largely confined to the hill tribes. Both cotton and silk are produced in the north-east. Domestic production and quality are not sufficient, so some yarns are imported. In recent years, the use of synthetic fibres has become more widespread. With the exception of the hill tribes, most weaving in Thailand is done on a frame loom with two or more heddles. *Ikat*, tapestry weave, and supplementary weft techniques are used to pattern traditional Thai cloth.

Articles woven include traditional clothing for both men and women. The traditional dress for a Thai woman is the *pha sin*, an ankle-length tubular sarong worn with a pleat in the front (Fig. 136). It is woven in three parts: the top, or *hua sin*, consists of a 10 cm band of cotton material attached to the *pha sin*, or body of the skirt. The latter may be plain, striped, or decorated with weft *ikat* designs. The *dtin sin*, or border, can be either a tightly woven striped band of cloth, called *jok*, or a strip of cloth, called *tdinjok*, which is intricately patterned in a supplementary weft embroidery weave and sewn by hand onto the *pha sin*. The *pha sin* is worn with a blouse or a long shawl, called a *pha sabai*, which is wound and draped over the upper torso. Men traditionally wear a long tubular cloth, called the *pha sarong*, which is folded over in front, then secured with a belt. The *pha sarong* is worn with a Western shirt or a jacket with a Chinese collar, called the *prarachatan*. The *prarachatan* has been officially designated as national dress for men and is usually worn with trousers for both formal and informal occasions. At one time, both men and women wore the *pha nung* or *jong krabeng*, an approximately 4 m long piece of cloth wrapped around the body and pulled through the legs to form knee-length pants resembling breeches (Fig. 137).[3] Other woven items include

Fig. 137. A nineteenth-century Thai lady and gentleman wearing the *pha nung* or *jong krabeng* pantaloons. The lady is also wearing a long pleated shawl, the *pha sabai*. From Frank Vincent, *The Land of the White Elephant* (1874).

pillows (*mawn*), sheets (*pha boo thi nawn*), blankets (*pha hom*), towels, mattress covers, and a general all-purpose checked cloth approximately 180 cm by 75 cm, called *pha khaw ma*, which is often used as a short *pha sin* by men working in the fields. It may also serve as a towel, a hammock, a baby carrier, a turban, or a shawl for a nursing mother.

Like the Burmese, the devoutly Buddhist Thais have the responsibility of seeing that the revered monkhood is adequately supplied with the basic necessities of life: saffron robes (which today may be purchased), pillows, towels, bedding, and banners for decorating the monastery or temple. Such gifts are presented by the faithful to the monastery on certain religious days.

North-east Thailand

North-east Thailand, or Esarn, a semi-arid plateau bounded by the Mekong River in the north and east and populated by people of Thai–Lao stock, is the most important area for weaving in Thailand.

Cotton planting begins in the north-east in May and June with the onset of the first monsoon rains, while silk rearing is a year-round activity. The cotton is picked in November and carded and ginned into tufts for spinning. Weaving steps up from January to June after the rice harvesting season, when the women are not as busy in the fields. To herald the weaving season, the villagers in some areas gather around a bonfire on a clear, chilly, full moon night for the *Long Kwuang* spinning ceremony. The women gossip and snack on local delicacies as they spin, while the young men make the most of the opportunity to court the unmarried girls with witty reparteé.[4]

Esarn is particularly famous for its silk and cotton weft *ikat*, called *mud mee* (Fig. 138). Traditional designs are handed down from mother to daughter. A skilled worker ties in the *ikat* pattern from memory with jute, banana fibres, or polyethelene string. Occasionally, an old *pha sin* may serve as a model for a particularly complex design. Today, apart from indigo, most dyes are derived from chemicals. The majority of *mud mee* motifs are strongly geometric, being based on a variety of line, wave, hook, and rhomb patterns ingeniously arranged and combined to create an almost unlimited repertoire of designs. Specific names for motifs owe their inspiration to the natural environment and may vary from place to place for an identical motif, which makes classification of designs difficult.

The simplest *mud mee* design is where odd threads of the weft (and sometimes the warp) have been wrapped at random to give small flecks of light colour after dyeing. This pattern may be referred to as *sai fon*, or 'falling rain'. Sometimes, a weft thread in this 'falling rain' pattern may be skilfully interwoven into the weft of a plaid design for added interest, such as in the *am prom*, a small plaid pattern from the Surin area. When small groups of threads are bound at intervals in regular or alternate rows, a pattern resembling dots and dashes is the result. This type of pattern is very popular on indigo-dyed cotton *mud mee*. On silk, this tying pattern is used to create a speckled texture within a major motif.[5]

In Thai *ikat*, great use is made of line patterns (Fig. 139). One pattern composed of short parallel lines is called *mee pai bai*, or *mee hol*, the bamboo leaf design (Plate 10). These lines may be extended to zig-zag back and forth down the cloth in a snake, or *nak*, pattern. A series of graduated 'V'-shaped lines may form the *mee son*, or pine motif. Large 'V'-shaped lines merging into diamond shapes form a frame or grid for many *mud mee* patterns. An 'X'-shaped design in various forms, often referred to as *kha beah* after a weaving implement, is a focal point in some repetitive *mud mee* patterns. Discontinuous diagonal lines in a diamond arrangement are referred to as *mee tak*. The cross, or *ga gabat*, is also widely seen on *mud mee*, as is a six- or eight-point star, which is a very popular motif in South-East Asian weaving. Cruciform chain patterns are also popular on everyday cotton *mud mee*.

One design, the sponge or *fawng nahm* pattern, is a *trompe l'oeil* in silk. Wavy lines in a contrasting colour on a 'shot' silk ground undulate and shimmer as the light catches the fabric so as to give a striking 'op art' effect. The cobweb or *yai maeng mum* pattern undulates around alternating cross and diamond motifs regularly spaced across the cloth and shows strong affinities with the sponge pattern.

Fig. 138. Indigo-coloured cotton *mud mee* in a traditional geometric design from the Nongkhai area. This fabric (warp 160 cm, weft 80–90 cm) is for *pha sin*.

Hook designs, called *mee kho*, are ubiquitous in Thai *mud mee*. They are usually arranged as dense interlocking zig-zag patterns across the cloth. Rows of double-headed hook shapes are sometimes aligned in such a way as to create a space for another motif. Hook patterns may appear as rearing confronting snakes between a pine tree or some other plant, or as a series of 'S' shapes in a linear arrangement. Swastika, rhomb, and key patterns, similar to those seen in Chinese art, are also incorporated into *mud mee* designs.

A very basic design in Thai *mud mee* is a small diamond-shaped pattern called *mee gung*, which involves the tying of groups of five or seven threads to create patterns. *Mee gung* designs are usually arranged diagonally across the cloth. They may be widely scattered or placed very close together in particular formations. A very basic version of this pattern is referred to as *yah*, or weed design, where small *gung* motifs resembling splashes are arranged in lines bounded by a small stripe in a two-coloured twisted weft thread.

Some *mee gung* are rendered in two or three colours to create quite complex motifs. The turtle (*tdao*) and pineapple (*sahp pa rot*) are variations of the *gung* design. A diamond-shaped *gung* surrounded by an open-work design in a lighter colour is referred to as the 'python skin' (*ngu loem*) pattern. Other well-known diamond-shaped patterns based on the *gung* motif include the water melon (*taeng moh*) design and the spider (*maeng mum*) motif. Other designs wrought within the confines of a diamond shape include the *khan bug beng*, which resembles a chandelier. It is inspired by the *bai sri* containers which are woven from banana leaves and filled with flowers and presented as offerings to a Buddhist temple. The squid (*phla meuk*) pattern resembles a snow flake, and is also very prevalent (Plate 11).[6]

Some *mud mee* designs are interrupted at regular intervals by the passage of a thread of a different colour or texture across the weft. This has the effect of muting the design and giving it a more linear flavour. Some motifs are arranged in vertical stripes separated by bands of ground colour and lines of two-colour twisted weft thread. Occasionally, a small band of supplementary weft decoration may

Fig. 139. Thai *mud mee* patterns.

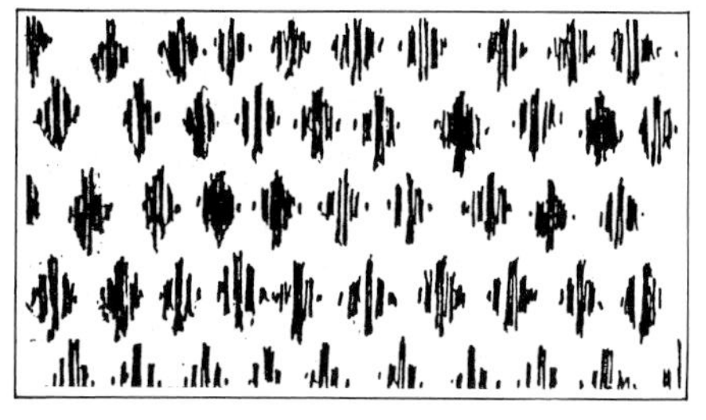

Simple *gung* pattern.

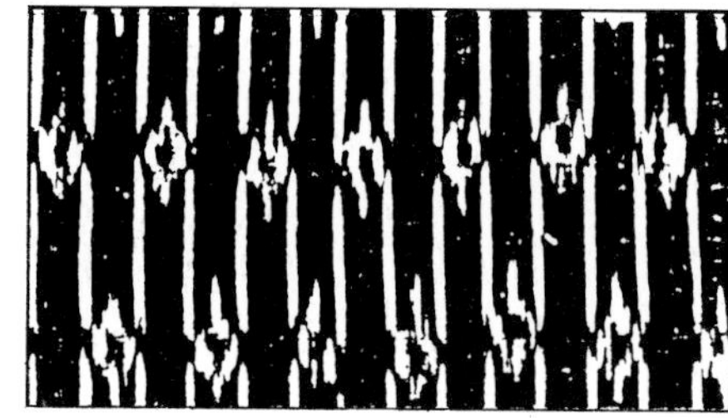

Simple *gung* pattern with stripes, sometimes called *yah* or weed design.

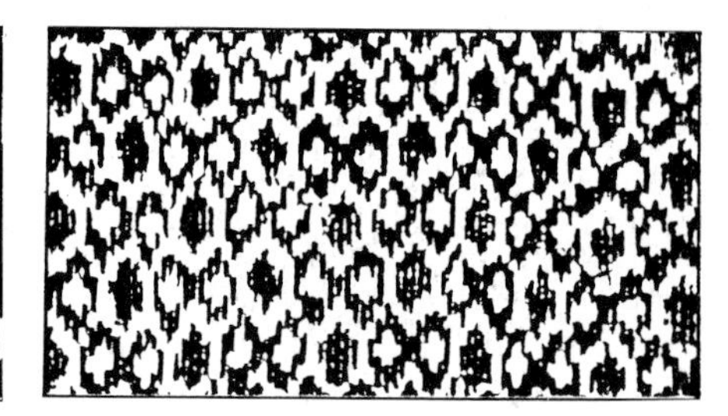

tdao (turtle)

ngu loem (python)

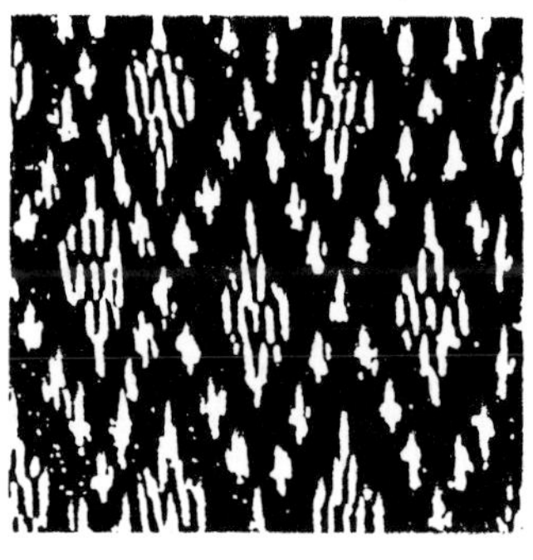

kao lam tdat (diamond)

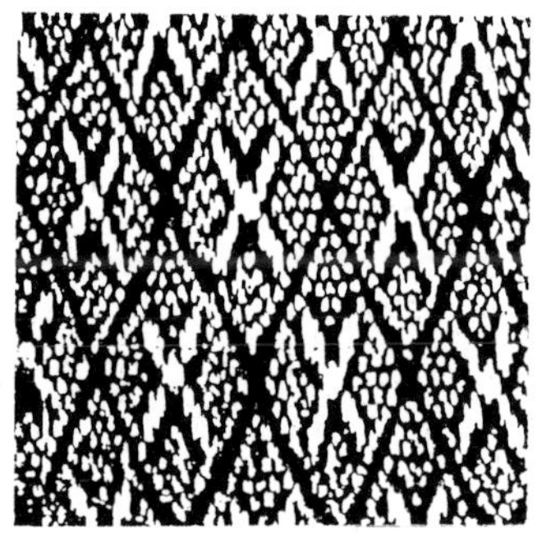

diamond with cross

kha beah

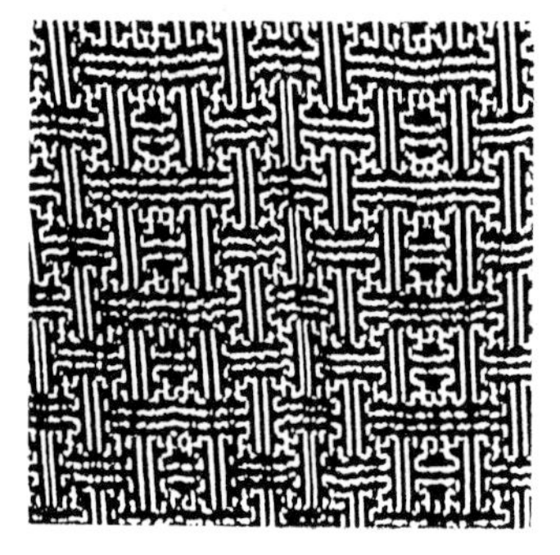

Chinese-inspired *mud mee* pattern.

mee kho (hook)

Two examples of *mee son* (pine and snake).

be added. This striped pattern, called *mee khan*, incorporates elongated linear versions of major motifs, such as the bamboo leaf, a zig-zag python, diamond-shaped *gung*, and glyph shapes, within the stripes. The bamboo leaf is probably the most distinctive and well known. It incorporates up to eleven colours within the design. *Mee khan* has a border pattern with a different motif within the stripes. In the north-east, the *mee khan* pattern is always worn with the stripes in a vertical alignment.

Buddhism, with its deep reverence for all living creatures, places no restrictions on representing living things in a realistic manner. There are a number of weavers from Surin, and more recently Khon Kaen, who specialize in depicting flora and fauna very realistically on *mud mee*. Birds, such as roosters and peacocks, may be shown along with elephants, snakes, butterflies, trees, and flowers. In most cases, these motifs form a border pattern along the bottom of the cloth.

Although *mud mee* designs are in themselves distinctive, it is the colour combinations selected by the weaver which modify and change the effect of the pattern. A silk fabric woven with a different-coloured warp and weft, gives off changing nuances of colour as the light strikes the fabric from different angles.

Fig. 139. Thai *mud mee* patterns. (*continued*)

Fawng nahm, sponge pattern, sometimes called wave pattern.

Yai maeng mum, cobweb design from Udon.

kalong pattern

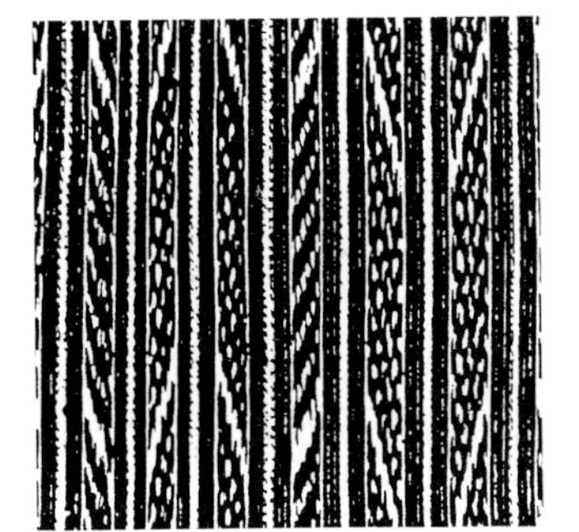

pai bai or *mee hol* (bamboo leaf)

Mee khan stripes with *nak* snakes.

Mee khan pattern with *ngu loem* python skin.

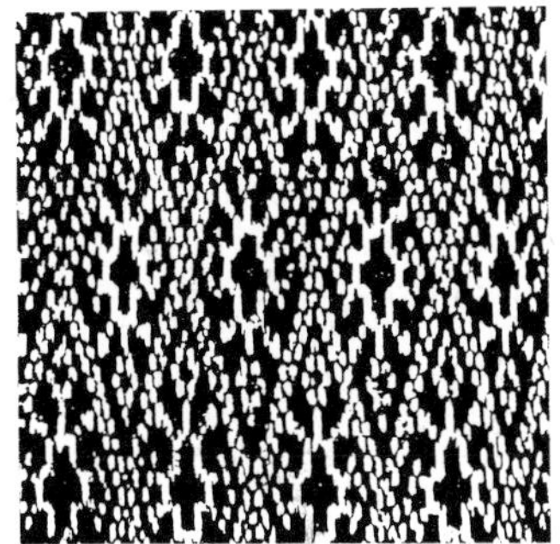

phla meuk (squid)

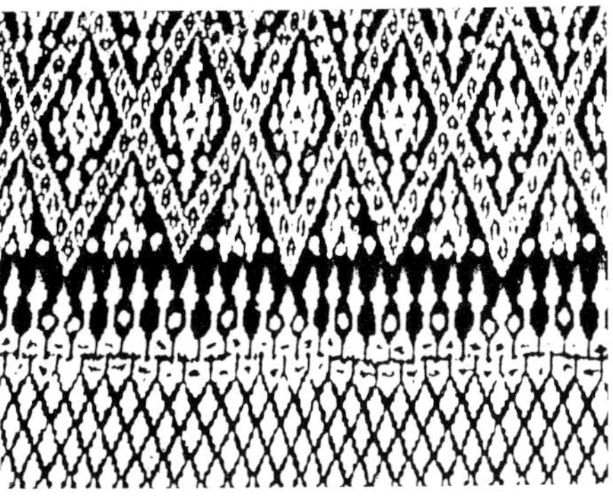

Khan bug beng, a design inspired by the *bai sri* religious floral offerings.

Peacock design from Surin.

Motifs appear and then seem to dissolve into the background with the change of light. This phenomenon is particularly pleasing on older silk textiles. Being coloured with natural dyes, these cloths are subtly understated, both in terms of design and colour. On the chemically dyed newer *ikat* cloths, the effect is much more theatrical and at times a little gaudy, with patterns that can be seen from afar in harsh bright colours.[7]

Mud mee is traditionally used for the women's *pha sin*. Today, it is also widely used for the men's *prarachatan* shirt or jacket. For the *pha sin*, the top-most part of the cloth may have a small *ikat* line extending the length of the cloth, while the border at the bottom may be somewhat more elaborate with a series of zig-zag, diamond, and *gung* patterns finished with a row or two of small, tapering vertical lines. Borders show the correct alignment of the major designs of the cloth. In former times, when etiquette and sumptuary laws were in force, it was considered ignorant to wear a *mud mee* design upside-down.

The most striking cloth in *mud mee* is called *poom som*, a magnificent 4 m length of cloth, formerly used to make the *pha nung* or *jong krabeng* pantaloons which were very popular for formal wear. Government officials during the reign of King Chulalongkorn (1868–1910) wore the *poom som* cloth in a striped design. The most beautiful of the *poom* cloth is *poom Khmer*, so named because it owes its origin to the Cambodian court. This cloth is usually in rich reddish brown hues enlivened with touches of yellow against a purplish black or dark green ground. The layout of this cloth obviously owes its inspiration to the Indian *patola* cloth. The centre field consists of finely balanced, repetitive floral and geometric designs arranged diagonally across the cloth. A closely patterned geometric or floral design extends the length of the selvages. The warp ends terminate in a series of border patterns arranged in bands. The outermost pattern ends in a series of tapering points supported by a pair of highly stylized *hong* birds or Thai swans.[8]

In addition to *mud mee* weft *ikat*, the supplementary weft *pha kit* technique is also important in the north-east (Fig. 140). The most traditional pieces are woven in a hand-spun unbleached cotton (Plate 12). Various colours are used for the supplementary weft; however, red and black are the most prevalent hues. Supplementary weft patterns are produced by setting up additional heddles or inserting pattern sticks into the warp.[9] Traditional *pha kit* designs are usually repetitive and are woven in horizontal bands. Motifs show strong affinities with those of *mud mee*. Many designs are strongly geometric and include hook and rhomb patterns, 'S' shapes, zig-zags, and six-sided stars. Frog and snake patterns are also portrayed in a geometric, stylized form. Flowers, plants, and *bai sri* offering trays are clearly recognizable in *pha kit* cloth. Horses with and without riders, caparisoned elephants, the goose or *hong* (swan) beside a plant, and the *rajah singh* (Thai lion) are also popular motifs in *pha kit*. A plain weave is used for the ground. On some cloth, a regular float weave, called *kit paow*, often done in red, may be used to finish the ends of the cloth. Supplementary yarns may be either of cotton or silk. They are thick and quite loosely woven into the cloth. On rectangular pieces, the ends may be fringed.

Fig. 140. A *pha kit* textile patterned with traditional geometric hook, zig-zag, and lozenge designs in red and black on unbleached cotton. Warp 117 cm, weft 42 cm.

Pha kit cloth has traditionally been used for a variety of domestic and religious purposes. Distinctive hard rectangular- and triangular-shaped cushions (*mawn kwan*) decorated with bands of *pha kit* weaving are probably the first items to catch the eye of a visitor to a north-eastern house (Fig. 141). Since people sit on the floor rather than on chairs for socializing and eating, these large, solid cushions provide support for the body. While sleeping, the pillow keeps the head, considered to be the most sacred part of the body, higher than the feet.[10]

The ability to weave a fine *pha kit* featuring a variety of traditional motifs was at one time regarded as an indicator that a girl was ready for

Fig. 141. Hard rectangular cushions (*mawn kwan*) from north-east Thailand which are decorated with bands of *pha kit* supplementary weft decoration. Collection of the Siam Society, Bangkok.

marriage.[11] As the marriage date approached, the bride-to-be would cut patterned strips of cloth and sew them into cushions, which would be presented to members of the groom's family on the wedding day. She was expected to present a *pha sin* to the groom's parents and to the wives of the groom's elder brothers. She would weave her own clothes and some for her future husband. Custom decreed that the bride should provide all the linen needed to set up house, such as a mattress cover, bedcovers, sheets, pillows, and towels. She would also need to prepare the bridal chamber, for in the north-east after marriage the groom usually lives at the bride's house for a number of months.

Triangular cushions, used by monks as back rests while chanting and delivering sermons, are presented, along with monk's robes, to the monastery at *Thodt Krathin*, which falls sometime in October or November at the end of the rainy season. At this time, it is customary for the faithful to present gifts to the monastery. When a young man enters the monkhood, it is a Buddhist tradition that he receive as gifts items deemed necessary for his stay in the monastery. His sponsors, family, and friends present these gifts in order to gain merit. The young man may come to the monastery in state wearing a family *pha nung* and a white shoulder cloth, followed by his sponsors bearing gifts, such as robes and pillows.[12]

These gifts might include a specially woven sheet or mattress cover, called *pha boo thi nawn*, measuring approximately 170–200 cm long by 60 cm wide. There is a 20–30 cm wide *pha kit* border in red and black with three to four design zones at one end of the sheet where the head is placed. These design zones are usually separated from each other by one to five stripes in alternating colours. Motifs tend to be strictly geometric.[13] There are certain designs, such as the *kit kaw dorg kig*, a yellow flower pattern, the *kit dorg peng*, a ceremonial offering design, the *kit kan kra kang*, or ceremonial vase motif, and the *tapao hlang gaw*, or sampan boat design, which are considered especially auspicious for ordination and merit-making ceremonies (Fig. 142).[14] *Pha boo thi nawn* sheets and *pha hom* blankets and smaller towels are also widely used at home as well as in the monastery.

At one time, there were particular designs reserved for special occasions for use by particular people. Some patterns were thought to be especially suitable for elderly people, while others were reserved for household guests. There were even patterns reserved for the son-in-law. Cloths made for important occasions, such as a house blessing and Songkran, the Thai New Year, also have their own auspicious patterns.[15]

Pha kit patterns are sometimes applied to 250 cm long *pha tung* or temple banners which are suspended from bamboo poles to mark the eight directions outside a Buddhist meeting hall or temple. These banners have bands of designs, such as elephants, horses, *bai sri* offering trays, and geometric motifs. Some banners have flat bamboo sticks inserted into the weft to form the shape of a pagoda. Some are further decorated with bells and coins along the warp fringes. The banners are flown at *Bun Prawaet*, sometimes called *Bun Mahathat*, a festival held during March–April, where the villagers, dressed in their best clothes, gather in the monastery compound to listen to a recitation of the *Vessantara* story, the most important *Jataka* tale concerning the Lord Buddha's previous lives, which personifies the Buddhist ideals of selflessness and detachment.[16] For this festival, the villagers often construct a gigantic white elephant from bamboo and white cloth and tow it in procession to the monastery.[17] The procession is led by a cloth banner painted with scenes from the *Vessantara* story.[18]

In addition to *mud mee* and *pha kit*, the north-east weaver makes plain weave utility cloth, called *pha fay* (cotton cloth), in blue or white to be cut and sewn into garments. Plain white cloth has other uses. At the time of death, the corpse is washed and new clothes are put on the deceased. A white cloth is laid over the coffin. Prior to the cremation, the cloth is passed three times over the coffin by two men. The clothes, mattress, and blanket of the deceased are placed near the pyre. At later mortuary rites after the cremation, a monk's robes, a pillow, a mat, two pieces of cloth, a pair of pants, an aluminium pot, an exercise book, a torch, candles, and a rice dessert are presented to the monkhood.[19] Pious elderly ladies dress in white when they attend the monastery.

Today, by using additional heddles, it is possible to produce heavier cloth in a variety of float weaves. Examples of these may be seen in the two-colour 'peppercorn' pattern, where the weft passes under and over three groups of two threads and then under or over a group of four to form a small diamond pattern. A larger example of a diamond formed by changing the direction of the twill pattern at regular intervals is called 'crystal ball' (*luk kaew*) in a twill weave. Thailand has been experimenting with alternating the colours in the warp and colour and size of yarn in the weft to create basket-like weaving patterns commonly referred to as 'turtle tracks'

Fig. 142. Thai *pha kit* supplementary weft patterns. (After Peetathawatchai.)

Kit kaw dorg kig, a yellow flower, used at ordination.

Kit dorg peng, representing the *bai sri* offering tray, used for merit-making ceremonies.

Kit kan kra kang, ceremonial vase, used at religious presentation.

Tapao hlang gaw, sampan boat in current, used at ordination.

Kit maeng ngao, scorpion-like insect, for household guests.

Kit dorg keow (*Murrya paniculata*). This design expresses respect for elders.

Kit kho, hook, for household guests.

Kit marg chab, wild grass flower, for household guests.

Kit garb yai, a design which expresses respect for elders.

Kit maeng ngord, scorpion, household use.

Kit taeng moh, water melon, household use.

Kit ka jae bor kai, key design, an auspicious pattern for household use.

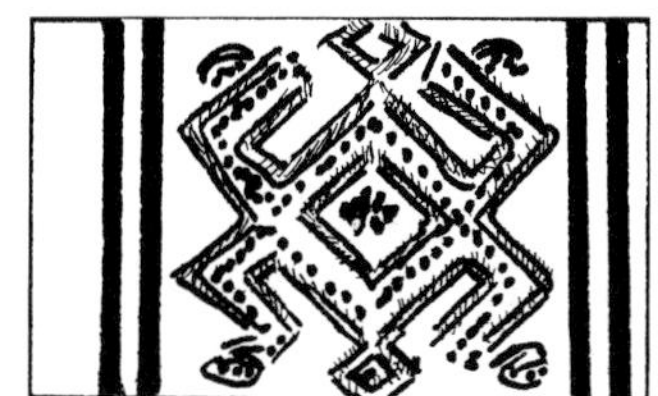

Kit oeng, frog, gifts for friends and relatives.

Kit chang, elephant, monastery presentations and house blessings.

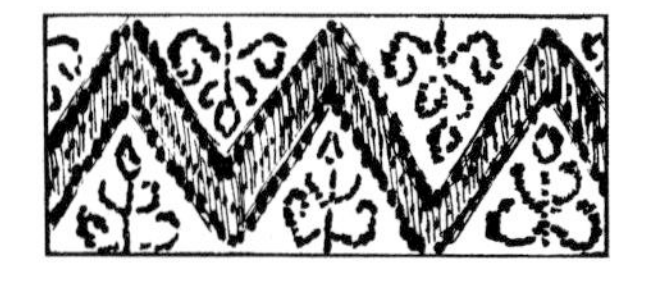

Kit dorg soey, used to decorate son-in-law's room.

Kit dorg chan, sandalwood flower, for Songkran New Year.

Kit ngu loem, python, household use.

Kit ma, horse and rider.

Kit phla meuk, squid.

Kit nak, snake.

Kit kho, hook.

(*glek tdao*). With encouragement from outside entrepreneurs, cloth in these new weaves is being produced in longer yardages for tailored Western-style clothing and furnishing materials.[20]

Although not as evident as previously, different parts of the north-east have their regional specialities. The northernmost provinces of Esarn specialize in weaving cotton textiles, such as plaid-patterned all-purpose *pha kaw ma* and *pha sin* patterned with white *ikat* on an indigo ground in bold geometric patterns, which incorporate hook, lozenge and zig-zag elements into some beautifully balanced designs. Serviceable cotton quilts, called *pha nuoem*, and lighter *pha hom* blankets are also made. A little *pha kit* and some Lao-style supplementary weft, embroidery weave *tdinjok* for sarong borders are also done in certain areas. Cotton comes mainly from Nan and Loei Provinces. Women today prefer to use machine-spun cotton yarn rather than spin it themselves. With the exception of indigo, most dyes used are derived from chemicals.

At Baan Chom Sii in Amphur Chieng Khan in Loei, which is the leading cotton producing area in Thailand, women specialize in weaving colourful checked and striped blankets in a twill weave on a four-harness loom. The villagers of Baan Na O, where a very prosperous weaving co-operative set up by the Accelerated Rural Development programme of the Ministry of Agriculture and Cooperatives has been established, weave both cotton and silk *mud mee* in geometric and animal patterns. Baan Na O also supplies surrounding villages with textiles.[21] Ban Chiang, in Udon Province, the cradle of Thai weaving, continues to produce textiles by specializing in traditional blue and white *ikat pha sin* and blankets. Baan Sri Than, 14 km from Udon Thani, is noted for its *jok* pillows.

In the Nongkhai area, a number of women from eleven villages produce *mud mee*, *pha kit*, and Lao-style *tdinjok* for the Village Weaver, a non-profit organization established in 1981 by the Sisters of the Good Shepherd, to promote the advancement of women in rural areas by actively sponsoring income-generating activities such as handicrafts. Yarns, beaters, and chemical dyes are provided by the organization. The weaver makes her own indigo dye (in the case of a blue colour), ties the *ikat*, and weaves the *pha kit* according to specified patterns. She is paid per metre of woven cloth. Much of the material woven is sold as *pha sin*, while some is made into bags, skirts, waistcoats, table mats, table runners, and wall hangings. The Village Weaver markets its textiles through interdenominational church-sponsored craft sales in Bangkok, catalogues, and mail orders.[22]

Other weavers in the Nongkhai areas may send their finished work to Nikki's House of Northeast Handicrafts, owned and operated by Khun Nisachol Poonark, a daughter of the north-east who comes from a long line of weavers. After growing up in the north-east and living in California as an American Field Service scholar, she trained as a teacher in the Philippines. On her return to Thailand, Khun Nisachol found that she did not like life in Bangkok. She moved to Udon and, while working at Udon Vocational College, became interested in promoting Thai textiles. She bought a farm just out of Nongkhai on which she built her showroom and workshop. Like the Village Weaver, she has a number of villagers from as far afield as Udon Thani weaving various types of cloth for her. Her sister supervises the weaving. Khun Nisachol employs one man to maintain the looms for the weavers. She uses Thai textiles to make a variety of items for sale to foreigners.[23]

The town of Renu Nakhon in Nakhon Phanom Province is a market centre for a number of Lao Phu–Thai people who have settled in the area. They specialize in weaving traditional cotton *pha sin*. Renu Nakhon textiles tend to be brightly coloured. *Mee khan* vertical bands containing white *mud mee* patterns separated by small puce and orange stripes on a dark indigo ground are particularly popular. The weavers sell their cloth to store owners who retail it to Thai tourists and buyers from elsewhere.[24]

Further to the south, villages in the Pu Wieng district specialize in *pha kit*, which is made into pillows. The villagers also weave *jok*, a narrow width of material patterned with small, repetitive supplementary weft patterns set within stripes of varying widths. The narrowest widths may be used as a border for *pha sin*. Pillows are also made from *jok*. In addition to making pillows and *pha sin* borders, the village of Baan Nong Hua Chang (just outside of Mahasarakham) has found a new use for *jok*. Some enterprising men of the village have begun making colourful purses, bags, and pocket books out of *jok* for the tourist trade.[25]

The busy market centre of Khon Kaen, the *de facto* capital of Esarn, lies at the heart of the silk weaving industry. The best silk comes from the Baan Hin Hoep area west of Khon Kaen. The small town of Chonnabot, some 50 km south of Khon Kaen, is particularly well known for its superb *mud mee* (Plate 13). In front of the shops along the main street, women may be seen winding silk yarn from bamboo spinning wheels onto rattan spools to the clacking of looms in the background. In recent years, Chonnabot has revived the art of making *poom som*, a 360 by 100 cm wide former royal cloth, in traditional designs using modern colours. *Poom som* cloth is particularly popular with Thai society ladies. With its extra yardage, it can be made either into a *pha sin* and neat-fitting Thai-style jacket, or a Western ensemble.

Roi Et, another busy market town approximately 120 km west of Khon Kaen, is also an important centre for fine silk *mud mee*. Within the area there are government silkworm production units and numerous weaving co-operatives. Possibly the most famous weaver in Roi Et is Pa Payom, whose traditional *mud mee* cloth is eagerly sought by connoisseurs of fine textiles, including members of the Thai royal family. All female members of Pa Payom's family are involved in weaving. A niece and granddaughters do the tying, while another family member is responsible for the dyeing, using chemical dyes. Daughters and in-laws do the weaving. Most of the *mud mee* is woven in a twill weave. Pa Payom was one of the first people in Thailand to add extra heddles to her loom to weave *mud mee* in a float weave. She also weaves metallic supplementary weft textiles on commission. Woven in traditional Thai motifs using gold thread imported from France, these are works of art. It takes a weaver about four to five months to weave a 350 cm length of cloth which retails at around US$1,000.[26] Just out of Roi Et, *en route* to Kalasin, a former pupil of Pa Payom and her two sisters have a flourishing weaving concern which specializes in metallic weaving that retails at prices even higher than those of Pa Payom.[27]

Fig. 143. A Lao Phu–Thai *pha pra wah* ceremonial shawl woven by Nang See Tun of Baan Pohn village, Somdet. Warp 220 cm, weft 52 cm.

At Baan Pohn, a Phu–Thai village out of Somdet, the women weave fine silk *pha beang* ceremonial shawls, called *pha pra wah*, with traditional Lao supplementary weft designs consisting of floral motifs set within diamonds and parallel bands on a red ground (Figs. 143 and 144). The women raise their own silk and dye it with chemical dyes. The different-coloured supplementary wefts are laid in by hand. It takes about one month to weave a shawl. Baan Pohn textiles are marketed through SUPPORT.[28]

Surin Province, close to the Kampuchean border, produces finely woven *mud mee* silk which is slightly stiff and scratchy in texture due to impurities in the local silk. Traditional colours are in the deep purple, golden yellow, and brown colour range, made until fairly recently with natural dyes.[29] Subtle plaids, such as *pha seku*, *am prom*, and *pha raberk* (Fig. 145), striped *mee khan* patterns, and realistic animals and birds are very characteristic of Surin *mud mee*. Some

Fig. 144. Lao Phu–Thai girls wearing their *pha pra wah* shawls. Photo courtesy of Mrs Nancy Charles, Bangkok.

Surin weavers can trace their ancestry back to Cambodia, so there are Khmer pattern influences in Surin's *mud mee*.[30] Baan Suai, a noted weaving village close to the Kampuchean border, specializes in producing traditional textiles, such as *poom Khmer*, tiny plaids and stripes, the *mee hol* bamboo leaf pattern, and *lai dok pikul*, which is a small, simple, repetitive floral motif woven in a supplementary weft.[31]

Pak Thong Chai, 32 km south of Nakhon Ratchasima (Korat), is a major weaving centre which produces silk for export by the Thai Silk Company founded by the legendary Jim Thompson. The company purchases raw silk from the north-east provinces of Khon Kaen and Buriram. Much of the silk is bleached and dyed on the company's premises. Since Thai silk is a little rough, it is used only for the weft. Silk for the warp is imported mainly from Japan. By using two different types of silk, the woven product can be standardized for export.. The factory also makes and repairs looms and prepares warps for direct placement on looms. Weaving is contracted out to reputable private workshops in the area. Some 300–400 women are known to be weaving indirectly for the Thai Silk Company. Weavers are paid according to the quality and length of cloth woven.[32]

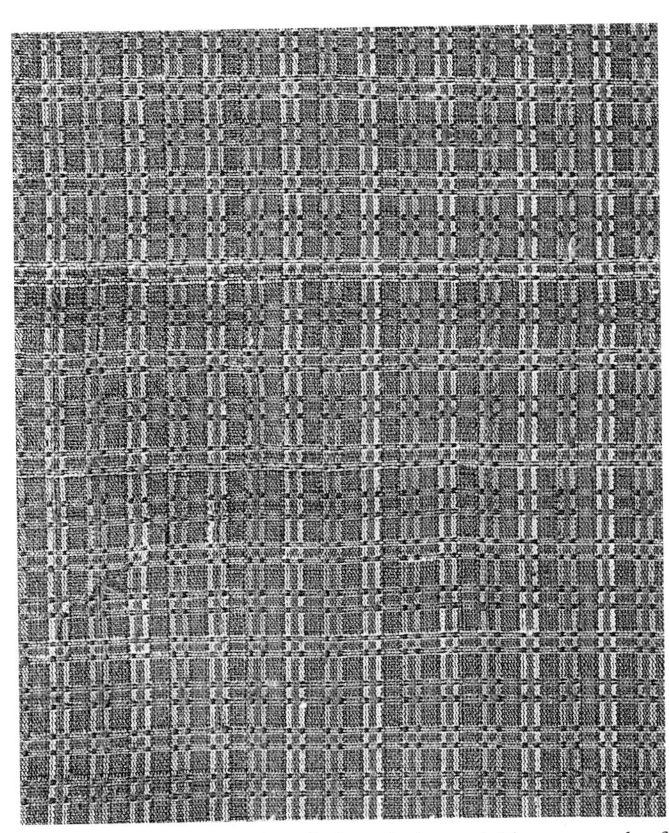

Fig. 145. An example of *pha raberk*, a plaid composed of subtly coloured warp and weft stripes highlighted by small squares of supplementary weft patterning. Surin. Collection of Khun Naiyanee Srikanthimarak, Bangkok.

Bangkok, the capital of Thailand, is also an important weaving centre for north-east silk. There are many small concerns which weave silk for entrepreneurs in Bangkok and abroad. One of the more well known is Sai Mai, which weaves top quality silk in various plys with intricate patterns for well-known fabric designers, such as Jack Larsen in New York (in association with the Thai Silk Company), Dux Furniture Factory of Sweden, and Pornpen of Bangkok. Sai Mai is owned and operated by Mr Ue Liang, who once had a cotton factory in southern China. Mr Ue Liang's move to Thailand after World War II coincided with Jim Thompson's efforts to resuscitate the dying silk industry. He and his wife, a local weaver, set up a silk factory in an old wooden house and surrounding sheds off a quiet, unpaved lane in the busy Yanawa district. Sai Mai uses raw silk from the provinces of Surin, Buriram, and Khon Kaen. Mr Ue Liang, his wife, and family supervise every stage of the weaving process. As head of the family, Mr Ue Liang takes charge of the inventory, while his three sons handle the degumming and dyeing of silk yarns. His wife takes care of the spinning, and the four daughters supervise the weavers and attend to the marketing. Sai Mai is particularly interested in new designs and techniques. Designers and craftsmen often come from abroad to work with the family on new patterns and processes.[33]

There are a number of progressive concerns who are most willing to apply modern technology to Thai silk weaving. H & M Silk Company of Bangkok has been experimenting with the discharge method of bleaching out colour to apply a lighter or brighter shade than the ground cloth. The results on handwoven silk have been most pleasing.[34] Supraphan Shinawatra, who has silk weaving factories in Chiangmai and retail outlets in Bangkok, has been developing machine-washable crease-resistant silk through the use of chemicals, which help bind the warp and weft together in such a way as to prevent slippage.[35]

Northern and Central Thailand

The old northern Thai capital of Chiangmai also has a history of weaving fine textiles. Today, weaving is largely organized as a cottage industry. There are large weaving concerns, such as S. Shinawatra and U. Piankusol, in and around the craft village of San Kamphaeng, who employ over 100 workers each to produce chemically dyed, handwoven iridescent silks in various plys for clothing and furnishings. A little *mud mee*, as well as fabrics in metallic yarns and screen-printed silks and cottons, are also produced. Like most commercial handwoven Thai silk, imported warp yarns are combined with a locally produced weft.[36]

The small town of Pa Sang, just south of Chiangmai, concentrates on weaving different coloured all-purpose plain cotton fabrics, brightly coloured cloth, and *pha sin* in stripes and plaids on hand-operated two-heddle frame looms with flying shuttles. In northern and central Thailand, stripes are worn horizontally on *pha sin* rather than vertically as in the north-east. Some hill tribe material is woven for casual Western clothing. A number of heavy cotton weaves are made for furnishings.

Some 60 km south of Pa Sang, in the Chom Thong district, is Rai Pai Ngarm, a private weaving concern founded in 1978. This firm proudly specializes in producing strong serviceable cotton cloth by time-hallowed traditional Thai weaving methods. Home-grown cotton is hand-spun onto bamboo bobbins with a bamboo and wood spinning wheel. All dyes used are natural, being made from the fruit, roots, and bark of locally available plants. Attached to the dyeing shed is a storage area where one can see piles of walnut shells, skins of mangosteen, sticks of sappan wood, baskets of limes, and burnt lime from eggshell, for indigo dyes. Under the servants' quarters, hanks of dyed yarn in soft, subtle blue, mauve, pink, beige, apricot, and green may be seen drying away from the sun, for these dyes, although colour-fast in water, fade a little on prolonged exposure to harsh sunlight. Under the main house are thirty traditional Thai frame looms with foot pedals. On these looms there is no warp beam, as the unwoven yarn is wound over the top of the frame. The finely carved, boat-shaped shuttles are hand-operated. Cloth is usually woven in 0.4 m by 25 m lengths in a solid colour plain weave or in stripes. Recently, Rai Pai Ngarm has added simple *ikat*, supplementary weft patterning, and float weaves to its repertoire. Cloth is made on commission or sold at their retail outlet in Chiangmai. For those of discerning taste in textiles, the beauty of Rai Pai Ngarm cloth lies in its wide range of soft, soothing colours, its nubbly texture, and its durability.[37]

The provinces of Prae and Nan have a large number of Thai Lu people who were forcibly moved into Thailand from Yunnan in China by King Rama I (1782–1809). They were brought to settle as farmers at the conclusion of the

Fig. 146. A young Thai woman from Nan forming zig-zag *nahm laay* patterns in a tapestry weave by interlacing different-coloured threads within their respective pattern areas along the weft.

Thai–Burmese wars.[38] Being Buddhist, they are culturally very similar to the Thais and have been largely assimilated into the northern Thai population. The Thai Lu are notable weavers. The women's traditional costume consists of a tubular *pha sin* with a series of blue and black horizontal stripes at the top, followed by a section of bright-coloured stripes in purple, pink, black, and occasionally red, turquoise, and yellow (Plate 14). Within this band is a series of zig-zag patterns in different colours woven in a tapestry weave which is locally called *nahm laay*, meaning 'flowing water' (Fig. 146). Black stripes on a blue ground, followed by a band of red, completes the *pha sin* decoration. The *pha sin* is worn with a neat-fitting, short black velvet jacket decorated with multicoloured embroidered stripes, strips of white braid, and silver jewellery (see Fig. 87). The Thai Lu also make stoles in a supplementary weft, patterned with bands of zig-zags and rhomb designs in bold primary colours (Fig. 147). Patterns similar to those on Thai Lu textiles may be seen on figures depicted in temple mural paintings at Wat Phu Min and Wat Nom Bua in Nan township.

There are a number of small workshops in Nan which make textiles patterned in the *pha nahm laay* tapestry weave on both traditional Thai and modern

Fig. 147. A small *pha beang* shawl patterned with brightly coloured geometric designs in an inlaid supplementary weft. Thai Lu. These small *pha beang* shawls are folded in four lengthwise and worn over one shoulder. Warp 137 cm, weft 32 cm. Collection of Khun Duangjitt Thavisri, Chiangmai.

floor looms. Bands of *pha nahm laay* are placed between brightly coloured horizontal stripes in a plain weave. With the aid of a porcupine quill, the weaver sends each colour back and forth in turn across sections of the weft to produce very distinctive zig-zag, lightning, and arrow patterns. Flowers are also worked in a tapestry weave along the borders of less traditional cloth. *Pha nahm laay* cloth, over recent years, has become very popular for Western-style clothing among fashion-conscious Thai ladies in Bangkok.[39]

The Thai Lu were not the only people to be displaced by Rama I. With the conquest of Laos by Thailand during the Thonburi period (1767–82), large numbers of Lao peoples were brought to settle in Thailand. The Lao Phuan people from the Phuan district, south-east of Luang Prabang, were settled in villages in Sukhothai, Uttradit, and Ratchburi Provinces. Other groups, such as the Lao Kang and Lao Ga, thought to have come from Vientiane, were moved to Uthai Thani. The Lao Song (also known as Thai Dam or Black Thai), who originally came from Dien Bien Phu in Vietnam, were settled in Ratchburi and at Hua Hin.[40]

In Sri Satchanalai district, Sukhothai Province, the Lao Phuan village of Hat Sieo, encouraged in the 1960s by Thai artist Uab Sanasen and his wife Visuta, has revived the art of weaving *tdinjok*—an intricately patterned, approximately 25 cm wide border which is attached to the traditional Lao woman's *pha sin* (Plate 15). The *pha sin* and *jok* border, collectively called *sin tdinjok*, was traditionally worn only on special occasions, such as Songkran, the New Year in April, and on Buddhist and state holidays where it would be worn to dance before honoured guests. *Tdinjok*, originally coloured by natural dyes, was never washed; when not in use it was stored inside out, away from light and insects, in a covered earthenware or glass jar.

Tdinjok is made on a two-harness floor loom strung with a black or red warp of approximately twenty-four threads to the centimetre. With the aid of a porcupine quill, the two-ply supplementary weft threads of brightly coloured cotton or silk are deftly woven into the warp. After each row has been laid in, two tabby shots in the warp colour are passed through the shed to hold the supplementary wefts firmly in place. At Hat Sieo, the weaver works on the right side of the cloth. It takes a skilled worker with good eyesight working, on average, about twenty days to complete a piece of *tdinjok*. There are nine basic patterns attributed to Hat Sieo *tdinjok*. They are all strongly geometric and symmetric in design and are named after flowers, banana leaves, drops of water, and hook patterns. The main motif, consisting of a wide band of lozenge shapes, may be repeated up to twenty-four to thirty times along the length of the cloth. Within this band motifs vary. The narrower surrounding bands containing small geometric and floral designs do not vary greatly from cloth to cloth.[41]

The *pha sin*, to which the *tdinjok* border is attached, may be made of plain or horizontally striped material. Some stripes are made with two to four twisted warp threads, called *han krarok* (or squirrel's-tail design), in two colours (usually in black combined with green, yellow, purple, or red). Some *pha sin* are patterned in small repetitive supplementary weft motifs arranged either in bands or in an all-over pattern. The best-known design is the *lai muk*, a small floral pattern, which is popular throughout Thailand.[42]

West of Chom Thong and south of Chiangmai in Chiangmai Province, is the Mae Chaem district, where a few women of Lao descent weave a special style of *tdinjok* which is characterized by a centre band of large diamond patterns in yellow enlivened by touches of red and white (Fig. 148). Yellow also

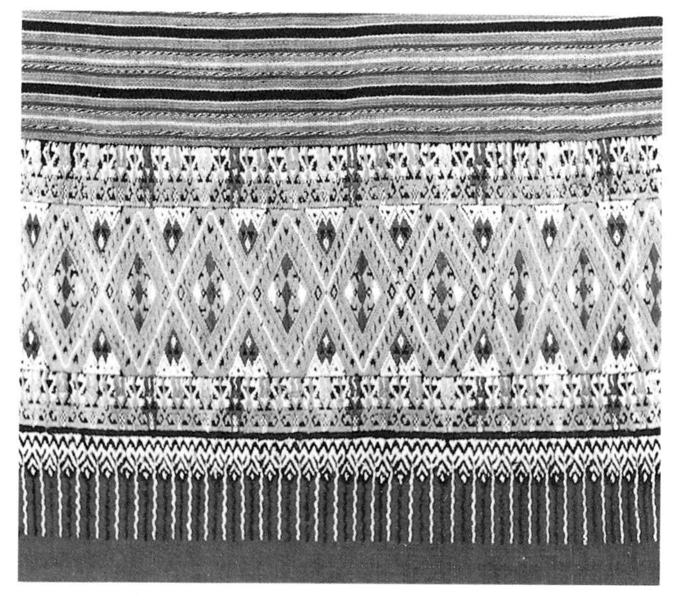

Fig. 148. *Tdinjok* border from Mae Chaem which features a centre band of large diamond patterns in yellow enlivened by touches of red and white. Width of *tdinjok* 13 cm.

Fig. 149. *Tdinjok* from the Lao Phuan village of Lap Lae is characterized by a wide band of diamond, wave, and zig-zag patterns in subtle green, yellow, and brown shades made from natural dyes. The *tdinjok* border is 28 cm wide.

predominates in the border areas. A few scattered motifs in soft blue, green, and mauve, as well as red and white, appear in the surrounding bands.[43]

The Chiangmai area, at one time, was also known to weave a distinctive type of *tdinjok*, with metallic threads mixed in with soft browns and yellows. The focal point in the diamond-shaped centre design and surrounding borders was highlighted with brighter coloured threads. With age, the threads have tarnished and the dyes have faded to leave a *tdinjok* with very interesting muted shades which offers quite a contrast to modern-day work. Such *tdinjok* is no longer made.

The Lao Phuan village of Lap Lae in Uttradit Province is noted for its wide *tdinjok* borders in subdued green, yellow, and natural brown dyes (Fig. 149). Diamond, wave, and zig-zag bands predominate. Small *hong* birds appear inside the large chain of diamonds in the centre, and as a small, scarcely discernable band in their own right. *Tdinjok* continues to be made in Lap Lae. Today, due to chemical dyes, the colours are quite bright.

The Faktha district of Uttradit, near the Lao border, has a tradition of producing fine textiles in both *mud mee* weft *ikat* and supplementary weft techniques. On *mud mee pha sin*, the *tdinjok* border is woven into the main body of the skirt rather than attached separately. Lozenge, hook, and zig-zag shapes in deep brownish reds, enlivened with touches of dark green, purple, and black, are typical of traditional *mud mee*. Newer pieces from this area, in red and purple on a pink ground, are very different in spirit. Chinese silk brocade and *pelangi*-patterned silk were at one time used to make the body of the *pha sin*. The top of the *pha sin* was often striped with supplementary weft decoration.[44]

There are a number of communities of Lao ancestry in Suphanburi, Chainart, and Uthai Thani Provinces in central Thailand, which make silk *mud mee* with a cotton *tdinjok* border. The silk threads on old central Thai *pha sin* tend to be somewhat coarse and uneven. The ground colour is usually red and is patterned with zig-zag and hook motifs in yellow, green, purple, and white. Designs include the *mee khan* stripe pattern and a variation called *mee san pao*, with interrupted stripes (Fig. 150). Sometimes, the stripes are separated by a small repetitive pattern in a supplementary weft. The *tdinjok* border attached to these textiles consists of a band of colourful, repetitive, geometric figures on a red cotton ground. The *tdinjok* border is not as wide, nor as intricately patterned as those from Hat Sieo and Lap Lae.[45]

The Lao Song people were first brought to Petchburi some 160 years ago to help build a palace for Rama I. Today, they may be seen in small

Fig. 150. Silk *mud mee* from central Thailand in the *mee san pao* pattern. The *tdinjok* is on a red cotton ground; 100 cm long by 130 cm.

Fig. 151. Some Lao Song people in Nakhon Pathom Province enjoying their traditional music and dance. Note that some of the women are wearing their distinctive traditional black and white *pha sin*. Photograph courtesy of the Church of Christ in Thailand, Bangkok.

villages in Nakhon Pathom, Petchburi, Suphanburi, and Ratchburi Provinces. They are distinguished by their sombre black costumes and swirling *coiffure*. The women wear a black blouse with silver buttons and a black handwoven *pha sin* with white or blue vertical pin-stripes (Fig. 151). It is worn with two pleats in front and is secured by a belt. It has a small closely woven band around the hem.[46] This *pha sin* is most unusual because it has a warp of bright red silk which is completely hidden by the dense indigo weft. The jacket, too, is sometimes surprising, for inside it may be covered with bright embroidery. Only at the owner's funeral is the embroidery displayed.[47] Apart from the *pha sin*, the Lao Song do very little weaving today. Instead, the women prefer to spend their time doing their unique appliqué embroidery. Under sponsorship from the Church of Christ in Thailand, they are finding a market for their products.

The art of *tdinjok* in the Ratchburi area has been revived by the inhabitants of Don Rhare village. The forebears of the present-day villagers, who were originally from northern Thailand, were moved to

Fig. 152. Lao Phuan *tdinjok* from Ratchburi Province. Width 20 cm. Collection of the Siam Society, Bangkok.

the area by Rama I. Up until World War II, they produced traditional *tdinjok*. Lack of thread led to abandonment of the craft. Through the enthusiastic efforts of villager Somboon Thomyot, who painstakingly transcribed many of the traditional *tdinjok* designs onto graph paper, the villagers have begun weaving again. The local government has provided the village with looms, while the SUPPORT foundation has assumed responsibility for the marketing.[48] Some Lao Phuan at Khu Bua in Ratchburi also make a distinctive *tdinjok* in pink and white, which may be lightly touched with green, yellow, and black (Fig. 152).

Southern Thailand

Southern Thailand is a peninsula which joins Thailand to Malaysia. Being connected to the fortunes of both countries, this region has had a most interesting past. From the seventh to the twelfth centuries, the various small states of Peninsular Thailand, such as Tambralinga (Nakhon Sri Thammarat) and Grahi (Chaiya), were part of the Buddhist Kingdom of Sri Vijaya. Historically, the town of Nakhon Sri Thammarat was the leading political entity of the south, and various Thai sovereigns relied on the rulers of that state to govern the southern principalities. (These at one time included the present-day states of northern Malaysia, which periodically acknowledged Thai suzerainty.) Although it is generally believed that local women have been involved in weaving since the days of Sri Vijaya, the modern art of supplementary weft weaving, called *pha yok*, was not known until around the eighteenth century.[49]

An 1811 uprising in Kedah (a present-day Malaysian state, formerly under the influence of Thailand), led the ruler of Nakhon Sri Thammarat to invade that state. On quelling the rebellion, he took back a number of skilled craftsmen, including weavers, as prisoners. The weavers were settled in the Tambon Mamuang Son Ton area of Nakhon Sri Thammarat. Malay-style looms were made and local women were drafted to learn the art of *pha yok*, which is very similar to Malay *songket* in that patterns are created by metallic and coloured supplementary weft yarns. Using local cotton and silk from the north-east, Nakhon Sri Thammarat was at one time famous for its delicate, gold-patterned textiles woven for the exclusive use of the ruler and high-ranking city officials. Unfortunately, the art of *pha yok* has virtually died out in Nakhon Sri Thammarat. At the time of this writing, only one family was still actively engaged in the craft. Classes given to women in the local penitentiary, in an effort to keep the art of *pha yok Nakhon* alive, met with limited success.[50]

Pha yok continues to be made in the small town of Phum Rieng, 200 km to the north, not far from the ancient town of Chaiya in Surat Thani Province. This village has a Muslim population, some of whom are thought to be descendants of the weavers from Kedah. Using north-east silk and imported gold yarns, *pha mai Phum Rieng* cloth used to be woven solely to provide the weaver's family with special clothing for weddings and religious festivals.[51]

A few years ago, the beauty of traditional *pha mai Phum Rieng* caught the eye of discerning ladies in Bangkok, and there is now a thriving industry to meet this need. Today, there are some 300 Malay and Thai floor looms in operation. On traditional cloth, three to five colours, together with gold or silver threads may be used in a single design. Well-known designs based on floral and geometric patterns

Fig. 153. *Pha yok* from Phum Rieng in southern Thailand in the *lai dok pikul* design in a white supplementary weft with touches of yellow on a green silk ground. This particular cloth was woven by a Khun Sittaya. This cloth is 100 cm by 350 cm, which is sufficient to make a *pha sin* and matching jacket. Note the similarities in cloth design and layout with *kain songket* textiles of neighbouring north-east Malaysia.

include *lai dok pikul* (Fig. 153), *yok dor ma li, lai yok bet, lai kuo lam dok, lai gan yeak*, and *lai ma leng wa* (Fig. 154). The patterns are set in the warp with additional pattern heddles placed behind the main heddles and suspended over the warp with a support hung from the frame. Up to twelve heddles may be used for simple patterns, such as *lai dok pikul*.[52] Many weavers are now imitating textiles from other parts of Thailand. To further promote weaving amongst the villagers, and to improve weaving techniques, a weaving centre has been set up by the provincial administration at Wat Phothram School in Phum Rieng. This project is financially supported by the Thai Social Welfare Council and the Women's Cultural Promotion Association of Surat Thani.[53]

South of Nakhon Sri Thammarat, towards the western side of the peninsula and 10 km out of Trang, is the small village of Baan Na Mun Sri. Weaving cloth from home-grown cotton has long been an important occupation of the women of this village. Much of the cloth woven is in a plain weave in a single colour, or in stripes and plaids. *Han krarok*, using a two-coloured, twisted thread and supplementary weft are important patterning techniques. Handwoven cloth continues to be important in rites of passage at Baan Na Mun Sri. A bride is expected to weave the groom a length of cloth which is presented to him to wear over his shoulder as he enters the bridal chamber. A cloth called *pha pan chang*, inscribed with admonitions to the living, is woven in a supplementary weft and placed on a coffin prior to burial.[54] Baan Na Mun Sri textiles are for local use. They do not seem to be retailed outside the village.

Ten kilometres out in Lake Songkhla lies the island of Ko Yor, whose population of approximately 4,000 is largely supported by weaving and fishing. Weaving is organized as a cottage industry using previously dyed cotton yarns purchased from Bangkok. The clacking of hand-operated looms with flying shuttles can be heard everywhere. Women and young men may be seen outside their houses apportioning yarns on warping frames and winding vast yardages of warp threads onto warp beams with raddles. Young girls and old ladies are constantly busy spinning yarns and twisting different coloured cotton and silk threads together in readiness for weaving two-tone striped *han krarok* cloth. Ko Yor specializes in weaving all-purpose cotton cloth in fairly subdued colours in small stripes and checks, sometimes enlivened with a simple supplementary weft pattern set into the warp

Fig. 154. Some popular textile patterns from Phum Rieng, southern Thailand.

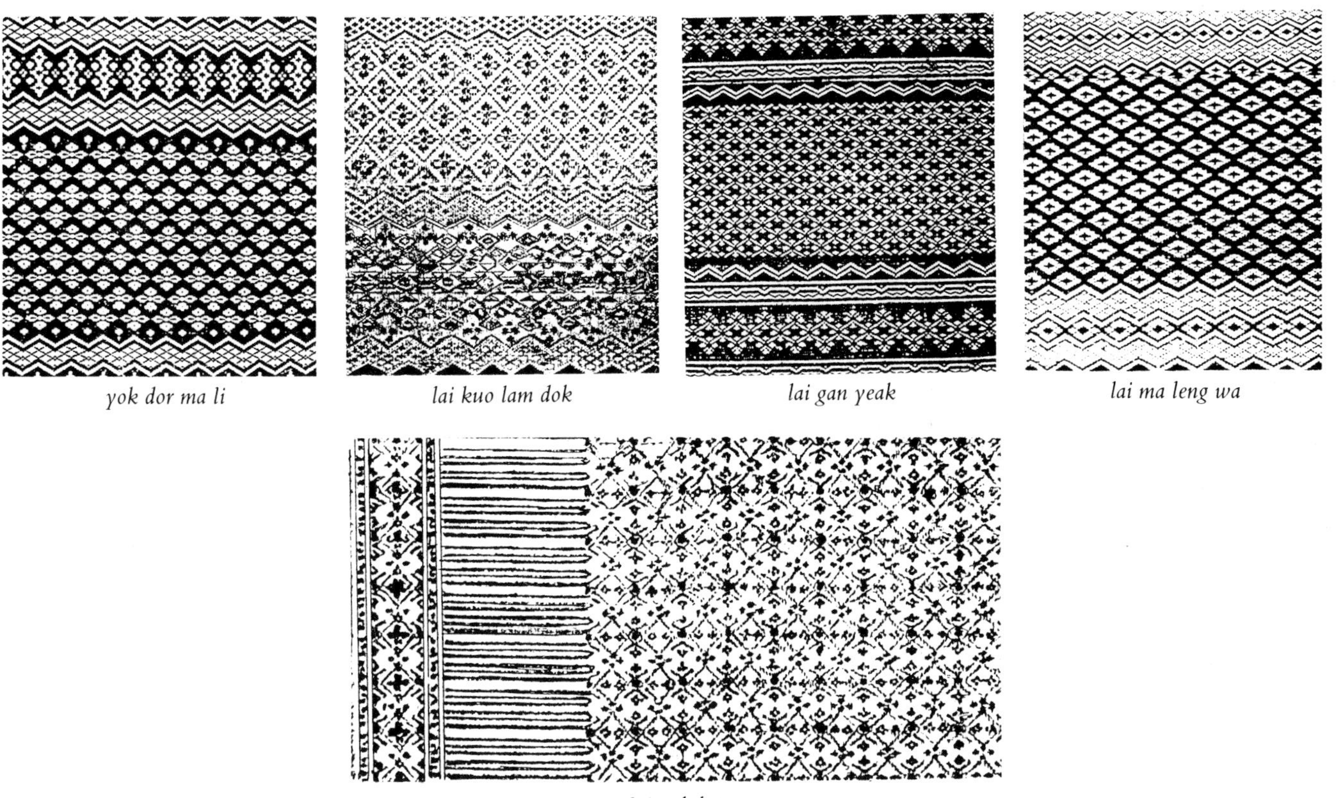

yok dor ma li — *lai kuo lam dok* — *lai gan yeak* — *lai ma leng wa*

lai yok bet

(The designs were kindly identified for the author by a group of weavers at Phum Rieng.)

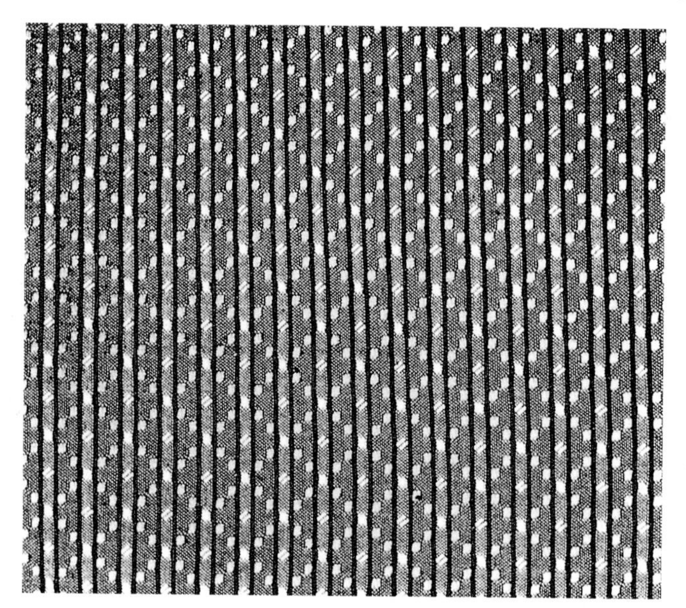

Fig. 155. Cotton cloth from the island of Ko Yor, southern Thailand, featuring warp stripes enlivened with a small diamond design in a supplementary weft.

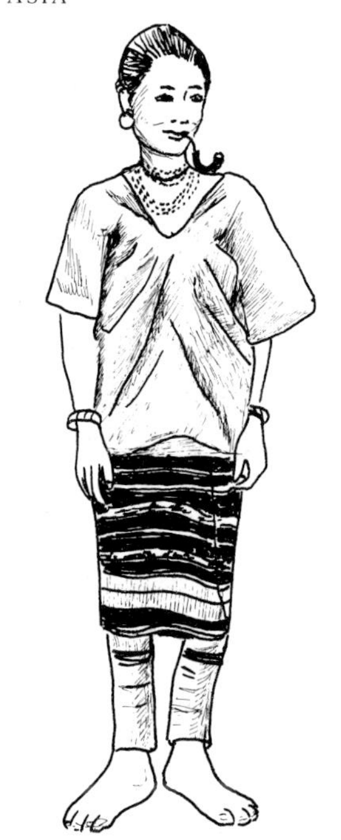

Fig. 156. Lawa woman in traditional dress.

with additional heddles (Fig. 155). Ko Yor cloth is popular throughout Thailand for its durability.[55]

Hill Tribe Weaving

Political upheavals in China, Laos, and Burma have caused many hill tribe people to migrate into northern Thailand. Over the past 100 years, large numbers of Karen, Akha, Lisu, Lahu, Hmong (Meo), and Mien (Yao) people have made their homes in the northern and western uplands of Thailand.

One hill tribe group who are the exception are the Lawa, an Austronesian group who may be found in the Mae Hong Son and Chiangrai Provinces. They claim to have been in Thailand for over 900 years and are possibly the most prolific weavers of all the hill tribe peoples. The Lawa grow their own short-staple cotton, which produces a warm, thick, nubbly cloth. Using a simple body-tension loom with a continuous warp, the women weave their short, tight, tubular sarongs which are made by joining two pieces of cloth together lengthwise. The sarongs are patterned with naturally dyed blue and red warp stripes of varying thickness, some of which enclose small dot and dash patterns of warp *ikat* in 'lightning' motifs. A sarong is worn with a white or blue sleeveless blouse, which is loose, bulky, and similar in cut to that of the Karen tribe. Leggings are also worn (Fig. 156).[56] The men dress in a manner similar to lowland Thai farmers in indigo-coloured shirts and loose-fitting pants.

In addition to clothes, the Lawa women weave elaborate ceremonial burial blankets that are strikingly different from the weaving of other hill tribes (Fig. 157). The colours, derived from natural dyes, are in muted pink or brown with narrow bands of grey, blue, and pale yellow. Small stripes of *ikat* decorate the outer selvages of the blankets. A band of zig-zag geometrical shapes in black and white woven in a supplementary weft provide a focal point for the central field. Prior to burial, the deceased lies in state with one blanket under the body and the other on top.[57] Weaving is regarded as one of the prime activities of women, for when a Lawa woman dies her loom is placed on her grave.[58]

The Karen, who number 200,000 in Thailand, have been coming over from Burma for the past 200 years. Like the Lawa, they are prolific weavers who produce most of their textile needs on a body-tension loom with a continuous warp. Their clothing and weaving techniques are similar to those of the Karen in Burma. Encouraged by various Christian missionary groups, the Karen of western and northern Thailand are now weaving their traditional striped warp *ikat* fabric in a wide range of soft hues in longer yardages. This thick, serviceable, cotton material is very popular among Westerners for bed-spreads, cushions, and casual clothing.[59]

The Akha, who may be readily identified by their colourful, towering silver head-dresses, live in the Mae Sai and Chiangrai areas of Thailand. Like the Karen, many Akha have immigrated recently into

Fig. 157. Lawa funeral blanket made from three strips of material sewn together. The central section is patterned with zig-zag patterns in a supplementary weft, while the side panels are decorated with warp stripes.

Fig. 158. An Akha woman wearing a blouse with a single halter strap, short skirt, and sash.

Thailand from Burma. The Akha continue to weave most of their own textiles. They grow most of their own cotton and whenever their hands are free, the women may be seen spinning their cotton with a wooden drop spindle. The Akha also grow hemp, which is used to make net fishing bags. The women weave with a distinctive foot treadle loom with a movable harness which is shifted along the warp as weaving progresses (see Fig. 36).[60]

The women first weave their cloth and then dye it indigo before it is made into clothing. The woman's costume consists of a long-sleeved jacket, a breast band with one halter strap, a short, kilt-like hipster skirt, a sash, and leggings (Fig. 158). Men wear a jacket and loose trousers. Both sexes make use of shoulder bags.[61] Akha weaving is restricted to a plain weave; *ikat* and other methods of patterning are not used. Instead, the Akha like to embellish their clothing with lines of embroidery interspersed between rows of brightly coloured geometric appliqué highlighted with seeds, buttons, shells, and silver bosses.[62] Plain, undyed cloth is worn by women to indicate a change in status. For the first part of the wedding ceremony, the bride appears in a plain white skirt before donning a full ceremonial costume. On attaining the status of a grandmother, a plain white skirt may also be worn.[63]

The Lahu, who number about 17,000 in Thailand, may be divided into four groups: the Lahu Na (Black Lahu), the Lahu Nyi (Red Lahu), the Lahu Shi (Yellow Lahu), and the Lahu Shelleh. Each group may be distinguished from the other by its distinctive dress.[64] Like the Akha, the Lahu Na and Lahu Shelleh weave a plain-coloured indigo cloth on a simple frame loom. Bright-coloured cloth used for banding and appliqué, is usually purchased from itinerant pedlars. The Lahu Nyi and Lahu Shi prefer to buy handwoven cloth from the Shans of Burma or the weavers of Chiangmai or Pa Sang. Recently, some Lahu groups have turned the weaving of shoulder bags with a body-tension loom into a major industry. In addition to their own distinctive bags, they are now copying those of other tribes. Traditional motifs, such as the eight-pointed star, the diamond, and zig-zag, are now being woven in a supplementary weft technique. Patterns are set with sticks and woven in different coloured cottons, acrylic wool, and metallic thread. Embroidery and silver bosses are sometimes added for further effect.[65]

The Lisu women of Thailand are noted for their colourful green and blue caftans with red sleeves, which they wear with Chinese-style pants and large black turbans (see Fig. 91). The Lisu, however, no longer weave extensively. At one time, they wove their costumes from cotton and hemp; today, they prefer to purchase their cloth from lowland markets and devote the lion's share of their time to doing appliqué and making tassels to adorn the clothing.[66]

The Hmong (Meo) people, like the Lisu, do not do as much weaving as before. The Blue Hmong continue to make their distinctive accordian-pleated batik-patterned skirts from hemp, which they weave themselves. Hemp is generally preferred over cotton for this skirt because its glossiness and greater weight gives extra bounce and swing to the skirt as the wearer moves. The Hmong loom is unique in that it combines features of both the body-tension and frame looms (see Fig. 33). In addition to batik, the Hmong decorate their clothing with intricate appliqué and cross-stitch embroidery.[67]

The Mien (Yao) people of Thailand no longer weave their own cloth. They prefer to purchase it from other tribeswomen, but they may dye it with indigo themselves. Instead of weaving, they spend their time decorating their elaborate pants with intricate cross-stitch embroidery in traditional designs (Fig. 159). They also do appliqué work.[68]

Fig. 159. Yao woman's indigo-coloured cotton pants embroidered with brightly coloured geometric patterns.

1. Ardis Wilwerth, 'Thai Textiles', *Newsletter, National Museum Volunteers*, Museum Volunteers, Bangkok, April 1985, p. 18.

2. Loretta Hoagland, 'Chitralada Shops—Handicrafts by Thai Artisans', *Look East*, Oriental Plaza Magazine, Bangkok, September 1985, p. 15; and Saowarop Panyacheewin, 'Now Soldiers' Wives Benefit from SUPPORT', *Bangkok Post*, 25 August 1985.

3. Cherie French, 'Thai Fashions Then and Now', in *Sawaddi Special Edition. A Cultural Guide to Thailand*, American Women's Club of Thailand, Bangkok, 1978?, pp. 127–30.

4. Vilmophan Peetathawatchai, *Esarn Cloth Design*, Faculty of Education, Khon Kaen University, Khon Kaen, Thailand, 1973, pp. 43–4.

5. H & M Thai Silk, *Mud Mee* (in Thai), Bangkok, n.d., p. 11.

6. Much of this information concerning the basic components of Thai *mud mee* designs was kindly given to the writer by Khun Nisa Sheanakul, collector and authority on Thai textiles. See also Peetathawatchai, *Esarn Cloth Design*, for photographs of the following designs: serpents, python, weeds, serpents and fir trees, waves, diamond pattern, and watermelon.

7. Virginia M. Di Crocco, 'Highlights from Ikat Isarn; A Lecture given by Khun Nisa Sheanakul', *Newsletter, National Museum Volunteers*, Museum Volunteers, Bangkok, August 1983, p. 20.

8. *Mud Mee*, op. cit., pp. 31–2.

9. See section on Supplementary Weft, Chapter 3, for a description of techniques used in north-east Thailand and Laos.

10. Over recent years, a few hotels in Thailand have begun using strips of *pha kit* alternating with teak panelling as wall covering. The results are most attractive: for example, the lobby of the Rossukhon Hotel in Khon Kaen has been furnished in this way.

11. Leedom H. Lefferts, 'A Collection of Northeast Thai Textiles: Content and Descriptions', typescript for the Department of Anthropology, Smithsonian Institution, Washington, DC, kindly made available to the writer by the Textile Museum, Washington, DC.

12. Stanley J. Tambiah, *Buddhism and the Spirit Cults in North-East Thailand*, Cambridge University Press, Cambridge, 1970, p. 106.

13. Merete Aagaard Henrikson, 'A Preliminary Note on Some Northern Thai Woven Patterns', in S. Egerod and P. Sorensen (eds.) *Lampang Reports*, Scandanavian Institute of Asian Studies, Special Publications No. 5, Bangkok, 1978, pp. 137 and 140.

14. The names of various motifs sometimes vary from place to place. In an effort to try and attain a little consistency, the present writer has adopted those used by Peetathawatchai in *Esarn Cloth Design*, pp. 49–51. Please note that the transcription spelling of motifs in some instances has been changed in keeping with the romanized system used.

15. Ibid.

16. Tambiah, op. cit., pp. 160–8.

17. William J. Klausner, *Reflections in a Log Pond*, 2nd edn., Suksit Siam, Bangkok, 1974, pp. 20–2.

18. Lefferts, op. cit., p. 3 of photograph section.

19. Tambiah, op. cit., pp. 180–8.

20. Much of this information has kindly been provided to the author by Mrs Rosemarie Wanchupela of ESARN Weaving, Bangkok.
21. Lefferts, op. cit., pp. 7–9.
22. Visit to Village Weaver and interview and tour of villages with Khun Suvan Boonthae, the Manager, 5 May 1986. See also catalogue *Village Weavers Handicrafts*, published by the Sisters of the Good Shepherd, Bangkok.
23. Visit and interview with Khun Nisachol Poonark, Nikki's House of Handicrafts, 5 May 1986; and Lefferts, op. cit., p. 22.
24. Lefferts, op. cit., p. 18.
25. Visit to Baan Nong Hua Chang with Khun Chantira Pisarnsin, Thai textile entrepreneur of Khon Kaen, 7 May 1986; also written up in Lefferts, op. cit., p. 16.
26. Interview with Pa Payom, 7 May 1986.
27. Lefferts, op. cit., p. 21.
28. Visit to Baan Pohn village, 7 May 1986.
29. Roxanna Brown, 'Collecting Surin Silks', *Living in Thailand*, August 1985, pp. 63–5.
30. Interviews with Khun Naiyanee Srikanthimarak, Bangkok, collector of Surin textiles, October 1985.
31. Visit to Baan Suai, 1 August 1986.
32. Visit to Pak Thong Chai, 31 July 1986; and Lefferts, op. cit., pp. 14–15.
33. Rangsit Vibhanand, 'Thai Silk, the Magic Material', *Bangkok Post*, 4 October 1981; and visit to Sai Mai, 13 September 1985.
34. Personal communication with Ms Susan McCauley, 4 January 1987.
35. Stephen Browne, 'The Origins of Thai Silk', *Arts of Asia*, September–October 1979, p. 100.
36. Visit to U. Piankusol, 8 September 1985; and S. Shinwatra, 26 July 1986.
37. Visit and interview with the owner of Rai Pai Nam, Khun Sawanee Chaisawvong, 27 July 1986.
38. Sonia Krug and Shirley Duboff, *The Kamthieng House, Its History and Collections*, Siam Society, Bangkok, 1982, p. 80.
39. Visit to weaving establishments in Nan township, 5 and 6 September 1985.
40. Clare S. Rosenfield and Mary Connelly Mabry, 'Discovering the Art of Teenjok', *Sawaddi*, American Women's Club of Thailand, Bangkok, September–October 1982, p. 24.
41. Ibid., p. 25; and Uab Sanasen, 'Notes on a Weaving Village', *Muang Boran Journal*, Vol. 16, No. 1, 1979, pp. 13–14.
42. Visit to Hat Sieo and tour of weaving establishments with Khun Sathorn Sorajpobsob Sunti, 29 July 1986.
43. Information kindly provided to the writer by Khun Duangjitt Thavisri, Chiangmai.
44. This information of Uttradit textiles is based on a study of examples in the collection of Khun Duangjitt Thavisri.
45. *Mud Mee*, op. cit., pp. 21, 27, and 29.
46. Marian MacAnnallen *Lao Song Handicrafts*, catalogue, Church of Christ in Thailand, Bangkok, n.d., p. 1.
47. Di Crocco, op. cit., p. 21.
48. Joyce Rainart, 'Village That Came Back To Life', *Bangkok Post*, 22 July 1985.
49. Vilmophan Peetathawatchai, *Folkcrafts of the South*, Housewives' Voluntary Foundation Committee, Bangkok, 1976, pp. 108, 113, and 115.
50. Ibid., p. 115.
51. Ibid., pp. 121 and 122.
52. Visit to Phum Rieng, 11 May 1986.
53. 'Silk Weaving Set Up For Villagers', *Bangkok Post*, 8 March 1985.
54. Peetathawatchai, *Folkcrafts of the South*, p. 124.
55. Visit to Ko Yor, 13 May 1986.
56. Gordon Young, *The Hill Tribes of Northern Thailand*, 5th edn., Siam Society, Bangkok, 1974, p. 57.
57. Margaret Campbell, *From the Hands of the Hills*, Media Transasia, Bangkok, 1978, 2nd edn., 1981, p. 90.
58. Ibid., p. 104.
59. Thai Tribal Crafts, 208 Bumrungrat Road, Chiangmai.
60. Paul Lewis and Elaine Lewis, *Peoples of the Golden Triangle*, Thames and Hudson, London, 1984, p. 206.
61. Anthony R. Walker, *Farmers in the Hills: Upland Peoples of Northern Thailand*, Chinese Association for Folklore, Taipei, 1981, p. 170.
62. Campbell, op. cit., pp. 118 and 130.
63. Ibid., p. 60; and Lewis and Lewis, op. cit., p. 230.
64. See Lewis and Lewis, op. cit., pp. 174–83.
65. Campbell, op. cit., p. 104.
66. Lewis and Lewis, op. cit., p. 244.
67. Campbell, op. cit., pp. 88 and 89.
68. Lewis and Lewis, op. cit., p. 138.

CHAPTER SEVEN

Indo-China

Laos

WITH the exception of Malaysia, the mountainous, land-locked kingdom of Laos shares a border with every country of mainland South-East Asia, as well as with China. Situated between the Mekong River in the west and the Annamite cordillera in the east, Laos has been a refuge for many diverse peoples displaced from their traditional homelands in South China, Vietnam, and elsewhere. The earliest inhabitants of the country were possibly the Mon–Khmer people, such as the Khmu, Tin, Lamet, and Loven, sometimes collectively referred to as the Lao Theng. They were followed by numerous Thai–Lao groups, who may be divided into lowland Lao (the politically dominant group—sometimes called Lao Lum) and tribal Thai. This latter category includes the Thai Lu from Sip Song Pan Na in China, the Thai Dam (Lao Song or Black Thai), the Thai Daeng (Red Thai), the Thai Kao (White Thai), the Lao Phuan, the Lao Neua, and the Phu Thai. These advancing groups occupied the lowland valleys, pushing the Mon–Khmer groups into the less hospitable mountain areas. Later arrivals include the Hmong and the Mien, who eke out a livelihood through swidden agriculture on the highland slopes. Laos is also home to a number of Tibeto-Burman people, such as the Akha and Lahu.[1]

As far as textiles are concerned, most groups have a weaving tradition. Over the course of time, each tribe has influenced the others in terms of design motifs and pattern layout to produce interesting 'Lao' textiles.

Indigo-dyed cotton textiles are worn by all ethnic groups for everyday clothing. Men wear indigo-coloured pants and shirts for working in the fields. The Lao women wear weft *ikat*-patterned sarongs with geometric designs in white on a blue ground, while the Hmong women clothe themselves in batik-patterned, indigo-dyed skirts.[2] Besides indigo, the Lao used to produce a wide range of bright colours from natural dyes. Black came from a tree bark, yellows from turmeric, reds from morinda and cinnabar, greens from banana stalks, and pinks, purples, and oranges from berries and flowers.[3] Most Lao weaving is done on a floor loom which is similar to the Thai loom. The Lao also delight in combining a number of intricate patterning techniques (such as weft *ikat*) with supplementary weft on a single cloth to produce designs of great visual complexity. Some hill tribe groups continue to use body-tension looms to weave cloth from hemp and wild cotton.[4]

Traditional dress for the Lao people is practically identical to that of the Thai. The Lao, are, in fact, credited with bringing their national dress to Thailand. For women, the traditional costume consists of a finely woven cotton or silk *pha sin* sarong, called *sin*. The *sin* is usually patterned with vertical stripes of weft *ikat* and finished with a *tdinjok* border. It may be worn either with a short-sleeved blouse or a silk stole, called a *pha beang*, which is wound around the breasts and over one shoulder. Lao women may wear two to three *sin* at the same time. The newest one is put underneath with the older ones on top for protection. If a guest drops in, the new one can easily be transferred to the top. Brighter colours and more flamboyant patterns are worn by young women. With the onset of middle age, small patterns and subdued colours are generally preferred.[5] For everyday wear, men dress in short trousers and a short-sleeved shirt. On festive occasions, the *pha nung* cloth wound around the body and through the legs may be worn with a European-style white coat.[6]

One outstanding group of weavers are the Lao Neua, who originally came from the old northern capital of Sam Neua in north-eastern Laos. Recent political upheavals in Laos have forced many of these people to leave their traditional homelands and go to the Thai border as refugees. A desperate need for money has forced the Lao Neua to part reluctantly with their treasured handwoven family heirlooms which, until the 1980s, were largely unknown to the outside world. Although the Lao Neua freely traded with their neighbours in items such as food, lacquer resin, and labour skills, textiles were never sold or bartered. Weaving was done strictly to meet the textile needs of the immediate family. Custom decreed that women could only wear what they themselves had woven. Every family was expected to have at least one spectacular textile woven by the inlaid supplementary weft process and patterned with abstract geometric designs.

When not in use, textiles were carefully stored in lidded stoneware jars as protection against sunlight, moisture, and insects.[7] Some antique textiles have been so immaculately cared for that, despite being up to 120 years old, they appear as new. Lao Neua textiles are usually woven with a cotton warp and a silk weft. Weavings consist of a *sin* sarong, *pha beang* shawls, blankets, and small decorative pieces for hangings (Plate 16).[8] The recent appearance of these beautiful, little-known weavings has caused some excitement amongst dealers and collectors of fine textiles.

The Lao Neua weave striking calf-to-ankle-length *sin* sarongs consisting of vertical bands of red and dark blue weft *ikat* patterned with elongated, hooked zig-zag designs (Plate 17).[9] The indigo-coloured band of weft *ikat* is of cotton. The Lao do not generally use the natural indigo colour for dyeing large areas of silk for they find that silk does not take the dye well and the indigo colour tends to corrode the threads.[10] The red weft *ikat* band, however, is of silk. Splashes of green, purple, and yellow may enliven the main design which is usually in white or blue. The bands of *ikat* are separated by broad strips of supplementary weft designs. These patterns, worked in silk thread, may be fairly simple, consisting of rows of small diamond shapes set in the loom with pattern heddles. Alternatively, they may be composed of intricate hook and rhomb designs and lines with serrated edges inserted into the weft with the aid of a quill. A variation of this basic sarong consists of three horizontal bands of *ikat* broken by three to four rows of supplementary weft patterns in chains of diamond, cross, and simple rosette designs. The weft patterns in white on an indigo ground are composed of simple geometric outlines separated from one another by a few rows of supplementary weft patterning.[11]

The *sin* is finished with a separately woven border. It may be a simple, tightly woven black band with a few coloured stripes, or it can be a *tdinjok* strip in a riot of colour patterned with birds and animals skilfully integrated into a geometric format. The *sin* is worn with a simple black blouse fastened down the front with two rows of butterfly clips. Blankets made from two lengths of cloth stitched together, may also be finished with a *jok* border. Blankets are often patterned with key and swastika motifs. They may be worn either as a mantle for extra warmth or serve as a cover for the dead prior to cremation.[12]

On special occasions, red-coloured *pha beang* shawls were worn (Fig. 160). These shawls are approximately 200 cm by 40 cm and consist of two distinct design areas. At one end, parallel bands of varying width contain rows of angular, swan-like *hong* birds or peacocks, along with diamond shapes.

Fig. 160. One end of a red *pha beang* Lao Neua shawl patterned with horizontal bands of interlocking *nak* (snakes), *hong* birds, vegetal patterns, and diamonds in an inlaid supplementary weft. Warp 300 cm, weft 53 cm.

On closer inspection, these diamonds appear to end in *nak*, or snake heads (Fig. 161).

Stylized renditions of animals and human figures may also appear in a larger central area. Animal figures include the *rajah singh*, or royal lion, and the elephant, often mounted by a highly stylized anthropomorphic figure with outstretched hands (Plate 18). Two-headed animals are quite common. *Lai kachere* hook motifs, swastika, and saw-tooth elements may subtly connect one motif to another, uniting all components into a well-integrated pattern. The other end of the cloth is decorated with rows of different designs arranged as an elaborate diamond pattern culminating in an 'eye' in the centre. This diamond motif (called *duang tda*) is a tantric symbol derived from Mahayana Buddhism. The form of Buddhism practised by the Lao Neua is somewhat different from that followed by the rest of the Lao who are of the Theravada creed. The strong Mahayanist overtones and animist practices of the Lao Neua religious beliefs are reflected in their weavings and the uses to which they are put. The *duang tda*, or third-eye motif, is thought to serve as a focus of concentration for gaining insight through meditation (Fig. 162).[13]

Fig. 161. Part of a red *pha beang* depicting a pair of *nak* (snakes) as a mirror image. Note how the use of colour helps to create an abstract dimension to the design.

Fig. 162. The other end of the shawl featured in Fig. 160 depicting a diamond pattern with the *duang tda* 'eye' in the centre.

The red *pha beang* can also serve as a head-dress during spirit invocation ceremonies for the sick and dying. The shawl is folded lengthwise in three parts, wound around the head of the priestess, and secured with a pin in such a way that the diamond shape lies squarely over the forehead. On festive occasions, two red *pha beang*, with long silk fringes, are carefully draped over the torso so that the fringes hang down the back and front. They are secured by an additional *pha beang* worn diagonally over the right shoulder and under the left arm. At these events, during games and dances, the young men delight in attracting the girls' attention by playfully tugging on the fringes of the *pha beang*.[14]

Some of the finest supplementary weft inlaid or brocade weaving may be seen on the indigo-coloured *pha beang*, which are comparatively rare. They are exceedingly subtle in design and were probably used as meditation pieces. They vary in size, and the design layout is a little different from that of the red *pha beang*. At each end, two to three bands of varying width contain zig-zag designs of great intricacy and provide a frame for a large central area consisting of sweeping diamond and zig-zag shapes of the utmost complexity. Zoomorphs, where present, hook in with other forms. White is the predominant colour, and varied textured effects on the surface are achieved by altering the thread count so that the discontinuous supplementary weft threads pass over and under each motif. Where it shows through, the ground cloth gives added depth and dimension to the design. Many designs are mirror images, but this is not always obvious due to variations in the supplementary wefts within each design area (Plate 18).

The placement of colour in antique Lao Neua supplementary weft weavings is unique. Colour is not used in the usual sense to highlight key features of a major motif; conversely, it is used to draw attention away from the main form. For example, part of an elephant's trunk may abruptly change from white to another colour. This makes the animal appear as a *rajah singh*. The head of an anthropomorph is sometimes in a colour different from the rest of the body so as to make the figure 'disappear' into the ground cloth. Patterns on the cloth appear as a series of ever-changing interwoven designs to tease and delight the mind as well as the eye. The viewer is led back and forth through a labyrinth of key, hook, and saw-tooth patterns. Creatures and geometric shapes appear and disappear as one's focus changes. Ms Patricia Cheesman, a leading scholar of Lao textiles, is of the opinion that:

> The Lao Neua use of tone and colour in their textiles stems from the Buddhist belief in impermanence, non-substantiality and suffrance. Where the expected sequence of colour does not occur the onlooker is forced to discard all preconceived ideas of balance and question the order of life itself.[15]

It is indeed unfortunate that these textiles are no longer made today. Because of the unsettled political situation in northern Laos, it is doubtful that such cloths have been woven at all in the last hundred years. Consequently, the art of creating such intricate designs has been lost. Various charitable organizations in Thailand have been attempting to revive weaving as a source of income amongst Lao Neua refugees. Today, these people produce various cloths and border strips featuring competently woven realistic renditions of animals, such as chickens and elephants, in bright colours. Their attempts at abstract designs, unfortunately, lack the subtlety of colour and design of their remarkable forebears.

The lowland Lao people in the Vientiane and Luang Prabang areas have a tradition of weaving fine silk *sin* and *pha beang* from local and imported silk yarns coloured with chemical dyes. Narrow, vertical stripes are typical of many sarongs from this area. Stripes may consist of a series of dark and light tones within the same colour range. Two-tone twisted warp threads and small black lines may break up bands of solid colours. Stripes may be formed by simple lines of supplementary weft patterning. The body of some silk *sin* may be patterned in small geometric and floral designs in a supplementary weft, while other *sin* may feature a scattering of French-inspired floral motifs woven in gold threads in a discontinuous supplementary weft (Fig. 163).

A separately woven *tdinjok* border is always added to the *sin*. It consists of one or two narrow decorative bands of simple repetitive designs, such as interlocking hooks and zig-zags tipped by small, triangular, vegetal motifs (Fig. 164). These smaller

Fig. 163. *Pha sin* from Vientiane is of green silk patterned with small yellow stripes in a yellow supplementary weft. It terminates in a colourful *tdinjok* border. This *pha sin* was woven in the mid-1920s. Warp 180 cm, weft 100 cm, width of *tdinjok* 13 cm. Collection of Khun Vanida, House of Handicrafts, Bangkok.

Fig. 164. A Lao *tdinjok* border from Vientiane depicting floral motifs within a diamond frame. It was made in the mid-1960s. Collection of Khun Vanida, House of Handicrafts, Bangkok.

Fig. 165. Some typical Lao *tdinjok* border patterns. Collection of Khun Vanida, House of Handicrafts, Bangkok.

bands frame a wider central area comprising a chain of diamond lozenges embellished with geometric and floral elements (Fig. 165). *Pha beang* shawls may also be patterned with bands of supplementary weft decoration in the form of small diamonds, birds, and animals woven in an array of bright colours.

Tdinjok borders woven in Laos today are generally not as subtly patterned, nor are the colours as carefully chosen as those made by Lao groups which have long settled in Thailand. Lao-style *tdinjok* is currently being made by newly arrived refugees to Thailand to serve as accessories, such as belts, sashes, and handbags, to go with Western clothing. *Pha beang* lengths of cloth are being woven for sale to Westerners as table runners and wall hangings.

The Pakse area to the south is noted for its fine, brightly coloured weft *ikat* silk *mud mee*. Designs range from repetitive stripes of zig-zag and lozenge shapes in various colours to wide horizontal bands containing a number of different motifs. One popular motif depicts sheaves of rice separated by large diamond shapes. The sheaves of rice sometimes alternate with small trees. Small anthropomorphs may be seen astride large 'X'-shaped motifs. A diamond-shaped, interlocking snake design is sometimes featured on *mud mee* from Pakse (Fig. 166). The Lao Phuan people of central Laos are also noted for their *mud mee* motifs which bear a strong resemblance to those of the central area of north-east Thailand.

To the north of Vientiane are the Thai Lu people, who were among the first Thai people to inhabit Laos. They are widely respected by other Lao groups for their skills in basketry, pottery, and weaving. Like their compatriots in Thailand, they have largely been assimilated into the mainstream of Thai–Lao culture. In the more remote areas, some Thai Lu continue to weave their colourful horizontal striped sarongs which are patterned with the odd band of tapestry weave in lightning patterns, amidst horizontal stripes. While they have adopted the weaving styles of their lowland Lao neighbours, they, in turn, have influenced both the lowland Lao and the Lao Neua.

Many of the Khmu people of the Luang Prabang area, like the Thai Lu, have been integrated into the lowland Lao way of life. They have adopted the Lao religion, language, and dress. It is generally thought that they acquired the skill of weaving from their Lao neighbours.[16] Women in the more remote areas continue to wear a sturdy plain weave indigo skirt which may be divided up by white stripes and simple supplementary weft patterning into horizontal and vertical design areas. The Hmong peoples of Laos, like members of their tribe in Thailand, continue to weave cotton and hemp for their batik-patterned skirts and baby carriers. The Lahu and Akha weave indigo cloth for clothing.

During the French rule and in the early post-independence years, Laos gradually drifted away

Fig. 166. Typical designs seen on weft *ikat*-patterned silk textiles from Pakse, Laos. From the collection of Khun Nisa Sheanakul.

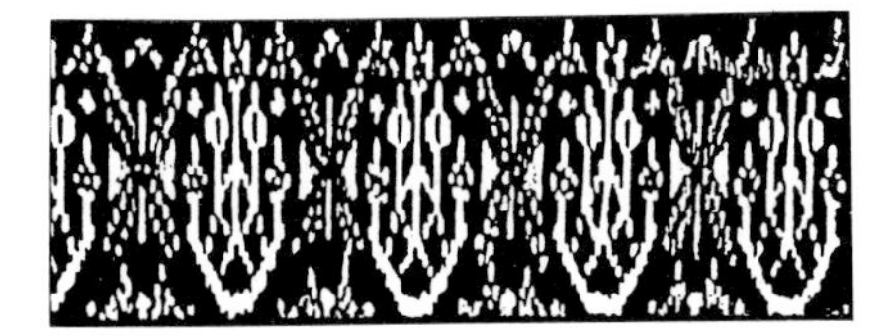

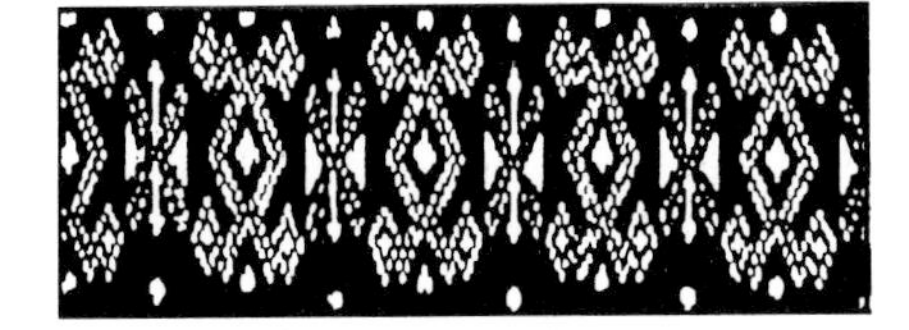

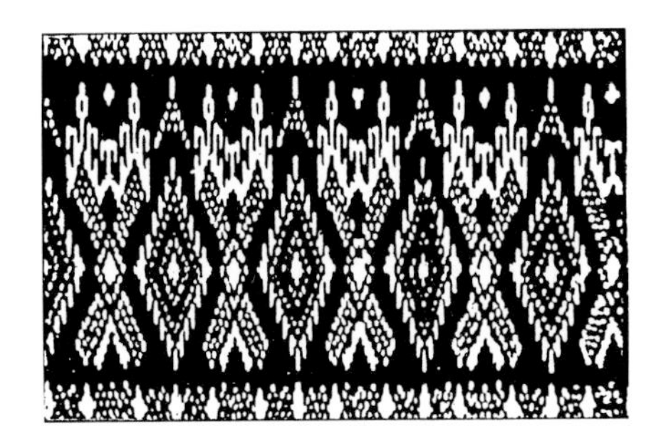

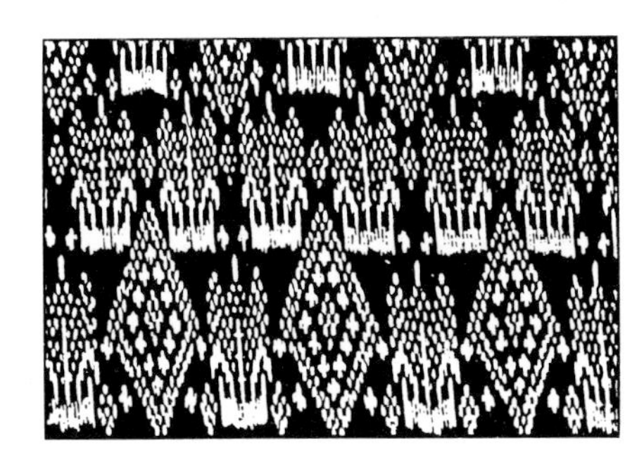

from the use of home-grown handwoven cloth in favour of cheap imported cloth, store purchased threads, and chemical dyes. With the closure of Laos's borders to the outside world since the 1970s, it has been practically impossible to import thread and dyes for weaving. As a result, there has been a return to the ancient methods of dyeing and weaving. It has been reported that in an effort to create national unity, the Government has discouraged individual styles of decoration in favour of a more 'national' style. This could eventually lead to the emergence of a number of interesting new designs.[17]

Kampuchea

Historically, Kampuchea (formerly Cambodia) is remembered for the glorious civilization which was centred on Angkor and flourished from the ninth through the fourteenth centuries. Thanks to the keen observations of Chou Ta Kuan, a Chinese emissary from the Mongol court who lived in Cambodia in 1296–7, we have a few snippets of information pertaining to life and ritual at the court of the *devaraja*, or Khmer god-kings. He took a particular interest in describing Khmer women, both royal and peasant, their hair, clothing, and rites at birth, puberty, and marriage. He also made note of rules pertaining to the use of cloth based on rank, and observed that at this point in time Cambodia produced only cotton, relying on Siam and India for imports of silk.

> A cloth of the king is worth three or four ounces of gold, and is very rich and fine. Although a few cloths are made in Cambodia, better ones come from Siam and Champa, and the finest from India.... Not only do the Khmer women lack skill with needle and thread for mending and sewing, they only know how to weave fabrics of cotton, not of silk. The spinning wheel is not even used! Women attach one end of the cloth to their belt and continue to work on the other end. As shuttles only bamboo tubes are used.... In recent times an influx of people from Siam have begun raising silkworms and growing mulberry trees; these people weave silk cloths, and know how to mend. When the cloths of the Khmer rip, they are taken to the Siamese to repair. Whether Siamese or Indian, the very finest cloths bear the continuous floral designs reserved for the king. Grand officers and princes can wear material with groups of spaced floral design. The simple Chinese can only wear material of two groups of floral design. This rule does not apply, however, to newcomers.[18]

On arrival in Cambodia in 1864, the French found the craft of cotton and silk weaving well established. Cotton was cultivated in the upper reaches of the Mekong, while silk was raised around Phnom Penh and to the south.[19] Women were reported as being weavers, while the men were mainly farmers or silversmiths. The Royal Cambodian Court supported a vast retinue of weavers who wove sumptuous silk textiles embellished with gold-patterned yarns in colours according to the days of the week. They also wove beautiful scarves for the dancers of the Royal Ballet troupe.

National dress for both Cambodian men and women traditionally is the *samphot*, a rectangle of cloth which is approximately twice the length of a sarong. It is worn wrapped around the hips, with one end being pulled between the legs before being tucked in at the waist to form a pair of loose knee-length pants. It is secured with a sash or belt. The *samphot* is worn with a jacket by men and a breast cloth by women. Sarongs, too, are widely worn for everyday wear. For country women, the sole item of clothing might be a sarong worn tucked in above the breast. Scarves and head-cloths are also widely worn.[20]

With colonialism came the importation of cheap, machine-produced cloth from Japan and other places. Cambodians purchased this for everyday wear, but for special occasions they continued to wear their special handwoven *samphot*. During the dry season, when they were not busy in the fields, women continued to weave *samphot* on their unique 5 m long looms, called *kei thbanh*, using both home-grown and imported silk yarns (see Fig. 40). Cambodian *samphot* are noted for their rich warm colours. Traditionally, each one was individually dyed with specially prepared natural substances which differed from region to region due to changes in the vegetation. A red colorant was made from sappan, anatto, and a variety of local species. Yellow was from the pounded seeds of *Gardenia grandifolia*, the heartwood of the jackfruit, and ginger root. A wide variety of blues, ranging from pale blue and violet through to deep royal blue and blue/black shades, could be achieved by adding various substances to indigo. Black was made from the viscous liquid of the lacquer tree, *Melanhorrea usitata*. The Cambodians seem to have made greater use of hot rather than cold water when preparing their dyes. They also had a good knowledge of mordants, such as alum, for their cloth was reported as being colour-fast.[21]

Samphot with a warp and weft in different colours is referred to as *samphot pamuong* and comes in a wide variety of different 'shot' silk colour combinations. Cloth which might be entirely of silk, or a mixture of cotton and silk, patterned in two or more differently coloured warp stripes is referred to as *samphot kanuih*. Plaid-patterned *samphot* are referred to as *samphot an lunh*. Red and yellow plaid combinations were at one time very popular. Colours and designs arranged in parallel weft stripes may be referred to as *samphot kosmos*.

The Cambodian loom has a transverse heddle, in addition to two parallel heddles for the ground weave. It is thus able to do a number of different float weaves. Patterns are set in the warp with extra pattern heddles supported by a cord suspended from the top of the loom frame. Up to eighteen heddles may lie between the reed and the two foundation heddles. As each pattern heddle is required, it is hooked to a special cord with a lever which connects with the extra treadle. Pressing down on the treadle raises the selected pattern heddle to create a shed. After the passage of the weft, another pattern heddle is selected to continue the design.[22] This method of patterning is used to create *samphot lobak*, the designs of which tend to be named after flowers and fruits. Ceremonial *samphot*, called *sarabap*, are patterned with supplementary gold and silver threads in floral patterns. These *samphot* were at one time popular with royalty. Dancers' sashes have also been patterned in this way. Where the gold design is confined to the border area, it is referred to as *samphot chorphnom*.[23]

The best known is the *samphot hol*, which is patterned by the weft *ikat* process (Plate 19). *Samphot hol* were traditionally woven in fine, smooth silk yarns in a weft-faced twill or satin weave. The softly iridescent colours are in the red/brown, yellow, and green range. The cloth layout of *samphot hol* has obviously been inspired by Indian *patola* cloth. The selvage border patterns are often filled with an array of rhomb patterns. The end border patterns are very elaborate, with two to three bands of close geometric designs ending in delicate, candelabra-like motifs. The centre field is sometimes composed of overall geometric rhomb patterns with 'S'-shaped snakes arranged to create a space for a central motif, which may be floral or geometric. The *bai sri* offering pedestal, the squid, and various floral patterns previously encountered in Thai weft *ikat*, may also be found on *samphot hol*. Geometric motifs in Cambodian *ikat* are often rendered in fine dot/dash staccato patterns to give an abstract dimension to the cloth. *Samphot hol* were highly regarded by Cambodia's neighbours. Malaysia, Thailand, and Burma readily acknowledge this particular textile as the inspiration for many of their weft *ikat* cloths.

The Cambodians also excelled in making large pictorial banners in weft *ikat*. These cloths were used to decorate the house for marriages, funerals, and other important events. Some of these are among the most elaborate and realistic portrayals ever seen in weft *ikat*. The most well known is a funeral banner with three to four repeats aligned horizontally across the cloth. The central band depicts one to three figures inside an open-sided pavilion crowned with a steep-sided, towering roof which is typical of South-East Asian Buddhist architecture. Between each pavilion may be a pair of figures set amongst small plants, trees, and, occasionally, birds. Below is usually a band featuring pairs of facing elephants, often with trunks upraised supporting a small parasol. Small human figures may appear between the elephants in this band. Above the pavilions may be more elephants or, more often, horses.[24] Scattered rosettes, rhomb, and star patterns dot the background. For the edification of the living, some banners have a saying from the Buddhist scriptures inscribed in Cambodian. Typical colours for these banners are red, brown, orange, and white. Cambodians no longer seem to be making these banners (Plate 20). Today, new examples are being made by Lao women in refugee camps in Thailand. Using a brighter and more varied palette, these copies are not as elaborate nor as sophisticated as the originals, but some of the better ones possess a simple, unfettered naïvety which makes them appealing to foreigners who eagerly snap them up for wall hangings.

The predominantly Muslim Cham–Malay people who live north-east of Phnom Penh in Kompong Cham Province, constitute about 1 per cent of Kampuchea's population. The women wear a neat-fitting, long-sleeved, open-neck tunic. Under this is worn a batik sarong. The men wear a shirt and a long, ankle-length robe. The Cham–Malay people at one time made use of a number of resist techniques to pattern their sarong. Batik, using a starch or wax resist, *tritik*, and *pelangi* are known. In recent years, the Cham have preferred to purchase imported Javanese batik rather than produce their own textiles.

Traditional Cambodian weaving came to an abrupt and untimely halt in the mid-1970s when the Khmer Rouge came to power. Apart from checked cloth for Khmer Rouge scarves, very little weaving was done during their four years of power. According to recent observers, the present Kampuchean Government is attempting to revive weaving as part of an effort to make the country self-sufficient in its textile requirements. The weaving of simple plaids and stripes has been revived. It is hoped that the art of weft *ikat* will eventually be resuscitated as part of Kampuchea's cultural heritage.

Vietnam

The country of Vietnam comprises a long coastal strip lying directly south of China. Bounded in the west by the Annamite chain of mountains, Vietnam is one of the most ethno-linguistically and culturally complex areas of South-East Asia. The population is divided into two groups: dwellers of the coastal plains, valleys, and deltas, and the tribal hill people. Lowlanders include the Vietnamese, who have been greatly influenced by Chinese civilization, and the Cham, who have incorporated many aspects of Indian civilization into their culture. The hill people of South Vietnam, collectively called Montagnards, number approximately one million and inhabit about 50 per cent of Vietnam's total area from the seventeenth parallel to Bien Hoa, some 32 km north of Ho Chi Minh City. They are divided into about three dozen tribes and subgroups, each with its distinct customs, mores, and religious beliefs. North Vietnam has an even larger number of minority people living in the north-west uplands. These include the Muong, Thai, Hmong, and Mien groups.

The Vietnamese, who comprise over 80 per cent of the total population, were under Chinese rule from 111 BC to AD 939, and are thus heirs to the great weaving traditions of China. It has been reported that during the Doi Tran Dynasty, from 1225 to the late fourteenth century, there was a great expansion in the trade of cotton, silk, and brocade, and a growth in crafts and trading guilds.[25] There were elaborate sumptuary laws regarding dress. The clothing colour of government officials was determined by rank and special clothes were worn for mourning.[26]

The Vietnamese at one time had numerous centres of sericulture, such as Phu Phong in the Mekong Delta, and Hai Duong and Nam Dinh in North Vietnam. These centres produced hand-reeled silk for a flourishing domestic silk industry.[27] The Vietnamese wove fine silk fabrics in a variety of intricate weaves and were known to spend much time on fine embroidery in colourful satin-stitch and gold couching. Unfortunately, French colonial policy encouraged the importation of luxury European fabrics, cheap silk from China and Japan, and cotton from India. These imports, along with the establishment of cotton mills at places such as Haiphong and Nam Dinh, led to a decline in the production of handwoven Vietnamese textiles.[28]

The Montagnards, apart from furnishing a few luxury items of trade, such as aromatic woods, elephant tusks, and rhinoceros horns, have generally remained culturally aloof from their lowland neighbours. They have not been greatly influenced by the traditions of India, China, and Europe.[29]

The Montagnard may be divided into two major linguistic groups, the Mon–Khmer and Malay–Polynesian. The Mon–Khmer affiliated groups of people inhabit the north and the highland areas of the south. Peoples living in the north include the

Bru, Pacoh, Katu, Jeh, and Bahnar, while those in the south include the Mnong, Stieng, and Maa. They are separated by a wedge of Malay–Polynesian speaking peoples, such as the Jarai, Rhade, and some Hre, who inhabit the central provinces of Pleiku, Plu, and Darlac.[30]

The short-staple cotton plant *Gossypium herbaceum* is the most important textile fibre crop in the highlands. In the past, the Mnong, Maa, and Kil tribes produced textiles from bark.[31] The Maa have also been known to make textiles from banana leaves.[32] The Hre people weave cloth from ramie and hemp.[33] Cotton is planted in the same fields as rice during the rainy season around July–August. It is harvested after the rice in the dry season when the bolls are ripe. Cotton is hand-processed in a small wooden gin and spun into thread on a simple hand-operated spinning wheel made from wood, bamboo, and rattan. After being wound into skeins, the cotton yarn may be left its natural colour or dyed blue or red, the two fundamental colours of Montagnard textiles. The blue colour comes from the indigo plant, which grows wild. Sappan wood, morinda root, annatto, and a number of lesser known local species, may be used to produce red/brown dyes. Yellow made from turmeric root and the heartwood of various trees may occasionally be used.

Warp threads are prepared for the loom by winding two balls of wool in and around four upright sticks, which are replaced by the breast and warp beams, heddle, and shed stick when placed on a simple body-tension loom. Montagnard body-tension looms use a continuous warp structure and are extremely simple. Only short lengths of cloth are woven on the loom used by the Jeh and Nong Mnong, where the warp is kept in place with the feet (see Fig. 30). Longer lengths are woven by people such as the Jarai and Rhade, who attach the warp beam to a tree or poles under the house. Most cloth produced is indigo in ground colour and is woven in a plain weave. Colourful warp stripes are the main form of decoration. Within the stripes there may be bands of finely woven, supplementary weft patterns made by carefully counting every warp thread and inserting a series of fine sticks over and under various warp thread combinations to produce small geometric designs (see Fig. 62).

Basic clothing for the Montagnard people is a loin cloth for men and a sarong for women, both of which can be worn with or without a jacket or blouse. Blankets may be worn by both sexes for extra warmth. A few tribes also weave bags.

The 20–25 cm wide loin cloth may be anywhere from 300 to 700 cm long. Its length varies with the tribe and the social position of the wearer. Longer ones may be worn by chiefs as a sign of rank. In the Pacoh tribe, the length of the loin cloth depends upon the wealth of the wearer. Men of property wear longer loin cloths than do those of lesser means.[34] Most tribesmen wear a longer loin cloth on ceremonial occasions. The style of draping the loin cloth also varies from tribe to tribe. For example, the Mnong male favours a fairly short 300 cm long loin cloth which is wrapped two times around the waist and drawn through the legs and up over the band formed at the waist so that the two ends fall in such a way as to leave flaps at the front and side of the body.[35] A Jarai male wears a much longer loin cloth which is wound several times around the waist before being drawn through the legs and arranged so that a wide folded flap falls over the right buttock, while the other end hangs down to the knees in front like an apron (Plate 21).[36]

Loin cloths may be plain or quite ornate. The Jarai loin cloth is woven in dark blue enclosed by red and white warp stripes. The fringe ends are beautifully finished with fine supplementary weft patterns worked in white on a red ground, which may be edged with a little twining above the fringes. Many tribes like to decorate the ends of their loin cloths with red pom-poms, copper coins, and small tubes of lead. Despite incursions of Western civilization through French missionary efforts, and the American presence in the 1960s and 1970s, many Montagnards have continued to wear their loin cloths with pride as a mark of tribal distinction. During the Vietnam War it was not unusual to see newly affluent tribal youths happily riding the latest Japanese motorcycles around the main towns in the central highlands with loin cloth ends flying in the wind.[37]

Women wear a knee-length rectangular piece of cloth approximately 50–70 cm long as a wrap-around skirt for working in the fields. It is tucked in or knotted at the waist. Sometimes, string or a silver belt is used to secure the skirt. Two lengths of cloth are sewn together to make an ankle-length skirt, which is worn on more formal occasions. Skirts are plain blue or black cloth; others are decorated with colourful horizontal warp stripes in red/maroon, white, and, occasionally, yellow. Like the loin cloth, some sarongs contain bands of two-tone supplementary weft designs. Some tribes like to embroider geometric and simple naturalistic designs in brightly coloured imported cotton and wool yarns. A warp fringe sometimes serves as added decoration down the sides of a sarong.

Blouses and jackets are simply tailored from rectangular lengths of cloth folded in half with an opening in the centre for the neck. Sleeves are created be sewing two cylindrical lengths of cloth to the shoulder openings of the jacket. The Rhade like to decorate their long-sleeved indigo jackets with small stripes along the sides and across the shoulders of the jacket. The top and bottom of the sleeves may also be banded.[38] The Jarai male wears a similar jacket but it is distinguished by a colourful square of red embroidery across the chest. Preng Mnong males decorate their jackets with fringes or a band of geometric designs in red and blue, while the Hre embroider both sides of their dark blue jackets with red and white designs.[39] Bru women sew rows of coins on their best blouses.[40]

Blankets are made by most Montagnard people by sewing two to four lengths of woven material together to produce a 200 cm length of cloth. These blankets are sometimes worn as cloaks and wraps. Most commonly, blankets are worn in a criss-cross fashion across the upper torso, so as to leave the arms free. Blankets serve as hammocks, baby carriers, and shrouds. Some of the most beautiful are decorated with geometric designs in a supplementary weft.

Montagnards also make shoulder bags from a piece of cloth folded in half and sewn along the selvages. To this is attached a strip of tightly woven white cotton to serve as a handle. Small personal items, such as tobacco boxes, flint, knives, and lucky charms, may be carried in the bags. Some Montagnard men prefer to carry such small personal items wrapped in the waistfold of their loin cloths. Various turbans are also worn by Montagnard tribesmen. These are traditionally made from lighter materials than the thick tribal homespun cotton. Such fabric is usually obtained from lowland merchants through barter.

The Montagnard design register is deeply rooted in Dongson-style neolithic geometric patterns based on the triangle, lozenge, zig-zag, and key and hook patterns. These have been refined over the centuries to produce elegant, beautifully balanced patterns (Fig. 167). Names given to these patterns show the Montagnard's strong attachment to their environment. Dot and dash patterns may go by names such as 'cucumber seeds', and 'palm leaves'. Small zig-zag triangular patterns are called 'buffalo teeth', 'jawbone of the cricket', or 'prickles'. Larger triangles include the 'hairs of the kapok', 'toucan's beak', and the 'rice pounder'.[41] In much Montagnard weaving, lozenges are the fundamental motif around which others are assembled. With respect to Mnong weaving, Huard and Maurice have suggested that small lozenges may represent the navel or womb, the female sex, or fertility in general. They are also of the opinion that a large lozenge could represent a buffalo pen, for the buffalo is of great ritual importance to the Montagnard.[42] They also identify a number of other designs, such as 'mottled python skin', 'caterpillar', and 'ant tracks' within a lozenge frame.

Boulbert has identified a number of designs in Maa weaving, such as 'caterpillar eyes', 'mouths of leeches', 'python skin', 'bird's feathers', and 'rice pounder' set within a lozenge frame.[43] Stylized renderings of birds such as the quail and dove, animals such as monkeys, and human figures occasionally appear in traditional designs. Cultural objects such as crossbows, stairs, and jars, and natural phenomena such as lightning and stars have all lent their names to textile patterns. Names for identical patterns vary from tribe to tribe, making a coherent classification of designs virtually impossible.

Like other peoples of South-East Asia, the Montagnards are also known to copy design elements from their neighbours. A number of distinctly Lao and Cambodian lozenge motifs have found their way into the Montagnard repertoire. The fact that the Montagnards are not adverse to design innovation was proved by the Vietnam War. Forced to flee from their tribal homelands, many women packed spindles and shuttles amongst their meagre belongings and set up their looms in makeshift refugee camps. In keeping with their practice of weaving what they saw around them, various types of helicopters (such as the 'Chinook' transport and the 'Huey' gunship), jet fighters, army trucks, bombs, Americans with dogs, and Japanese wrist watches (copied from US Navy catalogues) began to appear on textiles along with traditional designs (Fig. 168).[44]

The tribal people of North Vietnam also have a weaving tradition. The most numerous are the Thai, who may be divided into a number of groups, such as Black, Red, and White according to the colour of the upper garment worn by the women. The Thai practise both wet and dry rice agriculture and live in the upland valleys and river bottoms along the banks of the Red and Black Rivers. They are organized into *muong*, or districts, under hereditary chiefs. Like most Thai, they are talented weavers and at one time produced all their textile needs from local cotton and silk dyed with natural dyes and woven on floor looms. In general, these groups are much closer to the cultural level of the lowland Vietnamese than the

Fig. 167. Some patterns seen on Montagnard textiles. (After Dournes, Boulbert, Huard and Maurice and *France Asie*.)

buffalo teeth

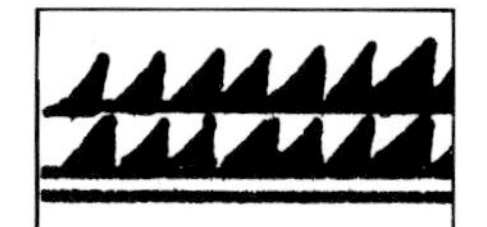

prickles

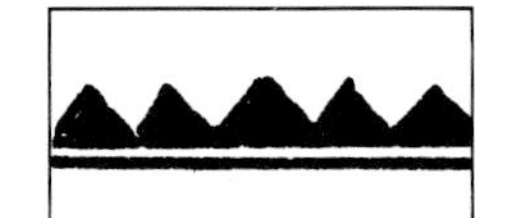

jaw-bone of
the cricket

kapok

spines

toucan's beak

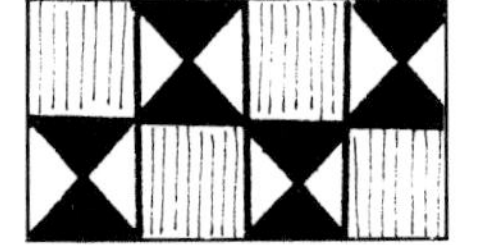

rice pounder patterns

bird's feathers,
Maa motif

LOZENGE MOTIFS

bird's feathers and caterpillar
eyes, Maa motif

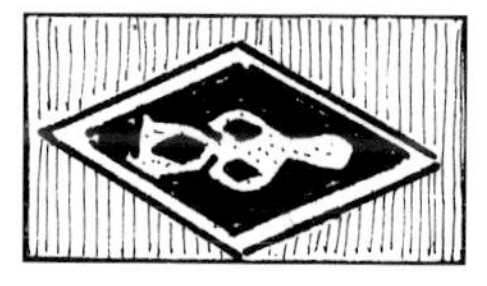

human figure in
lozenge,
Mnong motif

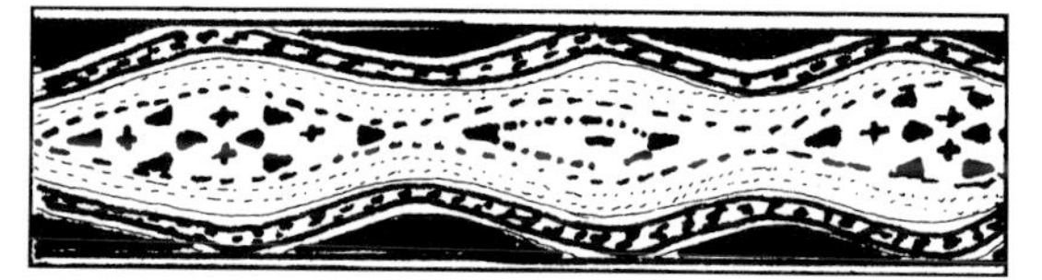

border on Jarai woman's skirt

rice pounder and lozenge motifs
on Jarai loin cloth

mouths of leeches,
Maa motif

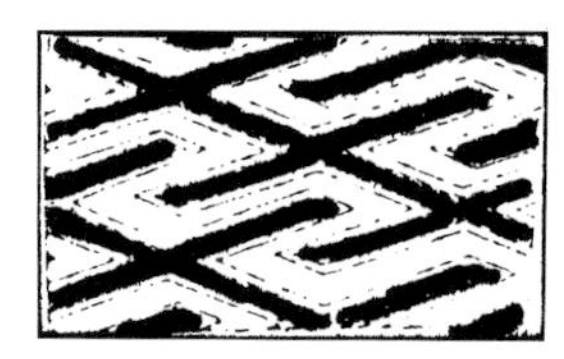

Mnong motif

Jarai loin cloth border designs

wing, Jarai

Mnong motif

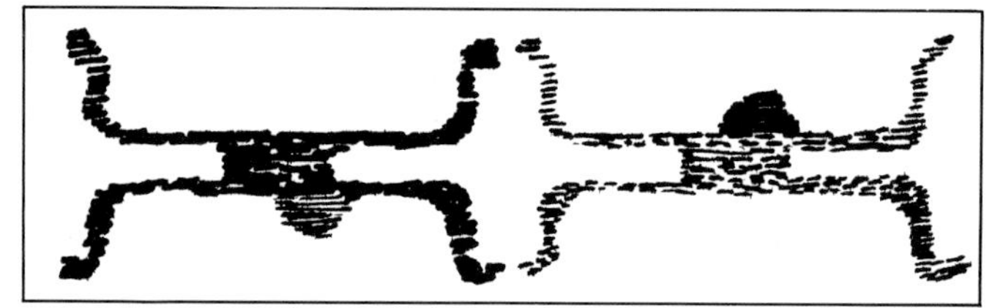

monkeys

heart motif,
Jarai

another heart
motif, Jarai

Fig. 168. Montagnard Vietnam War textile patterns. (After Hickey.)

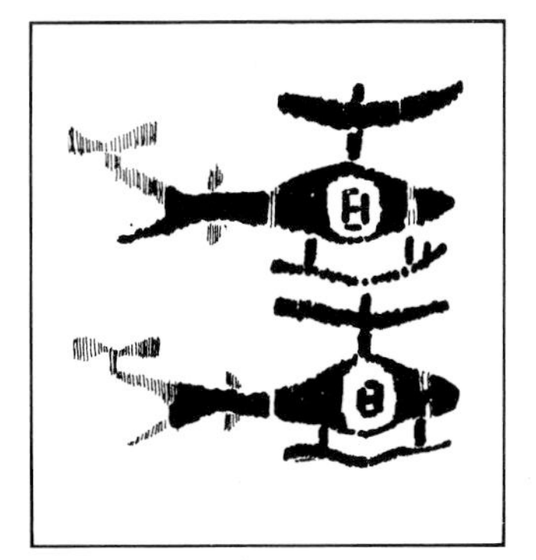

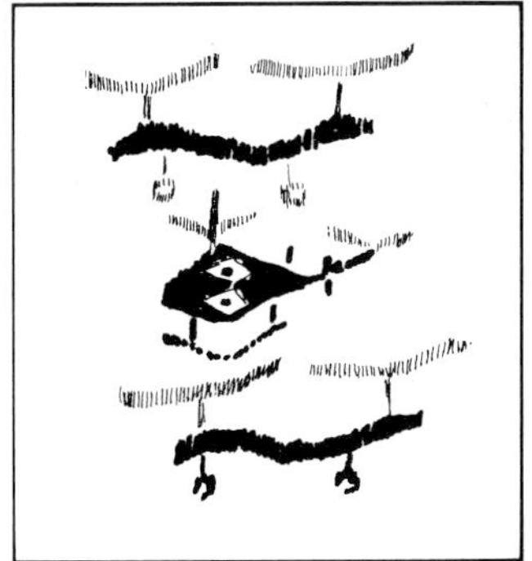

Montagnards and have a long history of contact with the Vietnamese. Over recent years, the Thai have been reported as purchasing much of their cloth from Vietnamese and Chinese merchants.[45] Being culturally close to the Vietnamese, the Muong people make use of an interesting loom set in a low frame. The heddles and beater are supported by cords to a 'crow', a vertical pole with one or two horizontal arms suspended above the warp.[46]

With the ending of the Vietnam War and the closure of that country's borders to much of the outside world, it is hard to know exactly what has happened in terms of hand weaving. Shortly after the fall of Vietnam, visiting journalists reported that post-war shortages had led to a dearth of manufactured cloth. Due to the length of cloth required for the *ao dai*, it is not as widely worn as before. It appears to have been largely replaced by a Chinese-style jacket, which was formerly only worn by Vietnamese farmers. The Government has been encouraging the development and continuation of handicrafts, such as silk and cotton weaving, embroidery, lacquer, ivory, bone, tortoise shell, and pottery production, the sale of which provides the funds to purchase machinery from Eastern Bloc countries for industrialization projects.[47]

The movement of large numbers of lowland people away from the crowded coastal cities into the 'New Economic Zones' in the highlands, and the Government's policy of bringing the Montagnards firmly under Vietnamese control will, in all probability, lead to great changes in the Montagnards' traditional way of life. Increased trade during the war years made many manufactured goods available. With them came such items as chemical dyes, mercerized threads, and cheap cotton clothing, all of which had an effect on Montagnard weaving and clothing styles. While the present austerities exist, the Montagnards will no doubt draw sustenance from the forests they know and love. The land they till will continue to furnish them with cotton and dye stuffs, and the women will continue to use the weaving skills acquired from their mothers to clothe their families with handwoven cloth, as they have done for generations.

1. Peter Kunstadter (ed.), *South East Asian Tribes, Minorities and Nations*, Vol. 2, Princeton University Press, New Jersey, 1967, pp. 236–8.
2. Patricia Cheesman, 'Laos: Indigo Dyed Fabrics', *Craft Australia Supplement*, Autumn/1, 1984, p. 87. The present writer has made extensive use of the published writings of Ms Cheesman for information on Laos.
3. Donna Mizzi, 'In Rescue of Historic Heirlooms', *Nation Review*, Bangkok, 29 May 1985.
4. Cheesman, 'Laos', p. 91.
5. Ibid., p. 90.
6. Erik Seidenfaden, *The Thai Peoples*, Siam Society, Bangkok, 1958, p. 72.
7. Mizzi, op. cit.
8. Patricia Cheesman, 'The Antique Weavings of the Lao Nuea', *Arts of Asia*, July–August 1982, p. 123.
9. According to Khun Vanida of House of Handicrafts, Bangkok, the Thai Deng, neighbours of the Lao Neua, also make this type of sarong. The supplementary weft patterns are not as intricate as those of the Lao Neua. Interview with Khun Vanida, 21 October 1985.
10. Cheesman, 'Laos', p. 88.
11. Cheesman, 'The Antique Weavings', p. 122. See photograph.
12. Ibid., p. 123.
13. Ibid.
14. Ibid.
15. Ibid., p. 124.
16. Frank M. LeBar, Gerald C. Hickey, and John K. Musgrave, *Ethnic Groups of Mainland South East Asia*, Human Relations Area Files Press, New Haven, 1964, p. 115.
17. Cheesman, 'Laos', p. 87.

18. Quoted in Elizabeth H. Moore, 'Meaning in Khmer Ritual', *Arts of Asia*, May–June 1981, p. 103.

19. See map from Jean Stoeckel, 'Etude sur le Tissage au Camboge', *Art et Archaeologie Khmers*, Vol. 1, No. 4, 1921–3, p. 396.

20. Thomas Fitzsimmons (ed.), *Cambodia*, Country Survey Series, Human Relations Area Files Press, New Haven, 1957, p. 41.

21. Stoeckel, op. cit., pp. 396–8; and Jean Galotti, 'Le Samphots du Camboge', *Art et Decoration*, Vol. 50, 1926, p. 6.

22. Mattiebelle Gittinger, 'An Introduction to the Body-Tension Looms and Simple Frame Looms of Southeast Asia', in Irene Emery and Patricia Fiske (eds.), *Looms and their Products: Irene Emery Roundtable on Museum Textiles, 1977 Proceedings*, Textile Museum, Washington, DC, 1979, p. 59.

23. Stoeckel, op. cit., p. 399; and Galotti, op. cit., p. 8.

24. These two animal mounts have an honoured place in Buddhism.

25. Kathleen Gough, *Ten Times More Beautiful: The Rebuilding of Vietnam*, Monthly Review Press, New York, 1978, p. 126.

26. Fitzsimmons, op. cit., p. 7.

27. Charles A. Fisher, *South East Asia*, Methuen, London, 1964, pp. 542 and 544; and E. H. G. Dobby, *Southeast Asia*, London, 11th edn., University of London Press, London, p. 22.

28. Because Vietnam (North Vietnam in particular) has been largely cut off from the outside world since 1950, it has not been possible to research weaving as done by the ethnic Vietnamese. It is possible that current austerities and the drive to become self-sufficient may offer some encouragement to the hand-weaving industry.

29. Gerald C. Hickey, *Sons of the Mountains, Ethnohistory of the Vietnamese Central Highlands to 1954*, Yale University Press, New Haven, 1982, Introduction, p. XV.

30. Robert Mole, *The Montagnards of South Vietnam*, Tuttle, Tokyo, 1970, p. 6.

31. R. Huard and A. Maurice, 'Les Mnong du Plateau Central Indochinois', *Institut Indochinois pour l'Etude de l'Homme, Bulletins et Travaux*, Hanoi, Vol. 2, 1939, p. 93.

32. Jean Boulbert, 'Modes et Techniques du Pays Maa', *Bulletin de l'Ecole Francais d'Extreme Orient*, Vol. 52, No. 2, 1965, p. 369.

33. Mole, op. cit., 207.

34. Ibid., p. 103.

35. Huard, op. cit., p. 100.

36. Jacques Dournes, 'Le Vetement Chez Les Jorai', *Objets et Mondes*, La Revue du Musee de l'Homme, Tome III, Fasc., 2, Ete, 1963, p. 111.

37. Hickey, op. cit., p. 169.

38. Ibid., photographs, p. 174.

39. Mole, op. cit., p. 207.

40. Ibid., p. 45.

41. Rene de Berval (ed.), *France Asie*, Special Edition, No. 12, 1955, p. 45.

42. Huard, op. cit., p. 111.

43. Boulbert, op. cit., plate XXV.

44. Hickey, op. cit., p. 249.

45. LeBar *et al.*, op. cit., pp. 221, 224, and 226.

46. Jeanne Cuisinier, *Les Muong, Geographie Humaine et Sociologie*, Institut d'Ethnologie, Paris, 1948, p. 219.

47. Gough, op. cit., p. 73.

CHAPTER EIGHT

Malaysia and Brunei

North-east Malaysia

TODAY, hand weaving in Peninsular Malaysia is largely confined to the north-eastern states of Kelantan and Trengganu. Because of their relative isolation from the lanes of commerce in the Straits of Malacca, these states have retained more of their traditional Malay culture than the west coast states of Malaysia, which over the past few centuries have been subject to considerable Chinese and Western influence.

The states of north-east Malaysia are heirs to the legacies of the former Hindu–Buddhist and Malay states that existed on the Thai–Malaya isthmus between the first and fifteenth centuries AD. Chinese annals dating back as far as the Liao Dynasty (AD 502–66) report the existence of a Malay Kingdom called Yang Yu Hsu (Langkasuka), which was located near present-day Patani in Thailand. A further report in AD 609 makes mention of a state, Tanah Merah, which was probably located in present-day Kelantan. The Indianized Kingdom of Funan in the third to sixth centuries AD, the Khmer at Ligor (Nakhon Sri Thammarat) in the second half of the tenth century, and the Kingdom of Sri Vijaya all exerted influence over southern Thailand and north-eastern Malaysia at one time or another.[1]

Contemporary eyewitness accounts attest to the fact that the northern Malay kingdoms, far from being mere trading posts, were thriving states with a tradition of royal patronage which actively sponsored the arts of drama, dance, metalwork, wood-carving, and weaving. The basis of prosperity of these states rested on a rich hinterland that provided ivory, tortoise shell, hardwoods, waxes, resins, rattan, and tin, which were traded for basic necessities such as pottery, rice, and sugar, and luxury items such as silver, gold, and skeins of Chinese silk for weaving.[2]

Up until the early years of the twentieth century, north-east Malaysia produced magnificent weft *ikat* cloth, called *kain cindai*, which was made from imported Chinese silk yarns and coloured with natural dyes (Fig. 169). Weft *ikat* in Malaysia is sometimes referred to as *kain limar* because the little dots and dashes which make up the majority of the designs resemble drops of a squeezed lemon. The Malays claim to have inherited the technique of weft *ikat* from Cambodia. In terms of patterns and cloth

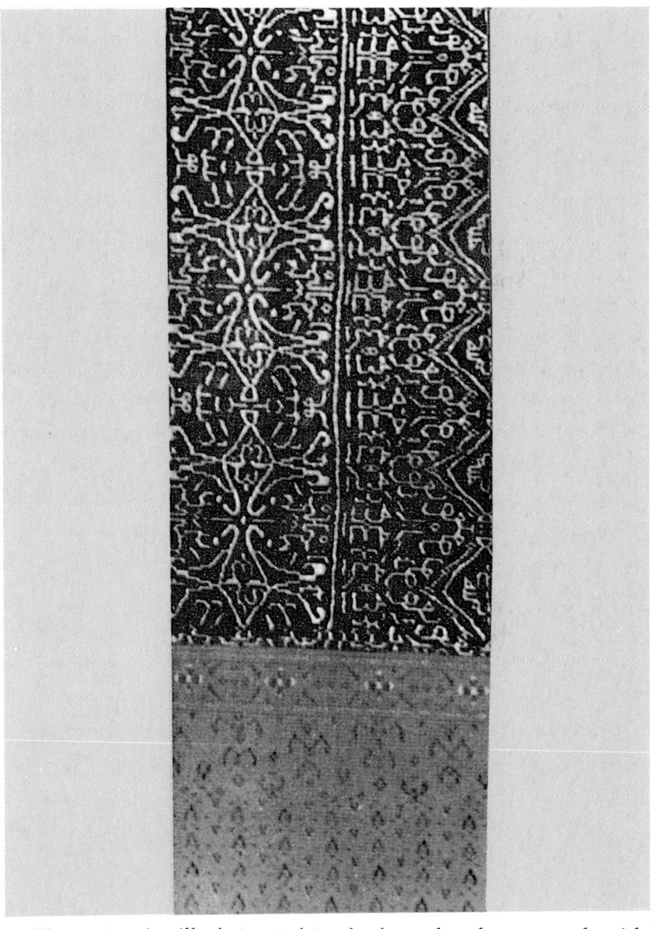

Fig. 169. A silk *kain cindai selendang* shawl patterned with weft *ikat* designs from north-east Malaysia. Warp 200 cm, weft 45 cm.

layout, *kain cindai* has been influenced by Indian *patola* cloth. Subtle stripes, arrowheads, zig-zags, *tumpul* patterns, and floral motifs are the most prevalent designs seen on *kain cindai*. Sometimes, the weft *ikat* technique was combined with the *kain songket* supplementary gold thread technique to produce an especially sumptuous cloth similar to that seen in Palembang, Sumatra (Fig. 170). The art of *pelangi* was at one time widely practised in north-east Malaysia.[3] Unfortunately, the art of weft *ikat* and *pelangi* seem to have been abandoned in favour of batik, which is now very popular throughout Malaysia. Fortunately, the National Museum, Muzium Seni Malaysia, and the Malaysian Handicraft Development Corporation (*Perbadanan Kemajuan Kraftangan Malaysia*) in Kuala Lumpur, and some museums abroad, have some magnificent examples of weft *ikat* which attest to the extremely high standard once attained.

The making of *kain telepuk*, or gilded cloth, is still practised by a few artisans in Malaysia. Dark-coloured woven cloth, which may be plain or have a subdued design, is first stamped with carved wooden blocks smeared with gum arabic. A gold-leaf pattern, cut from the same design block, is then applied and this adheres to the glue.[4] In the state of Perak, a few artisans continue to practise the art of *tekat*, or embroidery, to decorate textiles for the bridal bed, prayer mats, and dish covers for ceremonial occasions. The most popular form is embroidery in raised relief. The pattern is cut out in cardboard and secured by stitching to the base material, which

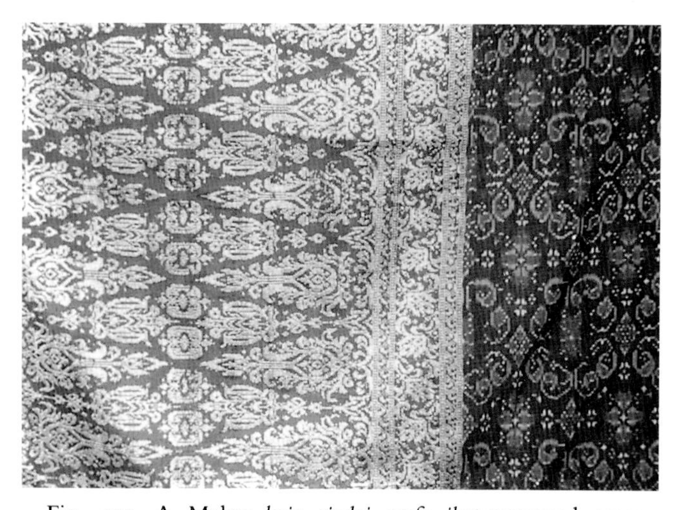

Fig. 170. A Malay *kain cindai* weft *ikat*-patterned sarong which has a *kepala* embellished with *songket* patterns in a gold supplementary weft. The floral border along the side of the *kepala* is called *bunga ati-ati*, while the zig-zag *tumpul* pattern is called *pucuk rebung lawi ayam* (cock's tail feathers), which alternates with the *bunga sarang celak* (cosmetic case) motif. Warp 240 cm, weft 95 cm. Collection of Kraftangan, Kuala Lumpur.

today is usually imported velvet. Gold thread is then carried back and forth across the design area to cover the cardboard pattern. Each loop is secured with a coloured thread along the edge of the cardboard. Sequins and small gold and silver discs may also be added. The Malays are of the opinion that the art of gold embroidery was probably introduced from India; however, the design motifs which feature foliage, flowers, and birds, obviously owe much to Chinese influences.[5]

The Malay love of gold patterning may be seen in the production of *kain songket*, which is considered the most traditional fabric of the Malay people. It is woven on an approximately 240 cm long by 100 cm wide Malay floor loom called a *kek tenun*. This loom is operated by two transverse foot pedals (see Fig. 37). Chemically dyed silk yarns from China and cotton yarns from Japan are used singly or in combination for the warp and weft of the fabric. Two-ply gold and silver threads from France or Japan are used for the supplementary weft patterns. To emphasize the richness of the gold thread, maroon, yellow, green, brown, and blue are used as the ground colours for *kain songket*.[6]

The loom has to be specially set up to weave *songket* patterns. Various warp threads are lifted and held by bundles of loops, called *ikat butang*, which are stored behind the main heddles until required (see Figs. 65 and 66). Malay *songket* patterns are set up so as to form overshots of five (or in the finest work, three) threads on the right side of the cloth. Working on the wrong side of the fabric, the weaver, after raising the appropriate warp threads, passes the shuttle containing the supplementary gold thread through the weft. If the gold thread is to be applied to a small area only, it is inserted with the fingers. The gold thread is held in place by two rows of plain weave.

Kain songket is used to make ceremonial attire for weddings and state functions. Full dress for the Malay male consists of a *destar*, a square silk scarf measuring approximately 70–80 cm on each side, which is folded into a peaked turban. The *destar* is worn with a long-sleeved tunic with an upright collar, called a *kemeja* or *kurung*, a wide sash, an approximately 180 cm by 80 cm knee-length sarong, or *sampin*, and long trousers (see Fig. 78). Together these items make up a suit consuming over 9 m of cloth. Formal wear for ladies consists of a calf-length silk tunic, or *baju kurung*, or a neat-fitting silk jacket, the *kebaya*, which reaches nearly to the knees. These are worn with an ankle-length sarong and a gold-patterned silk stole, or *selendang*.[7]

Kain songket was at one time woven by court artisans for the exclusive use of local royalty. Formerly, every sultan and major chief maintained his own group of weavers to produce sumptuous royal garments. Some sultans took a very active interest in weaving and were known to present weavers with sketches of designs they wished to have woven.[8] On occasion they would visit the weavers at work under the *astana*, or royal palace, to monitor progress. Some present-day weavers proudly boast that their forebears were master weavers at one of the royal Malay courts.

Today, *songket* weaving is confined to the east coast states of Kelantan, Trengganu, and, to a lesser extent, Pahang, where it is organized as a cottage industry. Women in the small kampongs, or villages, on the outskirts of the state capitals of Kota Bharu and Trengganu find time, in between attending to family and household chores, to sit at their looms and weave.

Weaving at the cottage industry level is organized by an entrepreneur who may or may not be a weaver. This person, usually a woman with some business acumen, acts as the middleman. She receives orders and supplies the weavers with materials, such as dyed silk yarns, beaters, and shuttles. The weaver usually has her own loom. The entrepreneur, if she is not a weaver, may engage the services of a specialist, a master weaver who is skilled in designing *songket* patterns and in setting up warping frames and looms. This person is paid to set up warps for weavers with designated patterns.[9]

Working about six to eight hours a day, it takes a weaver about a month to complete a 100 cm by 180 cm sarong with a warp count of 1,680 warp threads patterned with an overall *songket* design.[10] Young unmarried women with few household responsibilities are able to spend more time on weaving, so it takes them less time to complete a length of cloth. A young weaver, however, may not be as skilled as an older weaver. Weavers are considered to be their most productive between the ages of twenty and thirty when they have acquired sufficient skill and experience in their teenage years to become competent weavers. By forty, most Malay women have big families to rear and are too busy to practise their craft. After fifty, many women find that with failing eyesight they are unable to weave as easily as before.[11]

On completion of a length of fabric, the weaver returns it to the entrepreneur. She receives a cash payment, which is based on the size of the cloth, the workmanship, and the intricacy of the design. She may also pick up a new set of warp threads pre-set with a *songket* design, and hanks of pre-dyed weft yarn to weave another piece of cloth.

A piece of *kain songket* is judged by the neatness and evenness of the weave. The selvages should be straight and firm, while the gold thread should be woven with an even tension. There should be no mistakes in the woven pattern. Aesthetically, the best cloth should show a harmonious marriage between the ground weave and the overlying gold thread design, with each component subtly blending to complement the other.

Patterns on *kain songket* are aligned in vertical, horizontal, diagonal, and zig-zag bands, or in alternating motifs across the cloth. Designs may be very dense or widely scattered. Traditional *songket* patterns are very intricate, with gold thread covering the whole fabric. Many sarong cloths have a panel, called a *kepala*, in the middle of the cloth which differs from the body (*badan*) flanking the central panel (see Fig. 77). Border patterns may line the selvages and the sides of the *kepala*. The *selendang* shawl has border patterns and a triangular *tumpul* pattern at each end. The *kain destar* head-cloth has border patterns surrounding the central area. Special motifs may fill the border corners.

Traditional *songket* patterns are usually named after local flora and fauna, such as flowers, birds, and butterflies (Fig. 171). Food, such as rice cakes, and cultural objects, such as *kris* (the Malay dagger), fans (*kipas*), and cosmetic cases (*bunga sarang celak*), have also lent their names to textile motifs. Simple repetitive diamond patterns include honeycomb (*bunga madu manis*) and rice cake (*bunga potong wajik*). Basic floral designs include the jasmine flower (*bunga melur*), mangosteen (*tampuk manggis*), persimmon (*tampuk buah kesemak*), hibiscus (*bunga raya*), the *tanjung* and *chenkeh* flowers, pumpkin seed (*biji peria*), star flower (*bunga bintang*), banana flower (*bunga pisang*), and sunflower (*bunga sinar matahari*).[12] More complicated patterns include *bunga mahkota*, *rantai bunga pecah lapan*, sea horse (*rantai unduk-unduk laut*), enclosed bay (*teluk berantai*), open flower (*bunga kembang semangkuk*), and dish cover (*bunga tudung saji*). These motifs are also seen on old *kain cindai* textiles.[13]

The *kepala* of traditional Malay cloth usually consists of two vertical rows of triangular *tumpul* designs called *pucuk rebung*, or bamboo shoot. A particularly flamboyant example is called *pucuk rebung lawi ayam*, or 'bamboo shoot resembling the cock's tail feathers' (Fig. 172). Other variations include *lawi itek* (duck's tail) and *pucuk rebung berbungan dalam* (flowers inside

Fig. 171. Malaysian *kain songket* patterns.

Fig. 172. *Tumpul* motifs. (After Kraftangan.)

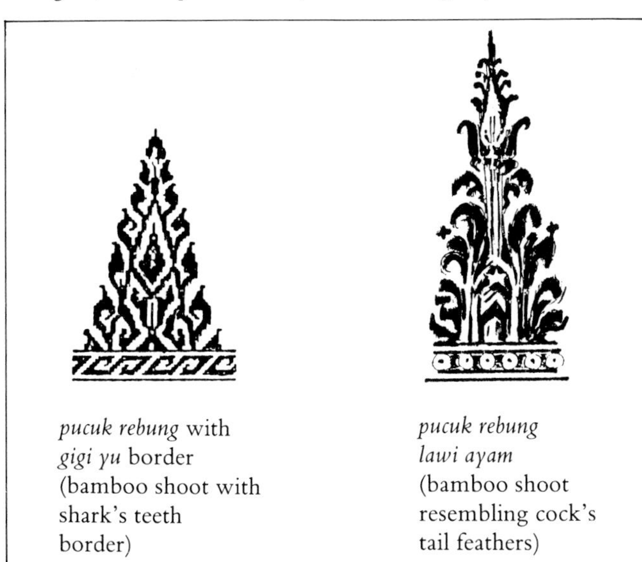

a bamboo shoot). Various floral and geometric motifs appear in the space between the rows of *tumpul* patterns on the *kepala*. Border patterns include *kendik tali* (a slanting design), *bunga sasok*, *bunga semangat* (rice offering), palace fence (*pagar astana*), shark's teeth (*gigi yu*), *bunga kerawan* (an open work pattern), and an 'S'-shaped hook design called *awan larat* (cloud motif) (Plate 22 and Fig. 173).[14]

The state of Kelantan has a long history of weaving. There are references to weaving which go as far back as 1610 AD. During the reign of Long Pandak (1794–1800), it has been reported that *songket* weaving was done at Kampung Gerung, Kampung Laut, and Kampung Atas Banggol. Since the reign of Sultan Mansur (1890–1910), the craft of weaving *kain songket* has been centred on Kampung Penambang, some 7 km out of Kota Bharu along the Kampung Pantai Cinta Berahi road. Today, there are five or six workshops which specialize in *songket* weaving. The most well known is that of Hajah Cik Minah bte Haji Omar, who was born in 1912. She is a fourth-generation weaver who had been weaving *songket* since the 1930s. She rose to national prominence in 1982 as a winner in the Contemporary Traditional Malay Handicraft contest organized by the National Museum. In addition to practising weaving, she is also an entrepreneur with her own workshop. She has over 100 women weaving for her in nearby kampongs.[15] Her son, Mohammad Hj. Hussain, is a well-known batik artist who specializes in soft filmy fabrics with floral patterns for an exclusive clientele.

Like Kelantan, the state of Trengganu has a history of *songket* weaving which dates back to the early eighteenth century. Today, there are many weavers in and around the state capital of Trengganu. The most well known in the town of Trengganu is the atelier of Tengku Ismail bin Tengku Su at Jalan Kenanga, which specializes in producing high-quality cloth in traditional designs. He has introduced the art of weaving in both gold and silver threads on a single piece of cloth. His highly skilled young weavers follow designs blocked out on graph paper. Tengku Ismail is also a collector of antique Malay weavings. It is from these that he seeks inspiration for new patterns. His *kain songket* is eagerly snapped up by members of the Malay aristocracy who not only admire and appreciate but have the means to pay for fine, labour-intensive cloth.[16]

Fig. 173. Malaysian *kain songket* border patterns. (After Kraftangan.)

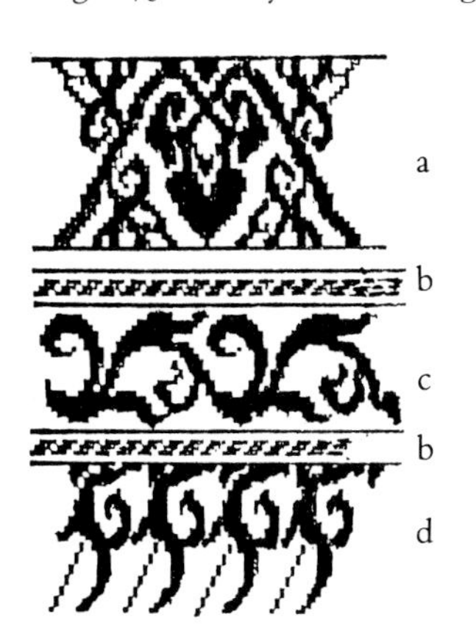

(a) *pagar astana* (palace fence)
(b) *kendik tali*
(c) *bunga kerawan*
(d) *gigi yu* (shark's teeth)

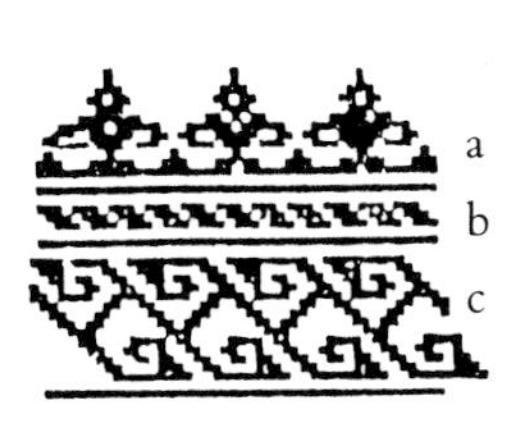

(a) *pagar astana* (palace fence)
(b) *kendik tali*
(c) *awan larat* (cloud motif)

paling putong

bunga semangat (rice offering)

There are about eighty weavers at Pulau Rusa, 4 km out of the state capital on the road to the airport, whose *songket* is marketed by a well-known entrepreneur, Hajah Atikah bte Mohamad. She has outlets for the sale of her weavers' products both in the kampong and in the MARA building in Trengganu.[17] There are also a number of other *kain songket* weaving establishments at Pasir Panjang, Bukit Besar, Chabang Tiga, Jalan Kuala Brang, Jalan Pulau Musang, Kampung Batu Buruk, Kampung Losong Haji Mat Shafie, and Kampung Haji Losong Haji Su in the vicinity of the town of Trengganu.[18]

A few entrepreneurs have expressed concern over the long-term future of *kain songket* in Malaysia. The cost of silk, dyes, and gold thread has risen sharply over the last few years, making the cost of a complete Malay ensemble at the time of writing in the vicinity of US$400, a price which only a shrinking number of Malay patrons are prepared to pay. Modern life since Independence has reduced full Malay dress occasions to a minimum, and Western dress is widely acceptable at most events. Since Malay looms currently produce *songket* in limited yardages, and by a slow meticulous process, opportunities for sale in overseas markets are limited. The current organization of weaving as done by part-time workers, makes it difficult to guarantee delivery dates and consistency of craftsmanship for potential large overseas orders.[19]

Weaving in Malaysia at the government level comes under the Malaysian Handicraft Development Corporation (usually referred to as Kraftangan), which was founded in 1979. This body was set up to develop traditional skills and handicrafts into commercially viable concerns by establishing handicraft training centres in states which had a particular craft tradition. Centres for traditional weaving have been set up in Trengganu, Kota Belud in Sabah, and in Kuching, Sarawak. Students are selected for one-and-a-half-year courses and are given small stipends during training. They are taught by experienced instructors. On graduation, with help from Kraftangan, students are given assistance in setting up their own workshops.

So that supplies are readily available to weavers, MARA (a government organization which works in co-operation with Kraftangan) is responsible for importing good-quality silk yarns, gold thread, and colour-fast dyes. Kraftangan also arranges for the washing and dyeing of yarns using modern mechanical equipment before selling them to weavers.

Under the sponsorship of Kraftangan, the art of weaving has been revived in the east coast state of Pahang, at Kampung Peramu Pekan. Interested women have been sent to Trengganu for training. On their return, they teach their fellow villagers the art of weaving colourful silk plaids using the *kek Siam*, or Thai loom. The Pahang weavers have been experimenting with simple *ikat* in random patterns and twisted weft threads. These new elements have been incorporated into their plaids with pleasing results.

Kraftangan, through Infokraf, its information section, mounts exhibitions and offers extensive advisory services to disseminate information on new and improved techniques of production, new product designs, quality control, and marketing information. Under its sponsorship, an International Craft Museum and a National Craft Museum have been set up to stimulate public interest and give recognition to the handicraft industry.[20]

Sarawak

The state of Sarawak, which occupies the north-west coastal strip of the vast island of Borneo, was once the private domain of the 'White Rajahs' of the Brooke family. Today, as part of Malaysia, Sarawak is home to a polyglot population of about one million, which includes Malays, Chinese, Ibans, Kayans, Kenyahs, Bidayus, and other indigenous groups, each of which has its own proud traditions and beautiful handicrafts made from the simplest of materials from the tropical rain forests which still cover much of the state. Handicrafts include wood-carving, basketry, mat making, weaving, embroidery, and beadwork.

The Iban people, who today live mainly in the Second and Third Divisions of Sarawak, have gained notoriety in history books as fierce head-hunters and pirates. In addition to their prowess in warfare, the Ibans are among the most artistic of the indigenous people of Borneo in terms of the diverse range of designs and patterns with which they embellish their crafts.

This skill and artistic sensitivity is nowhere more evident than in their weaving. The Iban are particularly noted for their finely crafted warp *ikat* in beige enlivened with touches of dark brown and bluish black against a brick red ground. Women weave on a simple body-tension loom with a continuous warp. Iban textiles at one time were woven from home-grown cotton and coloured with natural dyes prepared from plants of the jungle.[21] In addition to warp *ikat*, the Iban also pattern their cloth by the tapestry weave technique and by two unique decorative processes—that of *pilih*, where lengths of supplementary weft form a decorative background to the main weave, and *sungkit*, a method of inlaying supplementary threads into the weft with the aid of a needle (Fig. 174). Twining and embroidery are also known. Further ornamentation may be added in the form of beads, shells, coins, and gold and silver threads.

Motifs depicted on Iban weaving draw their inspiration from the immediate environment and serve as a visual record of tribal beliefs and values. In earlier days, designs were passed from mother to daughter. Motifs include readily recognizable humanoid figures (*engkaramba*) with up-raised arms, out-turned legs, and startled facial expressions (Fig. 175). Formerly, such *engkaramba* could only be made by wives and daughters of chiefs who had acquired a certain degree of skill and expertise in weaving warp *ikat*. *Engkaramba* designs are believed to prevent harm to crops and individuals.[22]

Zoomorphs include life-like representations of the crocodile, which is considered to be a protector of the rice fields and possibly a relative of man (Fig. 176).[23] Snakes, too, are treated with awe, for the Iban regard them as a possible place of habitation for *antu*, or spirits. Frogs are depicted on Iban textiles in a motif resembling a humanoid form. This is possibly due to a belief that *Salampai*, a female spirit who appears in the guise of a frog, is the maker of men and is thought to be in the vicinity when a child is born. Tigers and shrews may appear in very stylized forms on textiles in the belief that they confer bravery on the wearer. Birds, considered to be bearers of good omen, may be depicted to secure the success of any ventures undertaken. Deer may be portrayed to bring success in a hunt. As a bearer of less auspicious omens, deer might also serve as a talisman against ill-fortune. The teeming insect life of the Borneo rain forest includes spiders, beetles, fireflies, leeches, scorpions, and centipedes; all have lent their names to textile motifs. Imaginative depictions of useful plants, such as rattan, betel vine, bamboo shoots, fern tops, and various seeds and flowers, also appear in the Iban design inventory. Various natural and cultural phenomena, such as the moon, clouds, weaving implements, and cooking utensils, have also given their names to textile patterns (Fig. 177).[24]

The portrayal of motifs on Iban textiles has been greatly influenced by Dongson culture and the artistic styles of the Late Chou bronze culture of China. On Iban textiles, specific motifs are combined with hook and rhomb elements to give a strongly geometric flavour to the centre field. Motifs are usually arranged as an interlocking series of repetitive side-by-side patterns extending the length of the cloth.

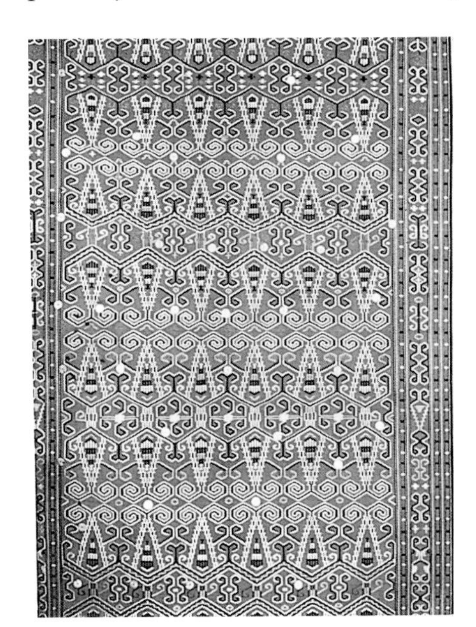

Fig. 174. An Iban textile with *sungkit*-patterned geometric designs which have been inlaid into the weft with the aid of a needle. Small iridescent shirt buttons have been added for extra embellishment. Warp 190 cm, weft 56 cm. Collection of Ramsey Ong, Kuching.

Fig. 175. *Engkaramba* humanoid figures depicted on an Iban warp *ikat pua*. Collection of the Sarawak Museum, Kuching.

Fig. 176. An Iban textile depicting crocodile and humanoid motifs which appear as part of the warp-faced ground weave against a background of red, green, yellow, and black supplementary weft threads inlaid by the unique *pilih* weaving process. Warp 164 cm, weft 64 cm.

Fig. 177. Some common Iban textile designs. (After Haddon and Start, and AGAS.)

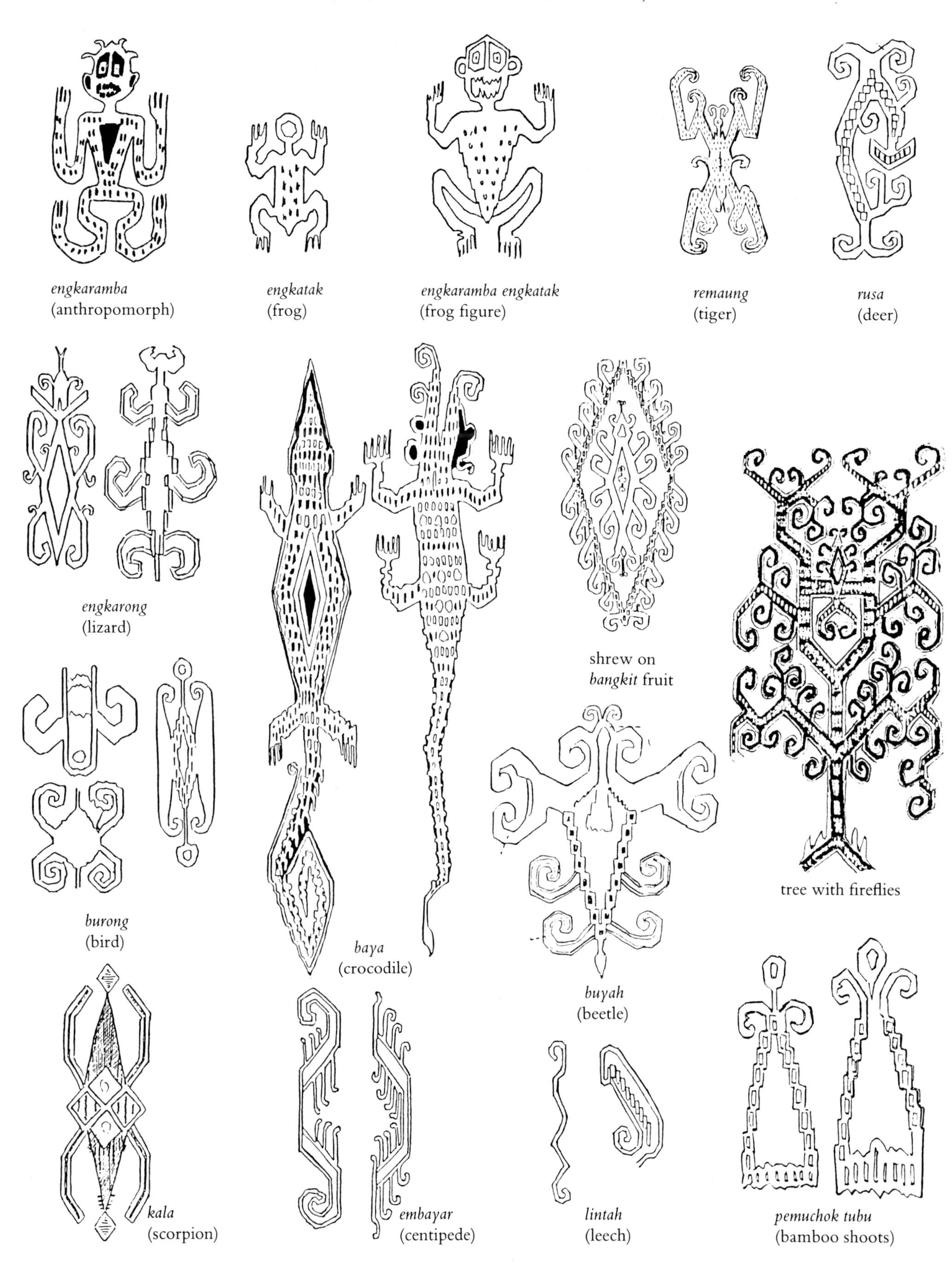

Many motifs have been cleverly modified by individual weavers so as to bring a spatial rhythm and subtle harmony to the overall design. Unlike the warp *ikat* from the neighbouring islands of Indonesia, the upper and lower halves of Iban cloths are not mirror images. This feature, along with the interlocking nature of the designs, makes Iban textiles unique.[25] Warp stripes in brightly coloured, chemically dyed yarns along the selvages, contrast nicely with the intricately patterned centre field, which is done in subdued natural dye tones.

The Iban traditionally spend a lot of time weaving their distinctive clothing. The basic garment for an Iban woman is the *bidang*, a short tubular skirt measuring approximately 110 cm by 60 cm, which is made from two lengths of cloth. For everyday wear, the *bidang* may be patterned with colourful warp stripes. Most *bidang* for ceremonial wear have a large central area delicately decorated with stylized geometric patterns in soft brown and buff shades. Over recent years, it has become popular to weave ceremonial *bidang* patterned with traditional motifs in gold and silver *songket* on a red ground.

Kalambi jackets, with or without sleeves, are worn on certain occasions by both men and women (Plate 23). They are tailored by folding a rectangle in half and cutting a hole for the neck and a slit for the front opening. Jackets vary in length from short boleros to knee-length coats. Every decorative technique in the Iban weaver's repertoire is used to embellish them. Some may be lined and further embellished on the inside with motifs which differ from those depicted on the outside. *Kalambi* for warriors were sometimes padded with kapok for extra protection. Motifs on warp *ikat*-patterned jackets are sometimes arranged in a linear fashion within a series of warp stripes, or are covered by a few repeats of humanoid and stylized bird motifs. The backs of some warp *ikat kalambi* are finished with a wide border worked in a different decorative technique, such as a brightly coloured tapestry weave or a supplementary weft technique (Fig. 178). The fringe is sometimes finished with twining. Beading may also be added for further decorative effect.

Fig. 178. Iban *kalambi* jacket patterned with linear warp *ikat* patterns and finished at the back with a border of brightly coloured geometric motifs in a slit tapestry weave. Length of back of jacket 60 cm. Collection of Dato and Datin Leo Moggie, Kuala Lumpur.

The loin cloth, or *sirat*, is the basic garment for men. It is a strip of cloth approximately 50 cm wide by 500 cm long. It is wound around the loins and through the legs so that the embroidered or supplementary weft-patterned ends (often weighted with beads) hang like an apron from the waist over the back and front of the body. Long, narrow sashes called *bedong* (about the same length as the loin cloth, but narrower), are finely patterned in a supplementary weft technique. They are worn by women over the shoulders as a stole, or in a criss-cross fashion over the breasts on formal occasions. *Dandong* shawls for men, which are a little wider than the *bedong*, vary greatly in length. They are patterned in warp *ikat* and sometimes serve as wrappers for ceremonial offerings.[26] Long shawls, called *pua menyanduk*, patterned with warp stripes and bands of warp *ikat* often serve as baby carriers.[27]

The largest and most impressive of the Iban textiles is the *pua kumbu*, a large blanket-sized textile made from two identical pieces of cloth tied together during dyeing, which form the upper and lower webs of the loom during weaving (Plate 24). On removal from the loom, the unwoven warp threads are cut at each end between the upper and lower web. The two pieces of cloth are joined together along a selvage with a lacing stitch to form the *pua*. The warp ends are reinforced with one or two rows of twining above the fringes.[28] A *pua kumbu* may be up to 250 cm long by 120 cm wide. Decorative motifs run the whole gamut of the Iban design repertoire. On the best *pua*, numerous motifs are skilfully combined to create a beautifully integrated composition. For added interest, border patterns may be different at the ends of the cloth.

In addition to being a textile of great visual beauty, the *pua kumbu* also serves to highlight the uniqueness of some of the social structures and religious beliefs of the Iban people. *Pua kumbu* traditionally have been highly regarded as items of wealth and important accoutrements to various festivals and rites of passage. At the time of marriage, parents of the bride present the groom's family with a *pua* as part of the dowry. The bride and groom sit under a *pua* canopy to receive blessings from elderly relatives. A *pua* also adorns the wedding chamber. Shortly after birth, a new-born child is carefully laid on a *pua*. It is also used for the first ceremonial bath. *Pua* are considered effective in warding off illness, and may be worn by a spirit medium when communicating with the supernatural on matters of importance to the tribe. At times of death, the *pua* shelters the body of the deceased as it lies in state. The *pua* also plays an important role in farming rituals by sheltering the *padong* structure containing various charms and offerings. In former times, the *pua* was used to receive the heads from a successful war party upon its return to the longhouse. At one time, a girl was not considered ready for marriage until she had woven a *pua*.[29]

Through Christian missionary activity, government development projects, improved communications, and mass education, change is coming to the Iban way of life. Except in the really remote areas, few women today grow their own cotton. Most prefer to buy processed yarns which are readily available in the bazaars of most small towns, which are now only a speedboat ride away from many longhouses. Chemical dyes are also becoming more widely used. Today, brightly patterned machine-woven fabrics are preferred by most women for sarong material over the *bidang*. Many young Iban are seeking employment in the towns, and many Iban women no longer know how to weave. An increase in tourism has created a demand for souvenir textiles which are quickly produced and cannot compare in design layout and workmanship with traditional cloth. Motifs on these new cloths range from severed human heads, river boats, helicopters, tea pots, and dogs rendered in a realistic style.[30] Despite these modern innovations, traditional textiles remain a source of pride to the Iban people. Traditional clothing continues to be widely worn at festivals such as Gawai Dayak, the Dayak New Year. It would be a rare Iban household which does not have at least one beautiful *pua kumbu* which was lovingly crafted by a female relative.

The Malay people who inhabit the coastal area of Sarawak have a tradition of weaving fine *kain songket* in 2 m lengths on a Malay loom. Today, there are only a few weavers active in the Mukah and Kuching areas.[31] One such person is Dayang Zakariah bte Haji Marhassan of Kuching. Although in her eighties, she continues to weave fine cloth in a variety of floral and geometric designs (Fig. 179). Her work is so fine that it takes over three months to weave a single sarong, which is very heavy because of the large amount of gold thread used. With encouragement from various women's organizations, such as the Sarawak Women's Institute, Dayang Zakariah has been passing on her craft to others.[32]

Another Malay craft that has recently been revived in Sarawak under the sponsorship of the Sarawak Women's Institute, is that of embroidering the *selayu*, the traditional head-cloth for Malay women. Using a flat gold needle and flat gold threads, women patiently embroider beautiful floral patterns on fine silk and muslin cloth. This craft has been revived in an effort to provide kampong women with a supplementary income. These *selayu* head-cloths find a ready market amongst Muslim women in Peninsular Malaysia.[33]

Kraftangan is also active in Sarawak in promoting the art of weaving as an occupation amongst young girls. Since 1978, Kraftangan has been selecting a small number of girls (currently seventeen) from applicants who have had nine years of basic education and have sat for their Lower Certificate of Education (the equivalent of a junior high school graduation diploma). The successful applicants undergo a one-and-a-half-year course in weaving, which also includes mat making. The students, who are usually Malay or Iban, specialize in either Malay or Iban weaving. To graduate, the Iban girl must be

Fig. 179. A very finely patterned *kain songket* sarong woven by Dayang Zakariah bte Haji Marhassan of Kuching. The *kepala* is heavily patterned with gold, while the *badan* has scattered motifs placed at regular intervals. Warp 216 cm, weft 90 cm. Photo courtesy of Sarawak Museum, Kuching.

able to weave a perfect *pua*, and the Malay girl a flawless *kain songket* sarong. On graduation, Kraftangan assists students to establish their own workshops by making materials and equipment available. Short refresher courses are offered from time to time. The Bimbangan Technical School, which is associated with Kraftangan, offers handicraft courses, including weaving, to people in various villages. Once the village begins production, Kraftangan assists in marketing the products.[34]

Sabah

Sabah, like Sarawak, has many indigenous groups, a few of whom are very competent weavers. Because of the state's close proximity to the Philippines, the textiles of Sabah show affinities with Filipino Muslim weaving, as well as with that of the Malays and other indigenous groups who inhabit the island of Borneo. Traditional weaving was done on a body-tension loom using home-grown cotton yarns coloured with natural vegetable dyes. Today, chemically dyed imported yarns are generally preferred by most weavers and, in addition to the body-tension loom, Malay floor looms are also used. Plain weave, tapestry weave, and supplementary weft weaving techniques are known. However, there is no known tradition of making *ikat* in Sabah.

The Kadazan, an agricultural people comprising one-third of Sabah's population of just over one million, are the largest indigenous group.[35] Traditional dress for the Kadazan woman is a black, calf-length handwoven sarong enlivened by a vertical and a cross-wise strip of southern Filipino-style *langkit* in brightly coloured geometric designs in a tapestry weave (Fig. 180). There may be matching embroidery in the four corners where the *langkit* strips intersect (Plate 25). Today, few people know how to make the intricate tapestry weave patterns and the *langkit* strips have been largely replaced by gold and silver braid. The sarong is worn with a matching black blouse cinched at the waist with a silver belt. A colourful Malay silk shawl is worn across the shoulders.

As a former Borneo head-hunter, the Kadazan male at one time wore the *cawat*, or loin cloth. Today, black loose-fitting pants and a long-sleeved jacket with a Chinese collar are generally preferred (see Fig. 180). Decoration is in the form of bands of gold and silver braid. A colourful Malay-style *destar* head-cloth, worn with a high point in front, completes the male costume.

The most traditional weaving in Sabah today is done by the Rungus, a subgroup of the Kadazan, who live in communal longhouses in the Kudat and Bengkoka peninsulas of northern Sabah. They weave a distinctive textile called *kain pudang*, a black, indigo-dyed, 45 cm wide cotton cloth subtly pat-

Fig. 180. Kadazan couple in traditional dress. Note that the woman's calf-length sarong is patterned with vertical and cross-wise strips of colourful geometric designs in a tapestry weave. Photograph courtesy of Victor Wah, Kota Kinabalu.

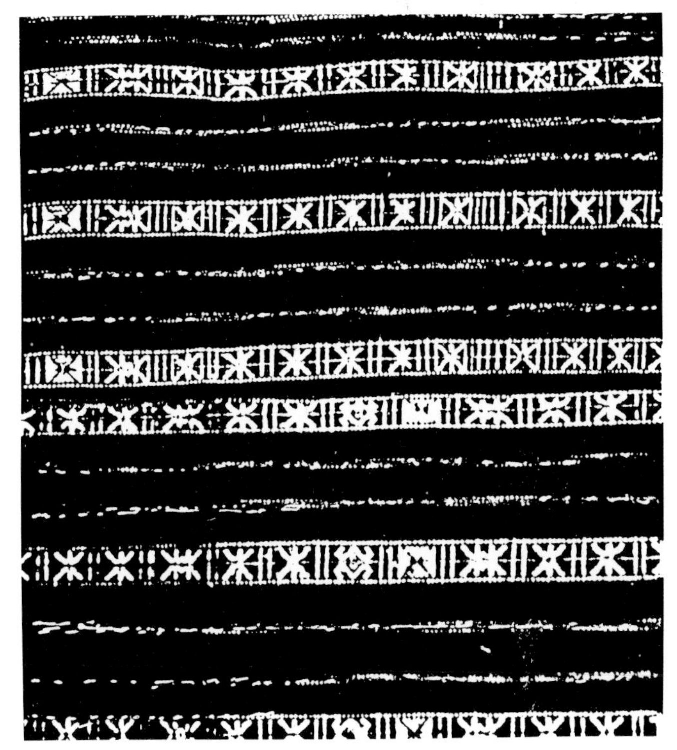

Fig. 181. A sample of a Rungus *kain pudang* textile which is patterned by a supplementary weft technique.

Fig. 182. A trio of Rungus women from northern Sabah in their traditional dress fashioned from *kain pudang* woven on a body-tension loom. Photograph courtesy of Victor Wah, Kota Kinabalu.

terned with 3 cm wide bands of repetitive geometric zig-zags, triangles, and rhombs in a supplementary weft (Fig. 181). These decorative bands may be edged with small pin-stripes of yellow and orange. *Kain pudang* is made into sarongs, breast cloths, and shawls. The Rungus brighten up their costume by wearing masses of coloured beads in the form of headbands, necklaces, and armbands. They also wear strips of brightly coloured handwoven cloth diagonally over the breast and shoulders (Fig. 182). Rungus men at one time wore indigo-coloured pants decorated with brightly woven cuffs and a colourful overlapping waistband in a tapestry weave.

The Rungus women weave with a body-tension loom. They are the only group in Sabah who still spin their thread from wild cotton and produce their dyes from local plants.[36]

The Dusun women (sometimes called Lotud) weave a cloth which is similar to *kain pudang*. It is black with red selvages decorated with small bands of white and orange pin-stripes. Two pieces are sewn together to make a short sarong, which is kept in place by a silver belt and rattan rings. It is worn with a black blouse and imported silk stole. On ceremonial occasions, the Dusun male wears an ankle-length sarong of handwoven striped material, called *kain mogah*, which is woven by Bajau women. It is teamed with a long-sleeved black satin jacket and silk shoulder cloth, and is worn with a *destar* folded in such a way that it has two 'ears' on top.

As shifting cultivators and gatherers of jungle produce, the Murut people of southern Sabah once made their clothes from bark and wild cotton. Today, they prefer to purchase plain woven cloth and decorate it with beads, embroidery, and appliqué.

Sabah is also home to large numbers of Bajau people. Comprising 15 per cent of the population, the Bajau are second in number to the Kadazan. Formerly a seafaring people from the southern Philippines, they have been drifting into Sabah since the nineteenth century. They have settled in the coastal areas of Semporna, Kota Belud, and Kudat. Today, they earn a living as rice farmers and breeders of ponies and buffalo. They are divided into a number of subgroups, such as the Illanun, Sulu, Obian, and Binadan.[37]

The women continue to weave on a very wide body-tension loom similar to that used in the southern Philippines. The steeply inclined warp is fastened to a roof beam by means of rattan wrapped around the warp beam. The breast beam rests in the weaver's lap and is held taut by the weaver as she leans back against a belt of pliable goat's skin. She braces her feet against a heavy beam attached to the side of the house.[38]

The most famous Bajau textile is the *destar*, a 100 cm square, heavily starched cotton head-cloth patterned in supplementary wefts of red, yellow, green, orange, and white imported yarns on a black ground (Plate 26). Metallic yarns are also used to highlight certain decorative elements. As with Malay *songket* weaving, the warp yarns are carefully counted and arranged so that designs will have overshots of five weft threads on the right side of the cloth. Since the designs are very intricate, the supplementary wefts may be both continuous and discontinuous. Coloured yarns are laid in by hand across the design area, rather than with a shuttle. It takes a woman working six hours a day for about one month to complete a *kain destar*.

Motifs on *kain destar* are strongly geometric and include the star-shaped *tuarah*, the *boras* swastika, the *saledap* cross, the *onsod* fence, the *sheku* parallel zig-

Fig. 183. Some popular patterns seen on Bajau *kain destar* from Sabah.

(Names of design motifs were given to the writer by the weavers at Kampung Merbau.)

zag, the arrowed snowflake *samping gapas*, and *anunan* floral designs (Fig. 183).[39] Because of their expert horsemanship, the Bajau are popularly known as the 'cowboys of Borneo', and horse and rider motifs are thus quite popular on modern weaving. The *destar* is worn by almost every indigenous group in Sabah, and the varying methods of draping and tying it serve as a handy reference for identifying the different ethnic groups of Sabah (Plate 27). In terms of design motif and cloth layout, the Bajau *kain destar* shows strong affinities with the Tausug *pis* headcloth of the southern Philippines.

The Bajau also weave *kain mogah*, a serviceable black cloth patterned with bands of muted red and orange bearing small squares and rectangle shapes worked in a supplementary weft. At one time, this cloth was woven from pineapple fibres or from tree cotton yarns. Today, weavers purchase cotton thread from Chinese shopkeepers at Kota Belud. It takes about two weeks to weave a 300 cm by 60 cm length of cloth. The Bajau only use a little of this cloth for personal use, usually to decorate rooms when they entertain, or for trimming a bride-groom's costume. The rest is sold to the Dusun people for sarong material. Only the Illanun group are now weaving *kain mogah*. Its popularity is declining due to competition from cheap imported cloth.[40]

Weaving in Sabah is not widely known, nor are handwoven textiles readily available. Even the colourful weekly markets, called *tamu*, which specialize in local produce, have very little in the way of local weaving. Kraftangan has set up a workshop to teach the art of weaving at its craft centre at Kota Belud, some 70 km north of Kota Kinabalu, the capital. Here, traditional Malay *songket* and Bajau weaving are taught to some thirty trainees. In 1982, a small workshop was established in the Illanun village of Kampung Merbau, 8 km out of Kota Belud. Here, Bajau-style weaving is done by thirteen women on Malay looms. In addition to traditional *destar* cloth, sarongs, table runners, and cushion covers are woven in new designs using Bajau supplementary weft techniques. A little all-purpose striped cloth, called *kain baraguru*, is also woven on a wide body-tension loom.[41]

Brunei

The tiny, oil-rich state of Negara Brunei Darussalam, sandwiched between Sarawak and Sabah, also weaves some distinctive *kain songket* on a Malay loom using imported cotton, silk, and gold and silver threads. Handwoven cloth in Brunei is classified according to its design rather than by the method of weaving. In addition to gold and silver yarns, silk threads of different colours are woven into the weft to highlight key elements of a *songket* design.

In keeping with the tenets of the Muslim faith, traditional *songket* patterns are based on floral and geometric elements. These are somewhat different from those found in Malaysia. *Jong sarat*, with an overall design, is the best-known *songket* pattern (Plate 28). There are nine variations to choose from. A less densely arranged *songket* pattern is called *kain berbagai bertabur* (Fig. 184). It consists of a number of parallel stripes and scattered or widely spaced, regularly aligned floral motifs. Some sixteen *bertabur* patterns are known. The most intricately patterned *songket* design is *si lubang bangsi*, which in the finest work is a mass of gold (Fig. 185). In Brunei *songket* work, the thread count for the supplementary weft overshot is not limited to three or five warp threads. It varies according to the design. A really fine example with a warp count of 3,400 yarns would take over two months to weave and would retail at about US$1,000 (in 1986). Brunei *songket* is worn for royal ceremonies and weddings. In the wedding ceremony, a piece of *jong sarat* is one of the gifts given to the bride from the groom.[42]

Apart from work done by royal artisans, hand weaving has always been a cottage industry carried out in the homes of women in villages, such as Kampung Sungei Kedaya, Kampung Bukit Salat, and Kampung Lorong Sikuna, within the vast riverside complex of Kampung Air lining the banks of Bandar Seri Begawan, Brunei's capital. *Songket* weaving is also practised in inland villages, such as Kampung Salam Bigar, Kampung Lambak, and Kampung Perpindahan Berakas. On relocating from weaving villages within Kampung Air, women have taken the craft of weaving with them.[43]

Weaving in Brunei was traditionally passed down from mother to daughter. However, soaring oil revenues in the 1970s led to a movement of young people to the capital in search of lucrative government jobs. Weaving, along with other crafts, was in danger of dying out. The Royal Brunei Government, anxious to preserve and promote traditional Malay culture, established the Pusat Latihan Kesenian dan Pertukangan Tangan, or Brunei Arts and Handicrafts Training Centre, in 1975. The Brunei National Museum was given responsibility for its organiz-

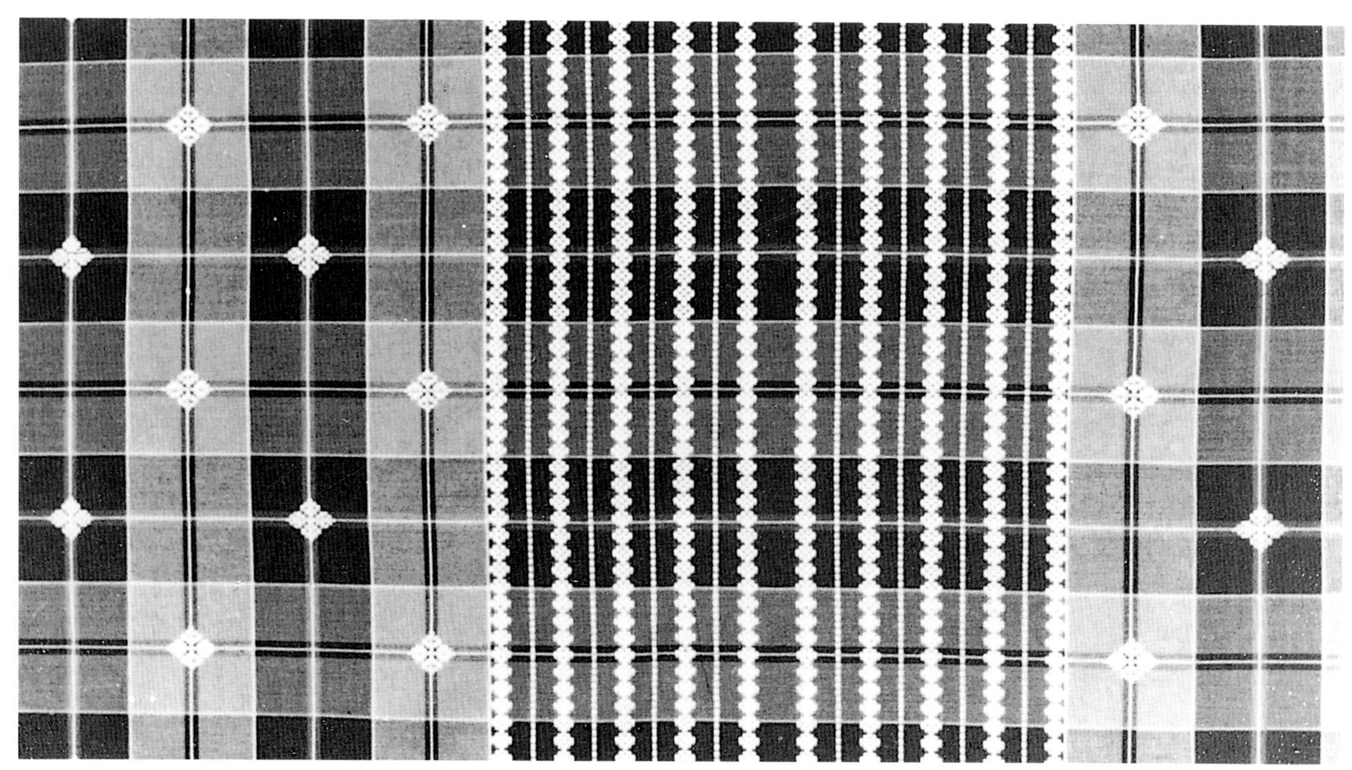

Fig. 184. A silk plaid sarong from Brunei called *kain berbagi bertabur*, which is highlighted with linear and scattered gold *songket* designs. Photograph courtesy of Matussin bin Omar, National Museum, Brunei.

ation and administration. At the time of writing, a total of fifty-five girls were enrolled in its three-year weaving courses held at the centre's new building on Jalan Residency. Under the experienced eye of the chief instructress, Hajah Kadriah, and eight teachers (all of whom are graduates from the centre), the girls are taught weaving from the very beginning. A new student first learns the names of the various tools and pieces of equipment and how to maintain them; then she is allowed to use them. Students begin by setting up the loom for weaving simple plaids. With a little practice, students are soon ready to set up the loom for complicated *songket* patterns. Before graduation at the end of the third year, a student is required to weave a piece of *si lubang bangsi*. While studying, all materials and equipment are provided by the centre. The students are given a monthly stipend of around US$115 per month and a percentage of the sale of products made by the handicraft centre.

While at the centre, trainees work on special weaving orders, many of which come from HM Sultan Hassanal Bolkiah, who is an enthusiastic supporter of the centre. For these special orders, the centre may also dye the yarns according to a client's specifications. The centre's textiles are so popular that they are sold as soon as they are made. The local demand for *songket* is so great that it is virtually impossible for the casual visitor to acquire a piece, for it is not sold in shops. Most work is ordered on commission. People go to a particular weaver and ask her weave a special piece in a particular design and colour within a certain price range. On graduation, weavers may work at the handicraft's workshop at Berakas or set up shop at home on their own. The centre offers assistance in the form of loans to purchase equipment and supplies. It also markets the finished products. With such a strong local demand for *songket*, weavers who are able to weave two sarong lengths a month, are assured of an income of US$200–400 a month depending on the intricacy of the design. To date, some 150 students have graduated from the centre.

Fig. 185. This silk *kain songket* sarong from Brunei is patterned in the *si lubang bangsi* design which has a warp count of approximately 3,400 yarns and takes over two months to weave. Photograph courtesy of Matussin bin Omar, National Museum, Brunei.

1. Norwani Nawawi, 'Malaysian Songket', unpublished MA thesis, Textiles/Fashion School of Art and Design, Manchester Polytechnic, England, 1985, pp. 4–8.

2. Ibid., p. 6.

3. B. A. V. Peacock, *Batek, Ikat, Pelangi and Other Traditional Textiles from Malaya*, exhibition catalogue, Urban Council, Hong Kong, 1977, pp. 26–8; and Malaysian Handicraft Development Corporation, 'Infokraf Malaysia', pamphlet (in Bahasa Malaysia), n.d.

4. Malaysian Handicraft Development Corporation, 'Seni Telepuk' (Gilded Cloth), pamphlet (in Bahasa Malaysia) n.d.

5. Malaysian Handicraft Development Corporation, 'The Handcrafted Textiles of Malaysia', pamphlet, n.d.

6. Interview with Raja Fuziah bte Raja Tun Uda, Director-General, Malaysian Handicraft Development Corporation, 21 May 1986.

7. Malaysian Handicraft Development Corporation, 'National Dress of Peninsula Malaysia', pamphlet (in English and Bahasa Malaysia), n.d.

8. Mubin Sheppard, *Living Crafts of Malaysia*, Times Books, Singapore, 1978, p. 55.

9. J. M. Gullick, 'Survey of Malay Weavers and Silversmiths in Kelantan in 1951', *Journal of the Malayan Branch of the Royal Asiatic Society*, Vol. 25, Pt. 1, 1952, pp. 134–44.

10. Interview with Hajah Atikah bte Mohamad, master weaver and entrepreneur, Pulau Rusa, Trengganu, 17 May 1986.

11. Gullick, op. cit., p. 144.

12. This information was gathered at a special *kain songket* exhibition sponsored by Infokraf, Kuala Lumpur, during the month of Ramadan, May 1986.

13. Malaysian Handicraft Development Corporation, *Serian Songkit* (in Bahasa Malaysia) n.d., Ch. 7, pp. 56–103.

14. Ibid.

15. Information from Infokraf Exhibition and from an interview with Mohammad Hj. Hussain, batik artist and son of Hajah Cik Minah, 15 May 1986.

16. Visit to the workshop of Tengku Ismail, 17 May 1986.

17. Interview with Hajah Atika bte Mohamad, 17 May 1986.

18. *Serian Songkit*, op. cit., pp. 108–10, has a list of names and addresses of some of the more well-known *songket* weavers currently at work in north-east Malaysia.

19. Sheppard, op. cit., p. 53.

20. Malaysian Handicraft Development Corporation, 'Objective, Policy, Programme and Activities', typescript, Kuala Lumpur, 1986; and interview with Raja Fuziah bte Raja Tun Uda, Director-General, Malaysian Handicraft Development Corporation, 21 May 1986.

21. Alfred Haddon and Laura E. Start, *Iban or Sea Dayak Fabrics and Their Patterns*, Cambridge University Press, London, 1936, reprinted Ruth Bean, Bedford, 1982, pp. 5, 19–22.

22. Ministry of Environment and Tourism, *Pua Kumbu*, AGAS, Kuching, n.d., p. 5.

23. Ibid., p. 7.

24. Haddon and Start, op. cit., pp. 125–34.

25. Michael Palmieri and Fatima Ferentinos, 'The Iban Textiles of Sarawak', in Joseph Fischer (ed.), *Threads of Tradition, Textiles of Indonesia and Sarawak*, exhibition catalogue, Lowie Museum of Anthropology and the University Art Museum, Berkeley, California, 1978, p. 73.

26. Anita Spertus and Jeff Holmgren, 'Borneo', in Mary Hunt Kahlenburg (ed.), *Textile Traditions of Indonesia*, exhibition catalogue, Los Angeles County Museum of Arts, Los Angeles, 1977, p. 41.

27. Palmieri and Ferentinos, op. cit., p. 78.

28. Haddon and Start, op. cit., p. 106.

29. Mattiebelle Gittinger, *Splendid Symbols, Textiles and Tradition in Indonesia*, Textile Museum, Washington, DC, 1979, pp. 217–18; and Palmieri and Ferentinos, op. cit., p. 73.

30. Spertus and Holmgren, op. cit., p. 42.

31. Lucas Chin, *Cultural Heritage of Sarawak*, Sarawak Museum, Kuching, 1980, p. 58.

32. Interview with Dayang Zakariah, 29 May 1986.

33. Visit and interview with Puan Fatimah, Sarawak Arts Council Shop, Kuching, 29 May 1986.

34. Interview with Puan Maisurwati, Kraftangan, Kuching, 29 May 1986.

35. Sabah Tourist Association, *A Guide to Sabah*, Vol. 4, Kota Kinabalu, p. 9.

36. Information from the National Museum of Sabah, Kota Kinabalu.

37. *A Guide to Sabah*, op. cit., p. 12.

38. John H. Alman, 'Bajau Weaving', *Sarawak Museum Journal*, Vol. 9, Nos. 15–16, 1960, p. 606.

39. Information given to the writer during a visit to Kampung Merbau, Kota Belud, 23 May 1986.

40. Alman, op. cit., pp. 603–4.

41. Visit to Kampung Merbau, 23 May 1986.

42. Interview with Haji Mohamad Yassin, Brunei Arts and Handicrafts Training Centre, 27 May 1986.

43. Shell Oil, Brunei, *Preserving a Proud Heritage* (in English and Bahasa Malaysia), Brunei, n.d.

CHAPTER NINE

The Philippines

THE PHILIPPINE ARCHIPELAGO encompasses over 7,000 islands inhabited by culturally diverse groups of people who have had a long history of producing textiles for clothing, household, and ceremonial use. Simple plant fibres, such as reeds, abaca, raffia, ramie, pineapple, and bark, along with wild and cultivated cotton and imported silk, have all been used by various peoples to create an array of textiles.

Weaving in the Philippines dates back to the beginning of the Filipino Iron Age *c.*200 BC. Clay, stone, and spindle whorls have been found at many archaeological sites.[1] Hand weaving in the Philippines today is largely confined to groups of people living in the northern and southern extremities who constitute about 4 per cent of the total Filipino population of fifty million.[2] Most weaving is still done on simple body-tension looms. Articles woven include blankets, loin cloths, wrap-around skirts, called *tapis*, sarongs, jackets, trousers, head-cloths, belts, and bags. Warp and weft *ikat*, supplementary warp and weft, tapestry weave, embroidery, and appliqué techniques are used to pattern handwoven cloth.

Northern Luzon

The mountainous area of northern Luzon, comprising the provinces of Apayao, Kalinga, Ifugao, Bontoc, and Benguet, is the habitat of a number of different groups or tribes which include the Isneg, Kalinga, Bontoc, Tinguian, Ifugao, Kankanay, Ibaloy, Gaddang, and Ilongot. These peoples, sometimes collectively called Igorot, are believed to be descendants of the original Malay–Indonesian inhabitants of the archipelago who came to the Philippines over 3,000 years ago. Subsequent immigrations have pushed them into the more remote mountain areas where they are known both for their slash-and-burn and terraced rice cultivation farming techniques.[3] In this mountain environment they have managed to retain the essence of their culture, which is rooted in a variety of animistic beliefs. As craftsmen, they are noted for their basketry, wood work, and textiles of bark and cotton.

Although textiles have always been highly valued in northern Luzon, not all tribal groups who live there have a weaving tradition. The practice of weaving appears to have been largely associated with groups of people who practised wet rice cultivation, such as the Kalinga, Tinguian, Kankanay, Bontoc, Gaddang, and Ifugao. There has always been a lively intra-area trade in textiles throughout northern Luzon. *Tapa* bark cloth was at one time produced by certain towns and widely circulated in the mountain areas. Tribes with a weaving tradition bartered their cloths to non-weavers. Textiles traded in other weaving areas were copied by local weavers. There was also an extensive trade in cloth between the highlands and lowlands, which has led to a diffusion of textiles in the mountain areas.

Until about 1900, the use of bark for clothing was widespread amongst the Bontoc, Ifugao, and others. Bark continues to be used amongst groups of Negrito and a few Ifugao; however, today cotton is the preferred yarn. This fibre used to be quite widely cultivated in the highlands and in the neighbouring Ilocos Norte Province. Today, much weaving is done using imported cotton yarns, and over recent years synthetic yarns have become very popular.

Traditional northern Philippine textiles are warp faced and woven with a double thread. Aesthetically, these textiles are subdued with blue, white, and red predominating, but enlivened with the odd touch of green or yellow. Warp stripes are the major form of decoration. On some textiles, stripes may enclose small woven, embroidered, or warp *ikat*-patterned figures in simple geometric forms which include stylized representations of natural phenomena, such as lightning, trees, snakes, lizards, frogs, fish, birds, and human figures, as well as cultural phenomena, such as axes, arrows, shields, houses, ricefields, and pounders.

The women of northern Luzon wear the *tapis*, a short, knee-length, wrap-around skirt approximately 140 cm by 75 cm, which is made from two pieces of cloth joined along one of the selvages. It is worn folded over in front and fastened with a belt below the navel. The warp ends are finished with a tightly woven striped band. A girdle or sash wound a couple of times around the waist, is worn with the *tapis* on ceremonial occasions. Men traditionally wear a loin cloth approximately 250 cm long by 25 cm wide, which might be quite plain or be striped with decorated borders and tassels at the ends of the cloth. Both men and women wear simply tailored jackets for special occasions and for protection from the sun while working in the fields. Clothing for ceremonial events is often embellished with stitchery, buttons, shells, pieces of brass, and tufts of yarn. Small woven bags and distinctive head-cloths are worn by various groups.

Some major tribal groups are distinguished by their dress. Traditional dress for the Ibaloy woman of Benguet Province is a wrap-around cotton *tapis*, called a *devit*, and a matching, long-sleeved *sa'day* jacket (Fig. 186). The outfit is patterned with alternating red and bluish black bands separated by small white pin-stripes. The lower part of the skirt, jacket, and sleeves is sometimes decorated with two or three overlapping layers of light cotton plaid fabric. This plaid material may form a sailor's collar and serve as trim for the neck and jacket openings. Formerly, only the wealthy could wear a multi-layered *devit*. In times of mourning, a blue and white *devit* was worn by widows.

Ibaloy men traditionally wear a white loin cloth with dark blue borders. A boy is given his first loin cloth when he is five or six years old. Large turbans wound two to three times around the head, are also worn by Ibaloy males. They are particularly popular with older people and may serve as a convenient place to store smoking equipment.[4] The Ibaloy acquired much of their cloth from the Kankanay, their neighbours to the north, and through trade with the lowland people of the Ilocos region.

Kankanay women wear a white *tapis* decorated with navy blue bands, while the men wear a dark blue loin cloth patterned with red stripes. Both men and women wear blankets with red and blue stripes containing anthropomorphs, snakes, and diamond patterns formed by a supplementary warp technique. Women occasionally wear white jackets edged with bands of striped material. A woman bearing her first child at one time wore a costume of plain white, undecorated cloth. The Kankanay also weave special burial jackets, called *losodan*, patterned with anthropomorphs in warp *ikat*.[5]

Ifugao women, who are among the most accomplished weavers of the northern Philippines, wear a number of different *tapis*. The most popular is the *intulu*, composed of blue and white stripes embroidered in blue and red with small geometric shapes along the seams.[6] They also wear a dark blue *tapis* patterned with red warp stripes and small, simple, supplementary warp designs, such as rice pounders and human figures, in dark blue on a speckled white ground. These are usually made from cotton. Some for sale in the village of Tam An, situated behind the Banaue Hotel, are made from a coarse thread produced from the bark of a local shrub called *dami*. The yarns, which are blue and white in colour, are combined with red acrylic wool to produce a cheap, loosely woven *tapis*, or blanket. Over the last few years, some Ifugao women at Barrio Amunganad, Banaue, have revived the art of producing simple warp *ikat*-patterned *tapis* coloured with vegetable dyes to produce yellowish

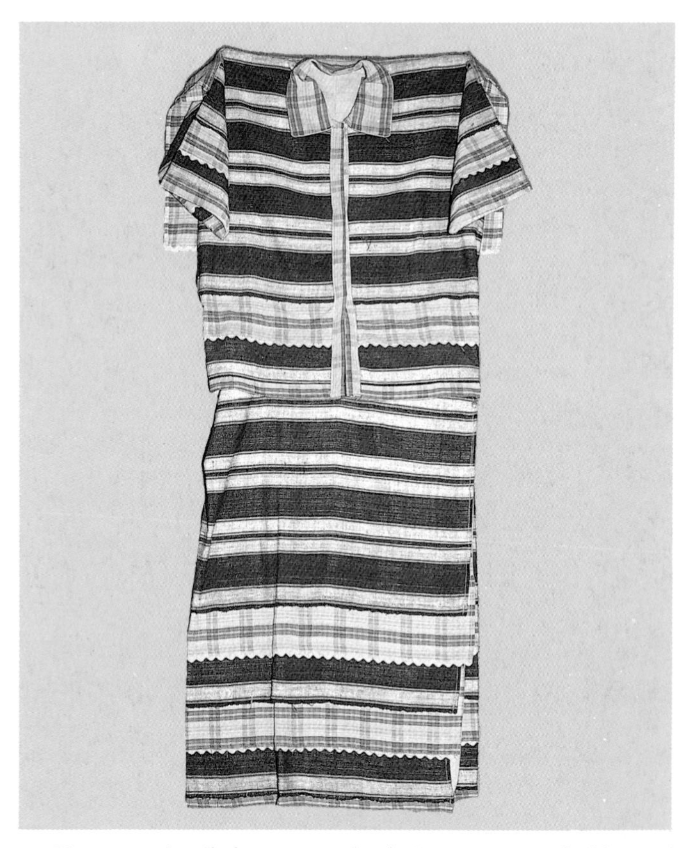

Fig. 186. An Ibaloy woman's *devit* wrap-around skirt and *sa'day* jacket from the Baguio area of Benguet Province. Note the plaid layers along the borders of the red and blue striped outfit which at one time indicated wealth and status. *Devit*, warp 130 cm, weft 75 cm; *sa'day*, length 55 cm.

brown, pinkish red, green, and black shades. The muted dye range is most pleasing but not entirely colour-fast. Blue or red sashes, called *mayad*, patterned in the centre and at the ends with bands of supplementary warp designs in red and yellow, are worn by Ifugao women on special occasions.

The Ifugao male wears a dark blue loin cloth, referred to as *binuhla'n*, which has a large red stripe in the centre. The ends are embroidered with geometric renditions of anthropomorphs, shuttles, and basket designs. A similar, but more intricately patterned loin cloth with almost identical designs in warp *ikat*, is referred to as a *pini'wa an nili'hha*. A poor man at one time was expected to wear a plain white loin cloth, called a *tina'nnong*. The Ifugao also weave special striped loin cloths, called *kupi'ling* and *taga'ktak*, which were once used by the well-to-do to clothe the deceased as they sat in state on the death chair.[7]

Bontoc women of Mountain Province weave a brightly coloured *tapis* made from three lengths of cloth (Fig. 187). The two side panels, which are identical, are patterned with red and dark blue/black warp stripes (Fig. 188). Some bands depict diamond, man, snake, and shield motifs worked in a supplementary warp on a white ground. Pin-stripes of yellow and green further break up the red and blue bands. The central strip of cloth usually consists of a red stripe in the middle, flanked on each side by a wide black band, followed by a slightly narrower white stripe. The last 30 cm at each end of the cloth is patterned with individual yellow, red, and green diamond motifs, followed by a continuous trellis work of diamond designs in different colours separated by parallel bands of zig-zag lines. The lateral edges are bounded by a 6 cm wide band of striped material. This style of *tapis* is also popular with other groups of mountain people.

To the north of Bontoc are the Kalinga people who weave brightly patterned garments, such as a red and blue striped *tapis* overlain with two or three bands of zig-zag and diamond shapes, embroidered predominantly in a yellow and green blanket-stitch and enlivened by touches of white (Fig. 189). Dangling diamond-shaped slivers of iridescent shell attached to small beads, are sometimes added for further embellishment (Fig. 190). The warp edges of the *tapis* are finished with rows of attenuated chain-

Fig. 187. A young Bontoc girl wearing a *tapis* wrap-around sarong patterned with red and blue warp stripes and supplementary warp designs of human figures and rice pounders in dark blue on a speckled white ground. Photograph courtesy of Eduardo Mansferré, Bontoc.

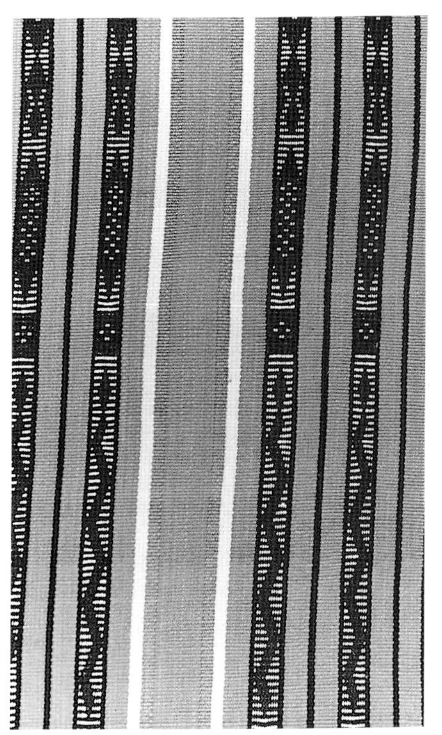

Fig. 188. Detail of supplementary warp patterning on the Bontoc *tapis* in Fig. 187.

Fig. 189. Kalinga woman in traditional *tapis*. Photograph courtesy of Eduardo Mansferré, Bontoc.

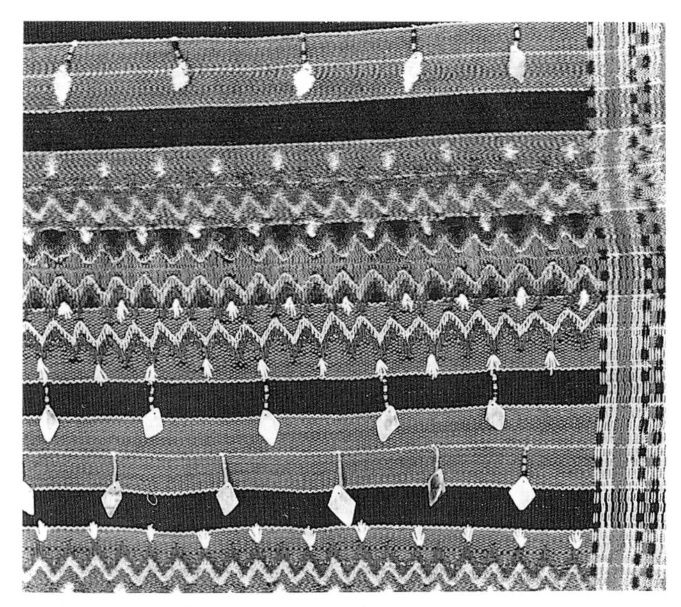

Fig. 190. Kalinga woman's red and blue striped *tapis* wrap-around skirt embellished with embroidery and slivers of iridescent shell.

Fig. 191. Kalinga man's loin cloth patterned with four red and yellow stripes on a blue-black ground. The ends are decorated with parallel bands of red, yellow, and white supplementary yarns arranged in rows of zig-zag and diamond shapes. Collection of the National Museum of Natural History, Smithsonian Institution, Washington, DC, No. 42396. Warp 240 cm, weft 32 cm.

stitch in yellow, blue, and red. This *tapis* was formerly worn topless with masses of beads. Today, it is worn with a Western-style blouse. The loin cloth for the Kalinga male consists of a blackish-blue ground patterned with four red and yellow stripes (Fig. 191). The ends are decorated with parallel bands of red, yellow, and white supplementary yarns arranged in rows of zig-zag and diamond shapes resembling a twill weave. Short striped jackets, which barely cover the breast, are worn on ceremonial occasions, along with mantles draped over one shoulder.

The Gaddang, an obscure nomadic people inhabiting the eastern area of northern Luzon, wear an outfit similar to that of the Kalinga, but the weaving is much finer. Their woven cloth has the highest thread count per centimetre in northern Luzon.[8] The Gaddang specialize in producing finely patterned striped material in narrow bands of pinkish red and dark blue, bisected by very narrow pin-stripes of white and, occasionally, yellow. Some stripes contain a chain of small diamond patterns (sometimes referred to as *mata* or 'eyes') which are created by a supplementary thread technique. Most ceremonial cloths are finished with masses of tiny white beads along the edges. Some beads are actually interwoven in the cloth. Beads are sometimes added to warp stripes to highlight the design. The loin cloth for the Gaddang is very narrow and the ends are often gathered into a brass ring before fanning out into a small beaded flap (Fig. 192). The warp fringes may be threaded with beads. Plaited cords with beaded pom-poms are used to fasten the very short jacket worn with the loin cloth (Fig. 193).

In addition to providing colourful distinctive clothing, textiles in northern Luzon are of tremendous ritual and social importance. Many circumstances involving economic exchanges, such as marriage payments, the acquisition of property, and the establishment of trading relationships, all traditionally involved the transfer of textiles. The acquisition and public display of numerous fine textiles was generally regarded as tangible proof of earthly attainments. Cotton textiles in northern Luzon have clearly established values based on type, rarity, and intricacy of design. A talented and diligent weaver at one time could make a tidy supplementary income producing ritual cloths.

Textiles were often used as offerings to spirits during special ceremonies to ensure their continued benevolence in the affairs of men. Shamans wear special blankets while performing sacred rituals. Some textiles are considered *de rigueur* for certain

Fig. 192. Gaddang male in full regalia with loin cloth, jacket, mantle, scarf, and head-cloth. Photograph courtesy of Eduardo Mansferré, Bontoc.

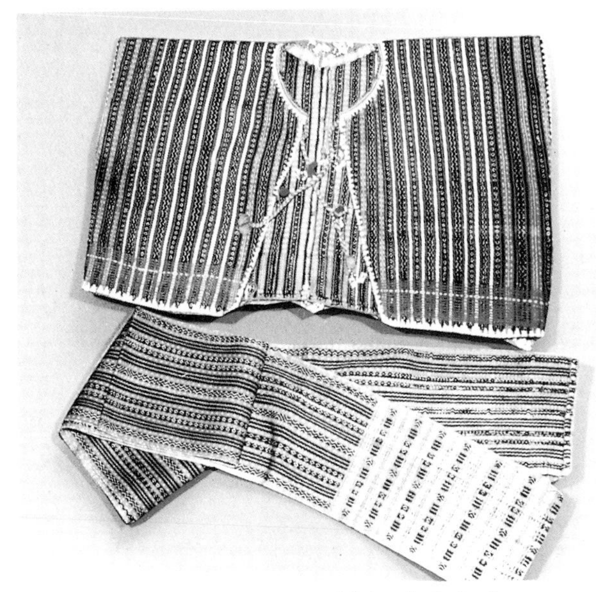

Fig. 193. Gaddang jacket and loin cloth finely patterned with stripes in geometric designs and finished with masses of tiny white beads. Loin cloth, warp 200 cm, weft 25 cm; length of jacket 35 cm. Collection of Ana M. Juco, Treasures, Manila.

ceremonies and a family will spare no effort or expense to acquire the appropriate textile, particularly where funerary rites are concerned. Custom decrees that every man should be buried in as many blankets as the family can afford.[9] Some of the finest weavings seen in northern Luzon today are in the form of funerary blankets.

The most sought-after blanket amongst the tribes of northern Luzon is one formerly produced by the Isinai, a Christian lowland people living in the towns of Aritas, Bambang, and Dupax in northern Nueva Vizcaya Province (Plate 29). This unique blanket is decorated with white *ikat* motifs on a blue ground enlivened with a horizontal red stripe at each end. It is made from two identical pieces of cloth joined down the centre to form a mirror image. Small blocks of *ikat* in the form of stylized representations of human figures, shields, fish, intertwining snakes, and other geometric forms constitute the main motifs. Imbued with special meaning, these motifs may be 'read' to depict the journey of man through life. These blankets have been widely imitated by the weavers of northern Luzon. For example, the Kankanay have been known to produce 'Isinai' blankets in a supplementary weft technique.[10]

The Ifugao make a highly prized warp *ikat* funerary blanket, called an *inla'dang*, for the more affluent members of their tribe (Fig. 194). It comprises a

Fig. 194. Detail of an Ifugao *inla'dang* warp *ikat* funeral blanket depicting *inan-a-ntak* (bean), *binula'ngon* (monkey), *hinikitan* (shuttle), *hinakhanlong* (dipper), and *linu'hhong* (rice pounder) motifs. Collection of the National Museum of Natural History, Smithsonian Institution, Washington, DC, No. 364748.

Fig. 195. Some common motifs seen on Ifugao weaving. (After Lambrecht and Reyes.)

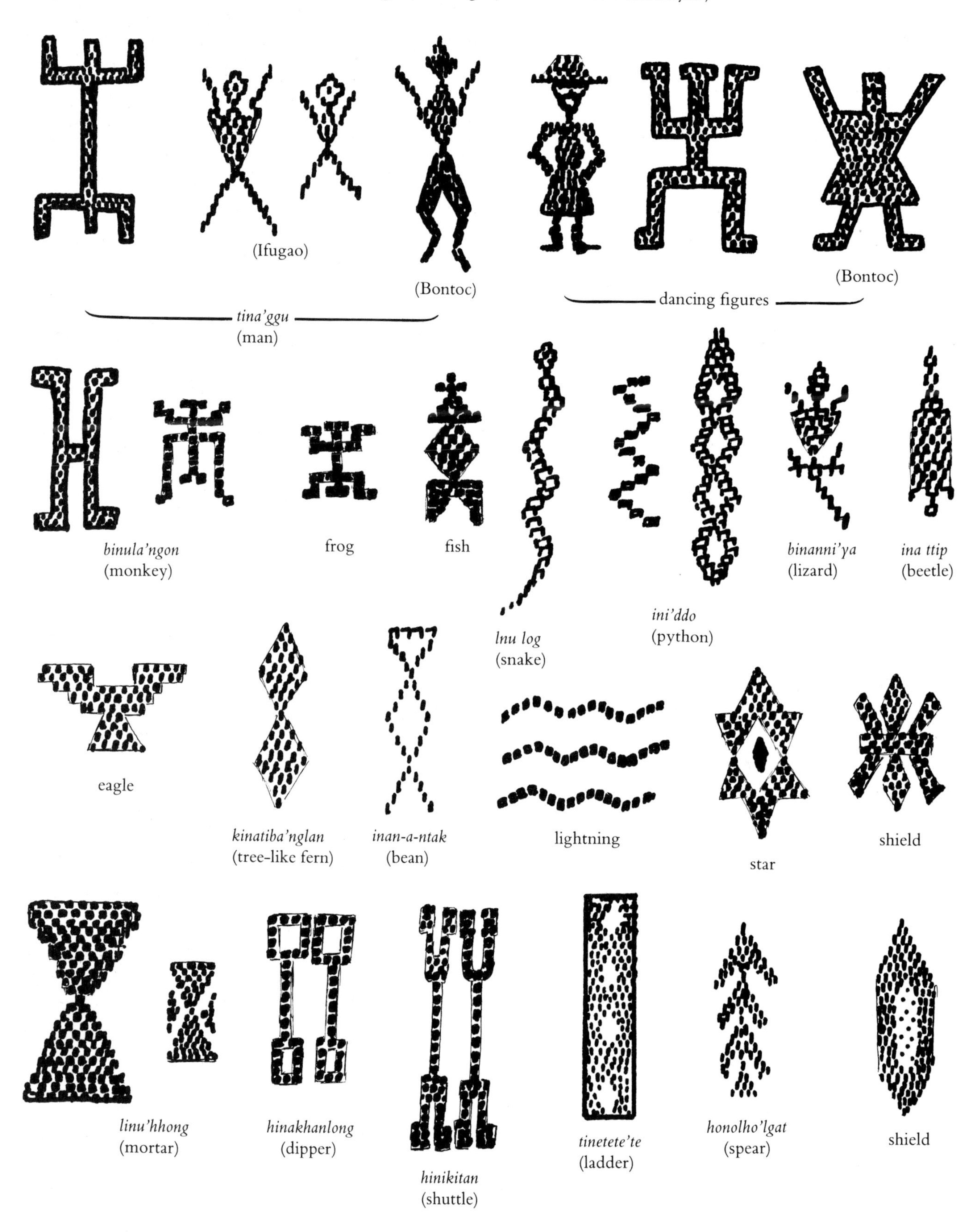

large central panel flanked by two smaller panels which are enclosed by orange and red stripes. Motifs in white on a blue ground include small stylized representations of human figures (*tina'ggu*), snakes (*lnu log*), monkeys (*binula'ngon*), beetles (*ina ttip*), lizards (*binanni'ya*), tree ferns (*kinatiba'nglan*), and mortars (*linu'hhong*) (Fig. 195). Like the motifs on Isinai blankets, these may be interpreted to formulate a story. *Inla'dang* blankets are occasionally worn for prestigious reasons by the living.[11]

The most common Ifugao funerary blanket is the *ga'mong*, which is approximately 240 cm in length and 150 cm in width, and is the outermost blanket to be placed on a corpse. It is also used to wrap the remains in the case of a secondary burial. One may also be placed on the death chair before the corpse is seated in state. It is also worn occasionally by the living (Fig. 196). The *ga'mong* is made from four loosely woven pieces of cloth joined together along the selvages (Fig. 197). The two central panels, called *ado'lna*, are in natural-coloured or white cotton. They are often patterned with one to three double bands of black stripes which terminate in a special design area at each end called a *pa'gpag*. Two alternating patterns, a trellis or *binogo'gan*, and diamond or *hinu'gli*, worked either by embroidery or in a supplementary weft technique, fill the *pa'gpag* area. These patterns are usually in blue enlivened by touches of red in the *hinu'gli* diamond area. The side panels (*bali'ngbing*) are in navy and red stripes. The navy bands may include small geometric patterns worked in a supplementary warp technique. Motifs are similar to those found on the *inla'dang* blanket (see Fig. 195). They include the *linu'hhong* (rice pounder), *tina'ggu* (human figure), *ini'ddo* (python), *binanni'ya* (lizard), *tinetete'te* (ladder), *ina ttip* (beetle), *honolho'lgat* (spear), *inan-a-ntak* (beans), shuttle (*hini-kitan*), and tree fern (*kinatiba'nglan*).[12] Prior to burial, pieces of the blanket may be torn off and distributed to relatives so that a spirit will not be tempted to 'steal' it.

The Ifugao also weave another blue and white striped blanket, called *baya'ong*, which has possible prototypes in Kankanay examples.[13] This blanket is generally associated with older people and is used as a cover, an upper garment, and a turban. A plainer, less prestigious blanket is the *ha'pe*, which is usually worn by young people. Both the *baya'ong* and *ha'pe* are worn by shamans when performing rituals. A black and white striped blanket, called the *kinto'g*, is sometimes used as a baby carrier by the Ifugao.[14]

The Tinguian, also called the Itneg, who live in the Abra region bordering Ilocos Norte in north-west Luzon, devote most of their textile artistry to weaving magnificent blankets patterned in a supplementary weft (Plate 30). The most stunning are those hung up as protective displays on ceremonial occasions. They may be patterned with pink/red and blue/black warp stripes, which contain geometric rhomb-like patterns thought to be stylized renditions of important items in the Tinguian environment,

Fig. 196. An Ifugao man wearing a *ga'mong* blanket. Photograph courtesy of Eduardo Mansferré, Bontoc.

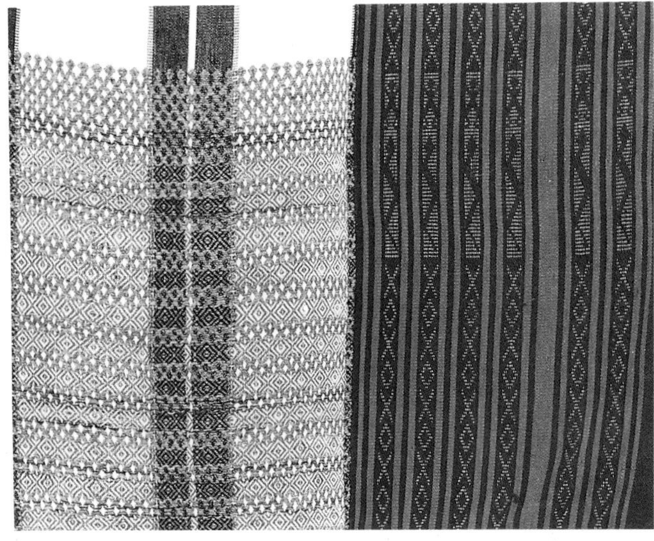

Fig. 197. Detail of an Ifugao *ga'mong* blanket showing the *ado'lna*, *pa'gpag*, and *bali'ngbing*. Warp 235 cm, weft 135 cm.

Fig. 198. Some popular motifs seen on Tinguian ceremonial blankets. (After Fay-Cooper Cole.)

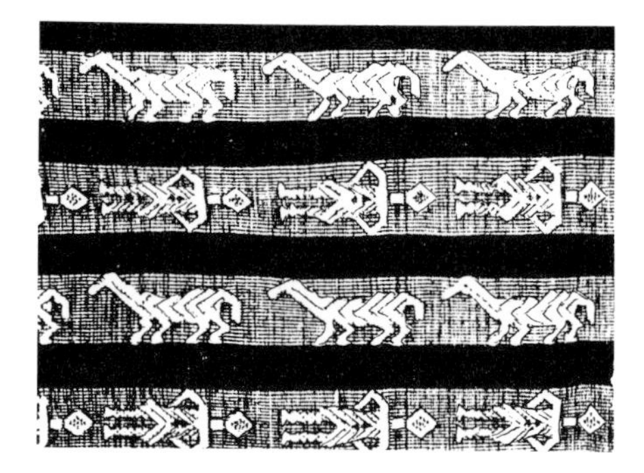

Horse and man motifs.

The pattern in the centre has been referred to as a star, a crab, or a frog. Flanking the central motif is a two-headed eagle which alternates with the small figure of a man.

A rhomb pattern sometimes called 'turtle' flanked by a small *mata*, or eye, motif. At the join between the two sections is a pineapple motif.

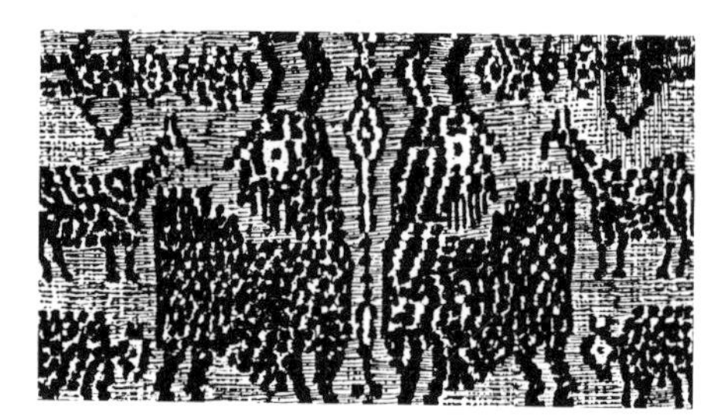

A deer, a horse, a caribou calf, and a man.

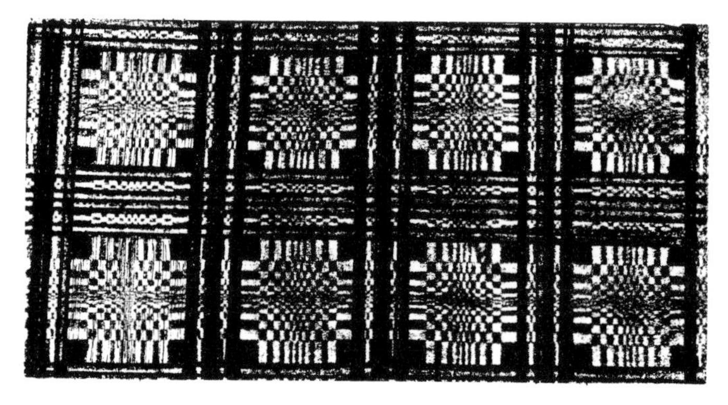

Mat design.

such as crabs, fish, frogs, pineapples, fish hooks, rice pounders, and bobbin winders. These are interspersed with realistic renditions of horses, deer, caribou, and birds, such as the chicken and the eagle (Fig. 198).[15] Human figures are also seen (Fig. 199). The most prevalent is a little standing figure of a man with arms akimbo. This figure is sometimes alternated with a horse on a striped or checked ground. Some blankets are cleverly patterned with a series of squares and rectangles of varying dimensions to create a very striking 'op art' effect. These blankets are now being made for the tourist trade in Piddig, Ilocos Norte.[16]

The Tinguian also weave a number of plainer blankets which are of tremendous ritual importance. A long white blanket, called a *tabing*, is used to screen off a corner of a room during special rituals to proclaim a family as one of wealth and importance (the *sayang* ceremony). In this newly created 'room', food and offerings are prepared for the 'black, deformed, and timid spirits', who may wish to attend the ceremony unobserved. At the *sayang* ceremony, certain women, when dancing along with spirit mediums, may wear a large silk blanket, the *pinapa*, which is decorated with yellow warp stripes. (The *pinapa* may also be placed over the feet of a corpse.)[17]

The *inalson*, a sacred blanket made of white cotton with red and blue warp stripes along the selvages, is worn by a spirit medium at the *gipas* ceremony held prior to the birth of a child.[18] This unassuming

Fig. 199. Tinguian blanket in pink and blue warp stripes patterned with a turtle motif. The warp ends are finished with whimsical effigies of women and dogs with small figures of chickens in the background. Collection of Roland Goh, Imperial House of Antiques, Baguio.

textile may be embellished with bands of honeycomb patterns in a supplementary weft along the warp ends (see Fig. 102).

The arrival of the Americans in the Philippines in 1899 led to the gradual development of communications in the mountain areas, which further stimulated the circulation of local textiles throughout northern Luzon. Public schools introduced industrial arts courses which included weaving instruction. Vigorous efforts were made to stimulate the sale of handicrafts. Vocational schools were established where weaving was taught and European-type looms were introduced. Students in these courses came from all over Luzon. On graduation, these women returned to their villages to organize classes and establish centres for the production of handicrafts.[19]

The city of Baguio, gateway to the mountain provinces, is an important centre which draws weavers from all over Luzon. The Easter School of Weaving, sponsored by the Episcopalian Church, offers employment and training to Bontoc and Ifugao women who weave textiles, including table-cloths, bags, and clothing, on traditional body-tension looms. The Episcopalian Church also sponsors the All Saints Weaving School in Bontoc, which specializes in weaving high-quality traditional textiles. Banaue Handicrafts is supported by the Roman Catholic Church. At its workshop in Baguio, under the supervision of Sister Vincent, some twenty girls weave table linens and furnishing materials patterned with traditional Ifugao motifs. Both cotton and synthetic yarns are used with hand-operated floor looms, which are able to produce longer and wider yardages than the traditional body-tension loom. The products are retailed locally and in Manila.[20] In Baguio, there is also Narda, a small handicraft concern which produces some very attractive handcrafted *ikat* patterned with skilful gradations of blocks of colour in the pinkish-red-purple, beige, blue, grey, and black dye ranges. On these loose-weave textiles, chemical dyes have been cleverly blended to achieve soft subtle shades. These new *ikat* are popular with tourists.[21]

Central Philippines

The making of cheap raffia cloth from the outer skin of the *buri* palm was at one time important in several isolated communities in the central Philippines. The best cloth is reported as coming from Isio in Occidental Negros, Leon, and Iloilo. A red dye from sappan wood and yellow from the tumeric root were used to produce coloured stripes in the *buri* cloth which was reported as being woven, in most cases, on a floor loom.[22] Raffia cloth continues to be woven for sails, mosquito nets, and cloth for packaging tobacco. It no longer appears to be woven for clothing.

Women from central Visayas use a floor loom to weave fine diaphanous *piña* cloth from pineapple leaf fibres and *jusi* fabric from a combination of pineapple and abaca yarns, with silk or cotton. Spanish writers in the mid-sixteenth century reported burials where the corpse was interred in multi-layered gauze shrouds which, in all probability, was *piña* cloth.[23] *Piña* and *jusi* continue to be very popular fabrics for lowland Filipinos. Caftans, shawls, blouses, and the large-sleeved Filipina dress, called the *terno*, are often made from these fabrics. The traditional Filipino man's shirt, the *barong tagalog*, is also made from local fibres. It is decorated with embroidery which is now usually done by machine (see Fig. 94). *Piña* cloth is also widely used to make vestments and altar cloths for the Roman Catholic Church. Some embroidered examples of *piña* cloth which have survived from the last century, display an intricacy which is almost beyond belief. On some, an extremely fine satin-stitch delineates floral motifs, while lace-like embroidery finishes the border areas. There may also be some *calado*, or open work, which involves pulling threads out of the ground cloth and replacing them with a decorative warp or weft.[24]

Ramie fibre mixed with cotton is a popular fabric for table linens. Much of this is now machine-produced in central Luzon, on the outskirts of

Manila. The Legazpi area in southern Luzon is noted for its abaca products in the form of mats and bags made by plaiting and weaving. These are sometimes embroidered with coloured raffia thread. This industry produces handicrafts primarily for the tourist trade.

Southern Philippines

There are two distinct weaving traditions in the southern Philippines, that of the animistic peoples who inhabit the upland areas of Mindanao, and that of the Muslim people who inhabit the lowlands of Mindanao, Palawan Island, and the Sulu Archipelago. The uplands of Mindanao are peopled by the T'boli, Mandaya, Bagobo, Manobo, Bukidnon, and Bila'an peoples who have a tradition of making warp *ikat*-patterned textiles from abaca fibre.

The Hill Peoples

The T'boli people near Lake Sebu in south-west Cotobato Province continue to produce *t'nalak*, their distinctive abaca cloth (Fig. 200). The preparation and weaving of abaca cloth is very much the preserve of women. Men, however, assist in felling the abaca plant and stripping the fibre into strands. After repeated combing and bleaching by the sun, women lay the warp fibres onto a 60 cm by approximately 300 cm bamboo tying frame. The women 'tie' in the design with pieces of waxed string. Tying is done from memory. No patterns are used. Symmetry and distances are measured with finger joints, the length of the index finger, the width of a hand span, etc. Traditional *t'nalak* is in three colours—deep reddish brown, black, and white. Black comes from the leaves of the *k'nalum* tree, while the red is from the roots of *sikarig* (*Morinda bracteata Roxb*). Creamy white is the natural colour of the abaca fibre. For dyeing, the fibres are placed in a 'steamer' composed of two earthenware pots placed rim-to-rim one on top of the other. Steam conveys the dye colours to the bundle of threads in the upper chamber. It may take up to three weeks to get the black shade required. After the initial black dyeing, the areas to be dyed red have their bindings removed and the fibres are steamed for a further two days. The last remaining knots covering the areas that are to remain white are then removed.[25]

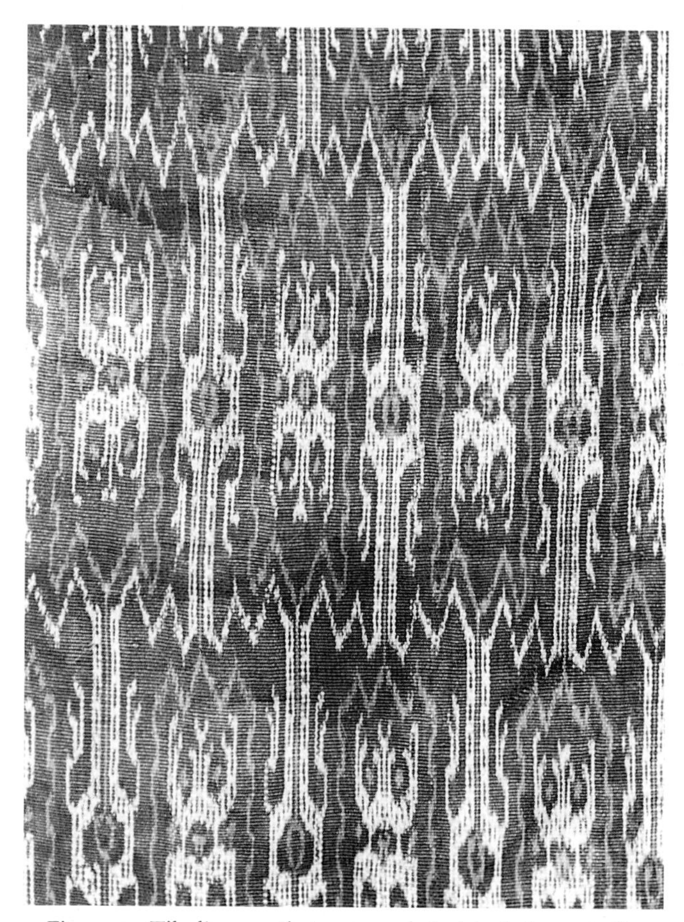

Fig. 200. T'boli warp *ikat*-patterned *t'nalak* cloth made from abaca. The motif depicted is *bakig*, or lizard. From Lake Sebu, South Cotobato Province. Warp 110 cm, weft 54 cm.

After washing and drying, the warp threads are transferred to a body-tension loom set up on one side of the house. Usually weaving is done in the evening when the coolness and moisture in the air impart a certain elasticity to the fibres. T'boli women revere *Baitpandi*, the Spirit of Weavers. While the weaver works, bells are hung on the edges of the unfinished cloth to frighten away evil spirits.[26] Children are not allowed near the loom while weaving is in progress lest they fall sick.

On completion, the cloth is removed from the loom. It is then softened by being pounded with a mallet and rubbed in ashes before being polished to a sheen with a bivalve shell. It takes up to two and a half months to weave 5.4 m of cloth.

The T'boli have an extensive inventory of traditional weaving designs composed of both geometric and stylized animal and human figures (Fig. 201). One very basic design is the *sigul*, or zig-zag arrangement of triangles or rhombs, arranged in a linear fashion across the cloth to provide the framework for an over-all pattern, such as the *kleng* (crab design), the *sa'ub* and *kofi* (bird) patterns, and the *g'mayaw* (mythical bird) design. Other patterns include the *tofi* (frog) design, the *bakig* (lizard), the *klung* (shield) motif, the *sawo* (based on the python), and the *bangala* design depicting a man (and sometimes a woman) in a house. These motifs originally

Fig. 201. Some well-known T'boli *t'nalak* patterns. (After Casal.)

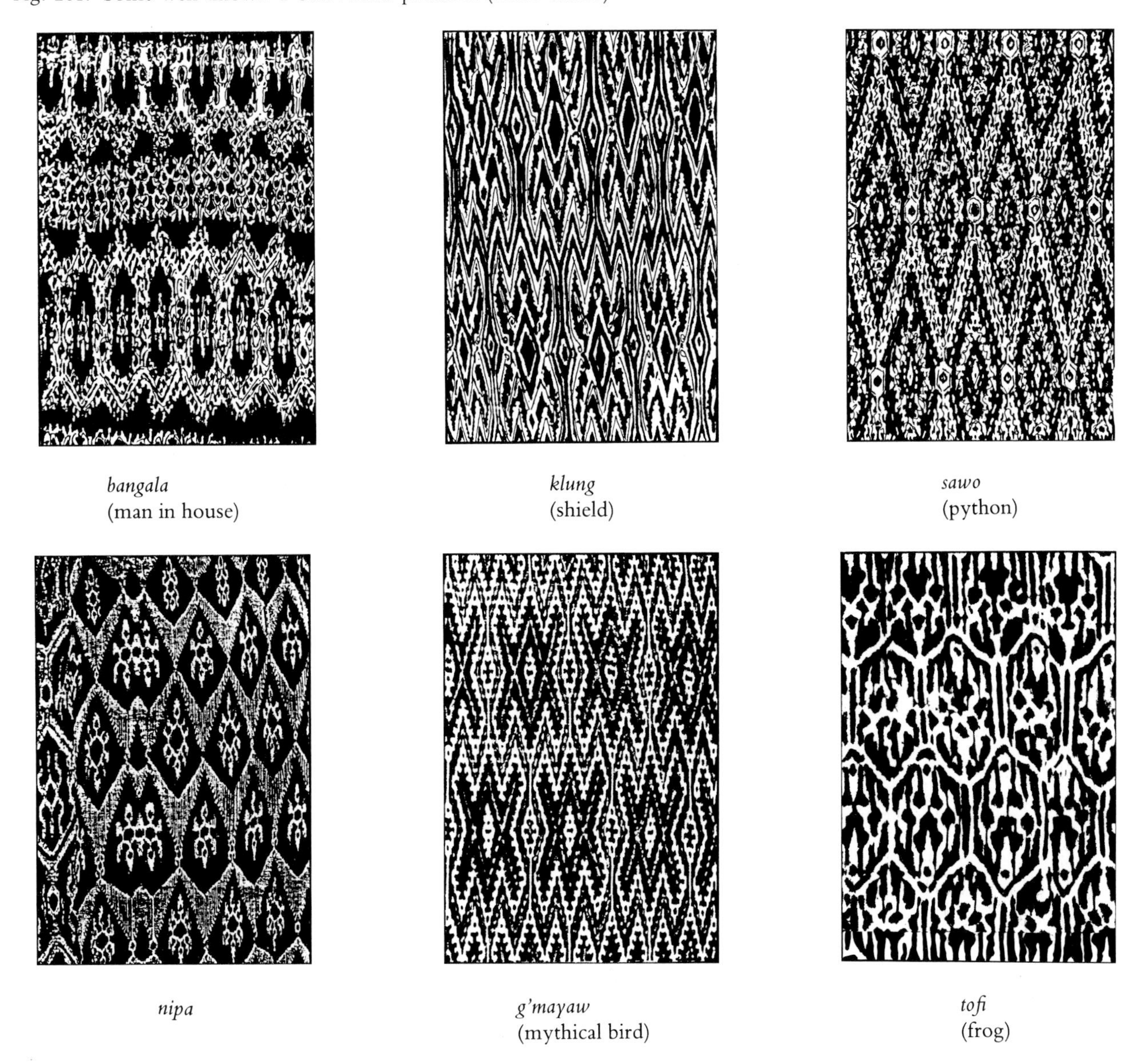

bangala (man in house) — *klung* (shield) — *sawo* (python) — *nipa* — *g'mayaw* (mythical bird) — *tofi* (frog)

had special meaning to the T'boli. Depictions of human figures could be either effigies of ancestors or for protection against evil. Birds often represented the souls of the departed. Lizards could be the reincarnation of a deity or an abode for the soul of an ancestor. Frogs symbolized rain, magic, and fertility.[27]

T'nalak cloth is still used to make clothing for ceremonial purposes. The traditional male costume, made entirely from abaca cloth, consists of a long-sleeved, waist-length jacket, and a pair of knee- or ankle-length trousers secured at the waist by a belt, with the excess falling like a skirt over the thighs. Women wear a short, three-panel skirt and fitting jacket. Today, cotton is generally preferred for everyday clothing.

Blankets, or *kumo*, made by sewing three lengths of cloth edge-to-edge, play an important role in ritual events. The T'boli believe that *t'nalak* cloth has special powers. Apart from making clothes, the cloth should not be cut. Cutting the cloth could result in sickness and even death. If the cloth is to be sold, a brass ring is attached to the cloth to appease the spirits. At the time of birth, a woman will be covered with a *t'nalak* blanket to ensure the safe delivery of her new-born child. The body of a deceased person is wrapped in a *t'nalak* blanket before being placed in a coffin hewn out of a tree trunk.[28]

Cloth is an important item of exchange at the time of marriage. On the day of the wedding, the walls and rafters of the bride's house are festooned with blankets. The groom, with the help of a female relative, 'claims' his bride from under a blanket. If the families are of substantial means, the marriage is followed by a series of nuptial feasts, called *mo'ninum*, which the bride and groom's relatives and friends take turns in hosting. Special buildings are constructed and 30–40 m long horizontal rows of bamboo poles are set up outside the bride's house. On these are hung the *kumo* blankets which form part of the dowry given to the groom by the bride's family. As part of the ritual, the groom's family and guests must 'penetrate' the *kumo* wall by passing under the blankets and on into the bride's house bearing reciprocal gifts. Usually a year is needed between each *mo'ninum* to give women time to weave sufficient blankets to provide a fitting display of wealth.[29]

The Mandaya people, who live in Davao Province of Mindanao, have carried the art of warp *ikat* to an extremely high level. They are particularly noted for their beautiful artistic representations of human and crocodile forms in red and black warp stripes (Fig. 202). These may be edged with delicate 'S'-shaped and wave-like curvilinear motifs. On the best Mandaya cloth, the natural light-coloured warp *ikat* patterns on a dark ground have a delicate lace-like quality. The Mandaya like to embellish their abaca *ikat* with embroidered geometric and curvilinear forms in yellow, blue, and white yarns. Motifs include hooks, crosses, and diamond shapes, which may be highlighted with glass beads. At one time, affluent Mandaya women wore red cotton blouses with black sleeves to distinguish them from the common women who could only wear brown or black abaca blouses. Formerly, only head-hunters who had taken at least ten heads were allowed to wear special red shirts with intricate design patterns. Today, such shirts are worn by village headmen.[30]

The Bagobo, who live in northern Davao, also pattern their warp *ikat* with pleasing repetitive rhomb designs and curvilinear decoration. In more recent times, the Bagobo have preferred to spend the greater part of their efforts on embroidery, bead, and shell work (Fig. 203). Warp-patterned abaca in narrow red and white stripes is made into bags and pants embellished with bands of embroidery (Plate 31). Motifs inspired by the natural environment include stylized renditions of dancing men, dwellings, zig-zags of lightning, 'X' designs, *binitoon* or star motifs, and plant forms. Embroidery threads are combined with sequins, glass beads, and beads carved from the shiny white seeds of Job's tears (*Coix lachryma*). Small brass bells may be added for an audio-visual effect. The women weave a sarong, called a *dagmay*, which is composed of three lengths of material. The middle section carries a number of readily recognizable geometric figures. Bagobo men's trousers have wide parallel bands of embroidery around the cuff, with plant-like forms

Fig. 202. Mandaya warp *ikat*-patterned abaca cloth depicting human and crocodile motifs within warp stripes. Warp 181 cm, weft 65 cm.

Fig. 203. Bagobo cotton blouse embroidered with geometric motifs. Length 45 cm. Collection of Ana M. Juco, Treasures, Manila.

embroidered on the legs (Fig. 204). Bagobo bags, with an intricately patterned overlapping front flap covered with masses of beads, are among the most beautiful in the Philippines. Handwoven jackets and colourful head scarves are also beautifully embroidered and embellished with beads.[31]

The Manobo, who occupy the coastal area and adjacent mountains of eastern Mindanao, weave cotton and abaca fibres into tubular skirts for women, and either loose-fitting pants or tight-fitting breeches for men. These are worn with red blouses with black sleeves by women, and neat-fitting jackets for men. Manobo woven cloth is somewhat lighter than that woven by neighbouring groups. Embroidery and appliqué are the major forms of embellishment. Embroidery designs are largely geometric and include diamonds, rectangles, squares, triangles, and the dancing man motif worked on a dark red or brown ground. Garments are finished with strips of blue, white, yellow, and red cotton cloth appliquéd around the sleeves and neck. Social status at one time was indicated by clothing, and certain motifs, called *binain*, were reserved for the wealthy. Warriors were given special privileges and, like the Bagobo, were allowed to wear special red shirts with geometric border designs.[32]

The Bukidnon of central Mindanao weave *jusi*, cotton, and abaca fibres with a simple body-tension loom. The making of warp *ikat* from abaca yarns is known, but greater effort is devoted to embroidery and appliqué. The women like to appliqué strips of red and white cloth to their distinctive, almost European-style 'Maria Clara' dress, consisting of a short blouse with large sleeves and a skirt. This is worn with a bow-like head-cloth and an embroidered red shawl. The men wear neat-fitting trousers which are sometimes elaborately embellished with embroidery and appliqué. The Bukidnon favour geometric designs, such as the 'X', star shapes, and curved triangular patterns.[33]

Fig. 204. Bagobo man's abaca trousers patterned with small warp *ikat* designs within warp stripes. The trouser legs are lavishly embroidered with beads and sequins. Collection of the National Museum of Natural History, Smithsonian Institution, Washington, DC, No. 286123.

The Bila'an, who inhabit the boundaries of Cotobato and South Cotobato Provinces, expend an even greater effort than the Bukidnon in decorating their handwoven cotton and abaca jackets and pants. Using cross-stitch and outline-stitch on a plain weave ground, the women spend hours embroidering stylized human figures with upturned arms, squared spirals, stepped triangles, eight-point stars, zig-zags, 'X' designs, and house and village motifs. For further embellishment, tiny discs of mother-of-pearl are stitched at regular intervals over parts of the cloth. Occasionally, coins and little brass bells decorate the edges of jackets.[34]

Muslim Weaving

The Muslim population, who constitute the largest cultural minority in the Philippines, are famous for their wood carving, metalwork, and handwoven cotton and silk textiles in bright glowing colours.

The Maranao, who live along the shores of Lake Lanao, and the Magindanao of Cotobato, make distinctive large wrap-around tubular cloths about 165 cm square, called *malong* (see Fig. 79). These are made by sewing two pieces of sarong cloth lengthwise along the selvages to make a square, which is then sewn up one side to make a tube. The *malong* is worn by women as a dress and by men over trousers as formal wear (Plate 32). This very versatile garment is worn for both everyday and ceremonial occasions and is the preferred garment for weddings and funerals. At a more mundane level, it serves as a baby carrier, a hammock for the injured, a raincoat, as bedding, a sack, or even a curtain.

The *malong* may be woven from either cotton or silk. For everyday wear, simple plaids and stripes are the preferred patterns. For ceremonial occasions, a more elaborate silk *malong* is worn. The most distinctive *malong* is probably the *landap*, which may be in one solid colour or be composed of two basic colours arranged in large alternating bands. Red, purple, and yellow are the most popular colours for *malong*. Yellow at one time was reserved for local royalty. Today, dark red shades are generally preferred by men and yellow by women. Colourful bands, called *langkit*, are separately woven in a tapestry weave by the Maranao on a small body-tension loom, and are added for decorative effect.[35] For the

Fig. 205. Some samples of traditional *langkit* stripes.

langkit, a 15 cm wide strip, called a *lakban*, is sewn vertically across the front of the *malong*, while a pair of even narrower strips, called *tobiran*, are sewn in parallel lines at right angles to the *lakban* (see Fig. 79).[36] *Langkit* may be attached either by machine or by embroidery to the *malong*. The Magindanao in their *malong* do not weave the *langkit* separately but incorporate special patterns into bands at the time of weaving.

Langkit bands are intricately patterned in three or four colours with *okir* designs based on abstract renditions of scroll, leaf, vine, and floral motifs (Fig. 205). Two of the more well-known *okir* designs are the *birdo*, or crawling vine motif, and *armalis*, a fern

Fig. 206. Some Maranao textile motifs. (After Reyes and Casino.)

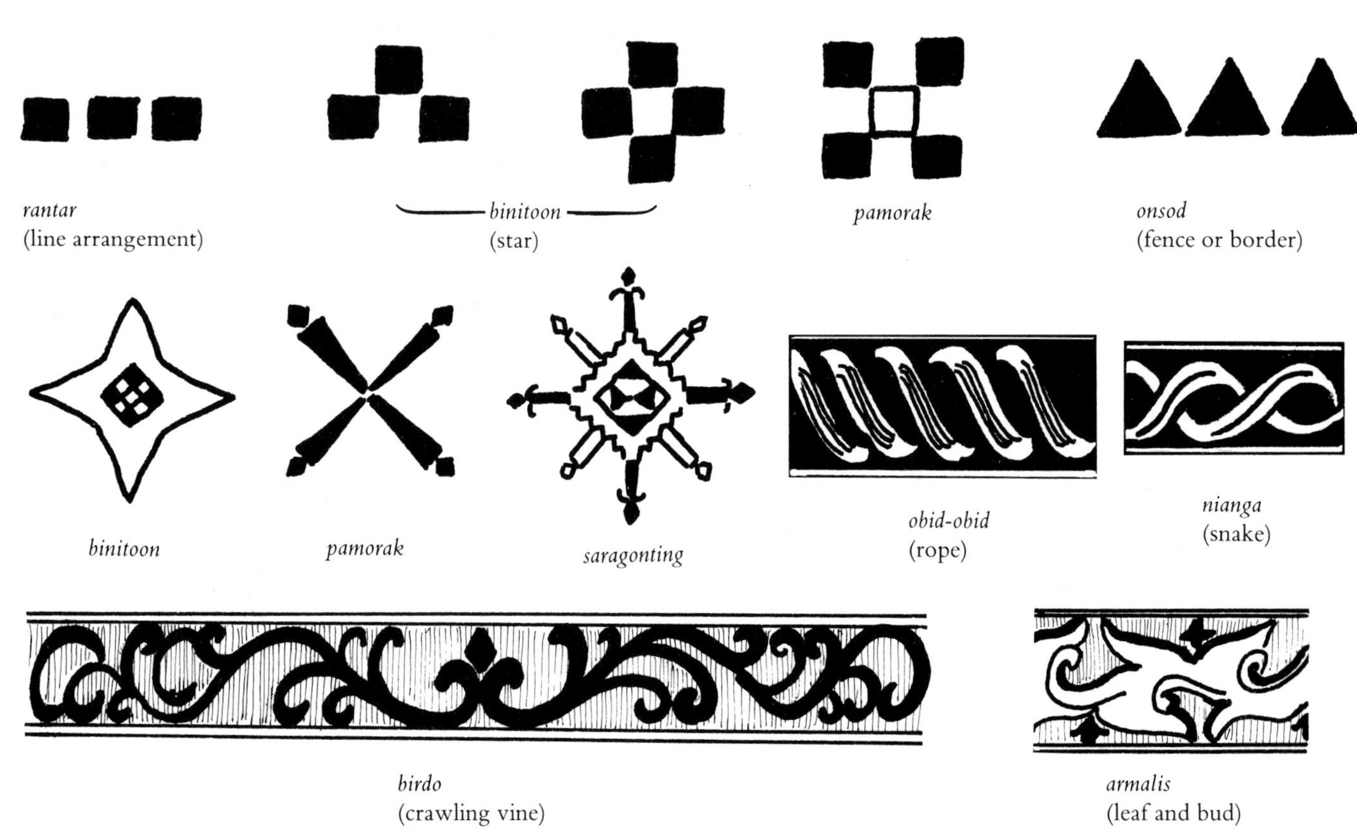

leaf and bud design. Despite an Islamic sanction against the use of representational forms, there are a number of stylized designs of birds and snakes which have survived in Maranao art (Fig. 206). They include the *obid-obid* rope design, the *nianga*, and the *magayoda*, based on the snake motif. Others could represent the *sarimanok*, the legendary rooster of Maranao art. A stylized crocodile also occasionally appears.

The *andon* is a brightly coloured *malong* patterned with *ikat* designs (Fig. 207). Both warp and weft *ikat* techniques are known to the Muslim peoples of the southern Philippines. *Ikat* motifs are sometimes placed between checks or arranged in a linear fashion between bands of warp stripes. *Malong* patterned with tiny *ikat* motifs between bands of many colours are referred to as *babalodan* (Plate 33). *Ikat* patterns are very simple and strongly geometric, being based on small triangles, squares, diagonal crosses, circles, and betel-nut shapes (see Fig. 206). Combining these basic shapes, the Maranao produce designs such as *rantar*, or squares in a line, *binitoon* (resembling a star), the 'X'-shaped *pamorak* motif, the *saragonting* (cross), and the *onsod* row of triangles.[37] These motifs show strong affinities with those of the tribal peoples of Mindanao.[38]

Some *malong* imitate *patola* cloth in terms of motifs and layout. The body of the cloth is patterned with floral geometric designs while the central area, or *kepala*, has two rows of facing *tumpul* patterns framing a series of small repetitive geometric motifs, such as diamonds, stars, and simple floral repeats. Sometimes, the *kepala* is worked in a supplementary weft in contrast to the weft *ikat* on the body of the cloth.

Some Maranao and Magindanao also weave colourful head-cloths, called *tubao*. The mode of tying at one time indicated the wearer's village of origin, his social status, and the importance of the occasion.[39]

The Yakan, who live on Basilan Island directly south-west of Mindanao, are also master weavers who specialize in weaving brightly coloured fabrics for clothing and ceremonial use. They weave in predyed mercerized cotton yarns on wide body-tension looms, which require a lot of strength to operate.

The men wear tight-fitting, handwoven trousers, called *sawal*, with a jacket and a 10–15 m long red sash, or *kandit* (formerly a loin cloth), wound many times around the waist to hold up the pants (Fig. 208). In former times, the *kandit* also served as protection against spear and dagger thrusts. One end may terminate in a tassel which covers the genital area. When travelling, the *kandit* serves as bedding or pillow. On formal occasions, instead of the *kandit*, a finely patterned silk *seputangan* scarf is worn as a sash or a turban. Clothing for women is practically the same except for the addition of a short, finely patterned wrap-around shirt which is worn over the trousers.[40]

For daily wear, stripes woven in a warp-faced plain weave are popular. The Yakan are noted for weaving extremely intricate repetitive geometric

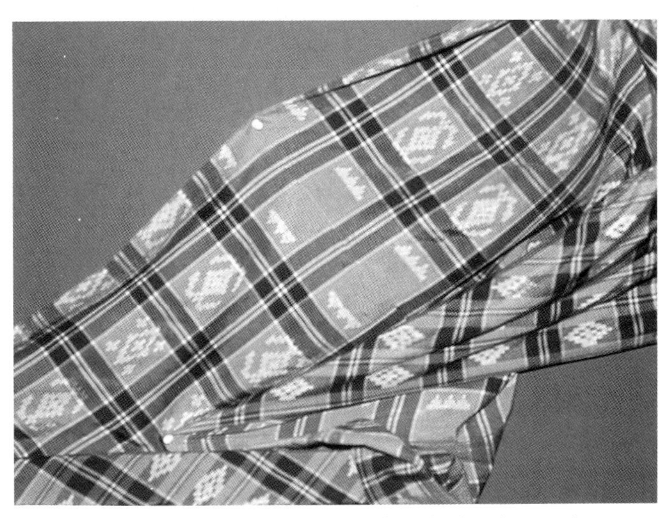

Fig. 207. A *malong andon* patterned with *ikat* designs between large plaids. Collection of the Cultural Center Museum, Manila.

Fig. 208. Two young Yakan men escorting a bridegroom to his wedding. The bridegroom is dressed in many layers of textiles. Note the tight-fitting pants patterned with fine geometric designs in a supplementary weft. Photo courtesy of Mrs Lelita Klainatorn, Bangkok.

patterns in a horizontal or diagonal alignment in a supplementary weft. The thread count per centimetre for these textiles is the highest in the Philippines.[41] Small shuttles are used to carry the slightly thicker, different-coloured supplementary weft threads through the warp. Some extremely fine supplementary weft patterns are laid in with a needle. Tapestry weave is also known to the Yakan, but they do not use this technique as much as their neighbours.

Yakan textile designs are strongly geometric (Fig. 209). One of the most popular designs is *bunga sama*, which is based on a diamond frame filled with varying patterns (Plate 34). About forty-five pattern sticks are required to weave this design.[42] Specific motifs include *bunga sama inalaman*, *bunga sama linalantupan*, *bunga sama kena kena*, and *bunga sama teed.* The colours of both the ground and supplementary weft are bold and bright. A finer, more intricate pattern, also based on the diamond, is the *sinaluan*, which consists of bands of small bisected and quartered lozenge shapes. This design, often woven in green, yellow, and white yarns, may be cut into strips to serve as trim for trousers and jackets. Other small repetitive designs include a saw-tooth pattern called *pussuk labbung*, a triangular–rectangular design called *kaban buddi*, a zig-zag design with triangles and small diamonds called *baggan ketan*, and *ukil lagbas*, a pattern composed of undulating lines (see Fig. 209).[43]

Important events, such as graduation or *pagtamat*, which takes place after a child learns to read the Koran, as well as wedding festivities, are carried out in full ceremonial dress by all participants. In addition to the usual clothing, extra shoulder cloths, waist cloths, and large turbans are worn. At the time of death, the funeral bier of the deceased is covered with a mound of textiles in the form of clothing, *seputangan*, and blankets (Fig. 210).

At one time, the weaving of cotton and silk

Fig. 209. Some typical Yakan supplementary weft patterns.

Sinaluan patterns combine *pussuk labbung* (saw-tooth), *kaban buddi* (triangles and rectangles), *baggan ketan* (a zig-zag design with triangles), and *ukil lagbas* (undulating lines to form distinctive linear designs in a very fine supplementary weft.)

(Names of patterns were supplied by Mrs Lelita Klainatorn, Bangkok.)

textiles from imported yarns was widespread amongst the Tausug of Jolo in the Sulu Archipelago at the southernmost tip of the Philippines. The weaving of the *kambut* sash, the *kandit* loin cloth, and an embroidered sarong called a *labur tiyahiran*, have virtually disappeared due to competition from cheap imported cloth (Plate 35).

The weaving of the *pis siyabit*, a 75–100 cm square head-cloth, continues to be made on a body-tension loom in a tapestry weave or supplementary weft by some families in the Parang area (Fig. 211). The cloth surface is divided into smaller squares and rectangles filled with diamond shapes, diagonal crosses, and zig-zag motifs, cleverly combined to produce a design effect similar to that of an oriental carpet. Pink, orange, and maroon shades, enlivened with touches of white and, occasionally, green predominate on silk cloth. Fine silk *pis* take about a month to weave.[44] The weaving of *pis* cloth in cotton has recently been revived to produce souvenirs for the tourist trade.

Fig. 211. Tausug cotton *pis siyabit* head-cloth patterned with traditional geometric designs in a tapestry weave. Warp and weft 83 cm.

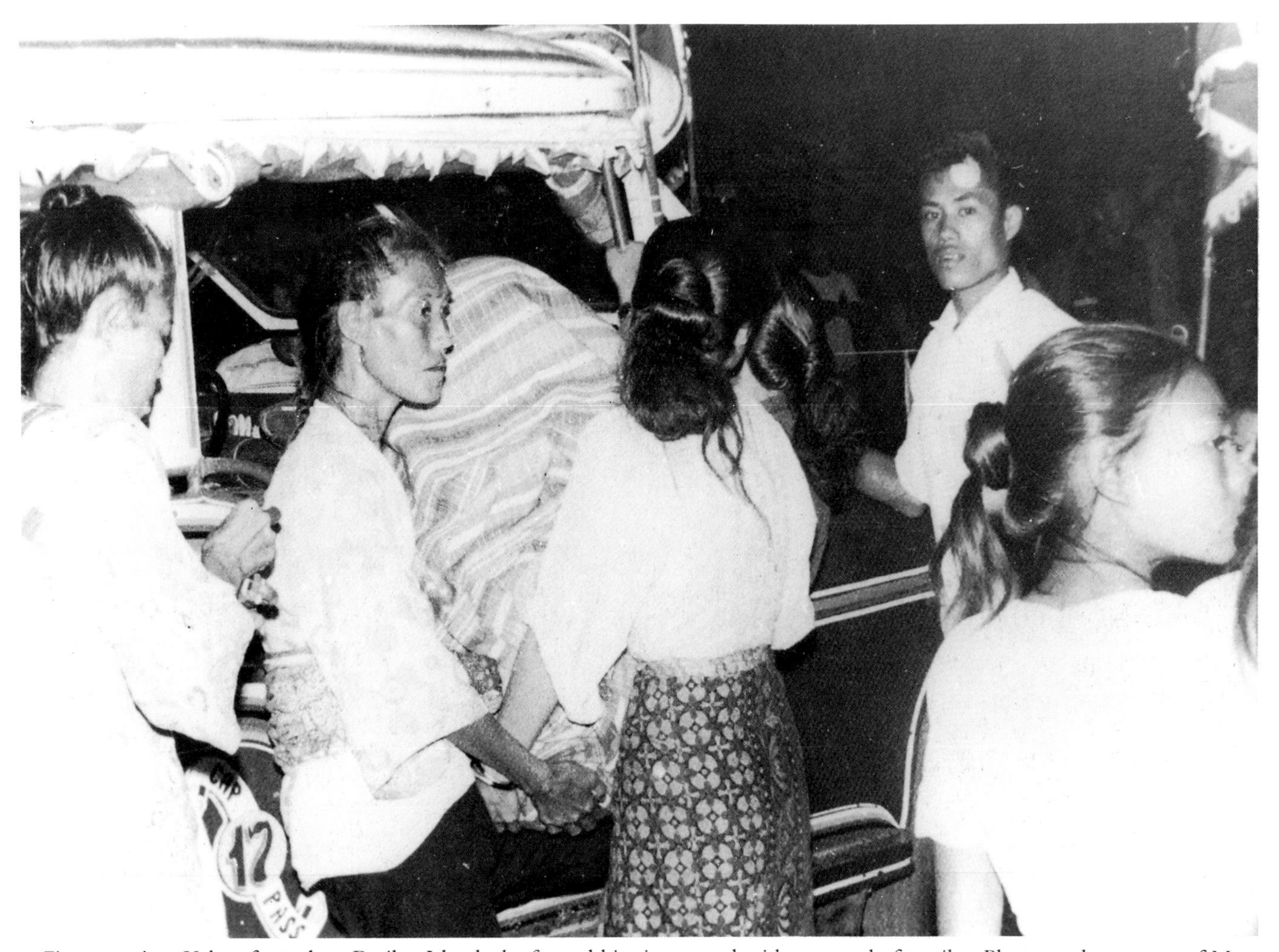

Fig. 210. At a Yakan funeral on Basilan Island, the funeral bier is covered with a mound of textiles. Photograph courtesy of Mrs Lelita Klainatorn, Bangkok.

1. Gabriel Casal *et al.*, *People and Art of the Philippines*, Museum of Cultural History, University of California, Los Angeles, 1981, p. 130.

2. Roberto de los Reyes, *Traditional Handicraft Art of the Philippines*, Casalinda Books, Manila, 1975, p. 3.

3. Casal *et al.*, op. cit., p. 184.

4. This information has been gathered from the Benguet Museum, Baguio, visited by the writer, 28 April 1986.

5. Casal *et al.*, op. cit., p. 237.

6. Francis Lambrecht, 'Ifugaw Weaving', *Folklore Studies*, Society of the Divine World, Tokyo, Vol. 17, 1958, p. 24.

7. Ibid., pp. 19–21 and 39.

8. Marian Pastor Roces, 'The Fabrics of Life', *Habi: The Allure of Philippine Weaves*, brochure, Museum Division of the Intramuros Administration, Manila, n.d., p. 29.

9. Casal *et al.*, op. cit., p. 224.

10. Lynne Klapecki, 'A Glimpse of Filipino Textiles', *Living in Thailand*, March 1982, p. 42; and Casal *et al.*, op. cit., p. 223.

11. Lambrecht, op. cit., pp. 35–7.

12. Ibid., pp. 10–15.

13. Casal *et al.*, op. cit., p. 223.

14. Lambrecht, op. cit., pp. 17–18.

15. Fay-Cooper Cole, *The Tinguian, Social, Religious and Economic Life of a Philippine Tribe*, Publication 209, Anthropological Series, Field Museum of Natural History, Chicago, 1922, pp. 434–5, and plates LXXI–LXXIV.

16. Roces, op. cit., p. 5.

17. Cole, op. cit., p. 313.

18. Ibid.

19. Casal *et al.*, op. cit., p. 224.

20. Visit to All Saints Weaving School, Bontoc, 28 April 1986; and Banaue Handicrafts, Baguio, 29 April 1986. It was not possible to visit the Easter School which was closed due to a strike on the part of the weavers!

21. Visit to Narda retail outlet, Hyatt Terraces Hotel Baguio, 29 April 1986.

22. Luther Parker, 'Primitive Looms and Weaving in the Philippines', *Philippine Craftsman*, Vol. 2, No. 6, 1913, pp. 376–7.

23. Pat Justiniani McReynolds, 'Sacred Cloth of Plant and Palm', *Arts of Asia*, July–August 1982, p. 99.

24. Roces, op. cit., p. 29.

25. 'T'boli Arts and Crafts: T'nalak', Santa Cruz Mission, pamphlet, South Cotobato, n.d.

26. McReynolds, op. cit., p. 97.

27. Gabriel Casal, *T'boli Art*, Ayala Museum, Makati, Manila, 1978, pp. 149–52.

28. 'T'boli Arts and Crafts: T'nalak', op. cit.

29. Casal, op. cit., pp. 77–80.

30. Reyes, op. cit., pp. 62–5.

31. Casal *et al.*, op. cit., p. 134; and Reyes, op. cit., pp. 55–9.

32. Reyes, op. cit., p. 61.

33. Ibid., p. 67.

34. Roces, op. cit., pp. 8 and 11.

35. Eric Casino, 'The Art of the Muslim Filipinos', in *Aspects of Philippine Culture*, Monograph No. 1, National Museum and United States Information Service, Manila, 1967, p. 19.

36. Casal *et al.*, op. cit., p. 134.

37. Reyes, op. cit., p. 45.

38. See chapter on Malaysia for a description of identical motifs as used on *destar* head-cloths made by the Bajau people of Sabah.

39. Casal *et al.*, op. cit., p. 137.

40. Andrew D. Sherfan, *The Yakans of Basilan: Another Unknown and Exotic Tribe*, Fotomatic, Cebu City, 1976, p. 207.

41. Roces, op. cit., p. 29.

42. Information kindly given to the writer by Mrs Lelita Klainatorn of Bangkok, formerly of Basilan.

43. Sherfan, op. cit., pp. 210–11.

44. David Szanton, 'Art in Sulu: A Survey', *Sulu Studies*, No. 2, p. 1973, pp. 32 and 65.

22. Malaysian *kain songket*-patterned man's silk sarong, the *sampin*. The *kepala* is composed of a *tumpul* pattern, *pucuk rebung lawi ayam*, which alternates with the *bunga sarang celak* motif. The border is composed of the *awan larat* cloud motif edged with *kendik tali*. Workshop of Hajah Cik Minah, Kampung Penambang, Kelantan. Warp 168 cm, weft 73 cm.

23. A bold, brightly coloured Iban *kalambi* jacket patterned with an *engkaramba* human figure formed by a combination of the tapestry weave and *sungkit* supplementary weft patterning techniques. Length of jacket 20 cm. Collection of Dato and Datin Leo Moggie, Kuala Lumpur.

24. Iban warp *ikat pua* textile from the Saratok area of Sarawak. Woven in a warp-faced weave from cotton yarns and coloured with natural dyes, this textile demonstrates the Iban weaver's ability to skilfully combine hook and rhomb elements into a tasteful integrated composition. The bold rhomb-shaped elements at one end of the cloth are a contrast to the delicate lacy composition at the other end. According to Mr Anking Kunding of the Sarawak Museum, the *ikat* stripes at the side of the blanket indicate that it was made by a woman of high status. This textile dates from the late 1920s–early 1930s. Warp 227 cm, weft 109 cm.

25. Kadazan woman's traditional black cotton sarong which is patterned in the front with a cross of *langkit* in brightly coloured geometric designs in a tapestry weave. There is a floral motif embroidered in each corner where the *langkit* strips intersect. Collection of Kraftangan, Kota Belud.

26. *Kain destar* from Kampung Merbau, Kota Belud, Sabah. This cloth is patterned in the centre field with a *gopas* border and *sinpul* motifs. The *samping gapas* star, the *saledap* cross placed at the four corners, the *sheku* zig-zag, the *onsod* fence and *bakagi* patterns, and the *anunan* floral designs enclose the centre field. Warp and weft 100 cm each.

27. A young Bajau male from Sabah dressed in his *kain destar* turban. Photograph courtesy of Victor Wah, Kota Kinabalu.

28. A *kain songket*-patterned sarong from Negara Brunei Darussalam. It is woven with supplementary gold yarns and coloured silk threads which are passed through the weft in small, flat claw-shaped shuttles. It is patterned in the *kain sukmaindra* design, which is one of the nine *jong sarat* patterns. Photograph courtesy of Matussin bin Omar, Brunei National Museum.

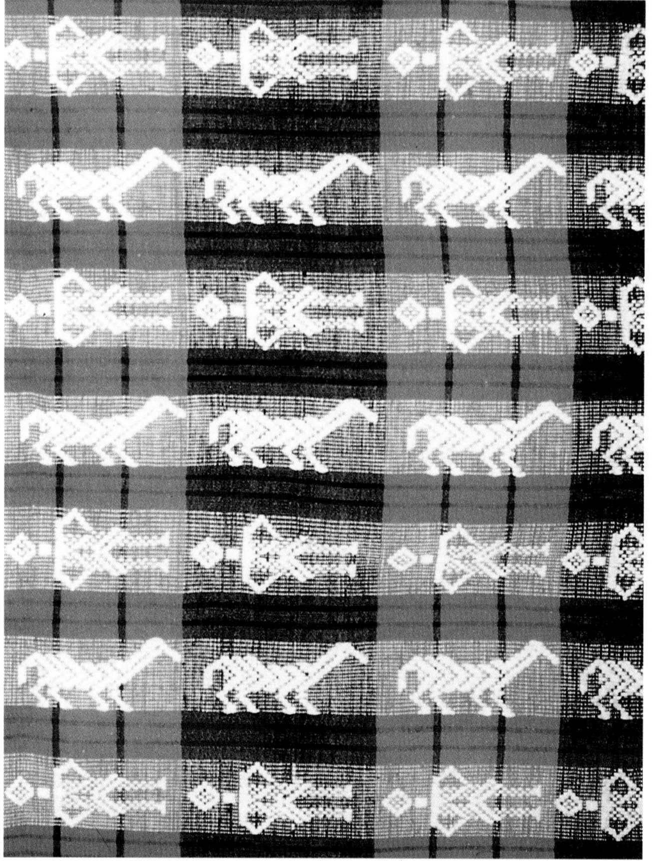

29. Isinai funerary blanket decorated with distinctive patterns in small blocks of white warp *ikat* on an indigo ground. These patterns are thought to represent stylized representations of human figures, local fauna, and cultural objects. The motifs at one time were 'read' by knowledgeable persons to depict the journey of a man through life. Isinai blankets were greatly sought after by the tribal peoples of northern Luzon and were often imitated by local weavers. Collection of the National Museum of Natural History, Smithsonian Institution, Washington, DC, No. 361025.

30. Tinguian checked cotton blanket patterned with alternating rows of horse and man motifs in a supplementary weft. Warp 198 cm, weft 115 cm. Collection of Roland Goh, Imperial House of Antiques, Baguio.

31. Bagobo abaca bag patterned with narrow warp stripes and embellished with small brass bells and coloured glass beads arranged in geometric patterns. Mindanao, southern Philippines. Collection of the National Museum of Natural History, Smithsonian Institution, Washington, DC, No. 286182.

32. Maranao *malong* worn by Ms Evelina Beninsig-Lu. Note the finely patterned colourful *langkit* strips in a tapestry weave. Warp and weft 165 cm each.

33. The *babalodan* is a *malong* which is patterned with bands of tiny weft *ikat* motifs arranged between multicoloured warp stripes. *Ikat* patterns are based on triangles, diamonds, and diagonal crosses to form simple designs, which include humanoid figures, turtles, crocodiles, birds, and cultural objects. These motifs show strong affinities with those of the tribal peoples of Mindanao. Magindanao cotton tube skirt, Mindanao. Warp 180 cm, weft 100 cm. Collection of the National Museum of Natural History, Smithsonian Institution, Washington, DC, No. 367146.

34. Yakan cotton textile from Basilan Island patterned with supplementary wefts in *bunga sama teed*, a complicated design to execute, requiring the placing of some seventy pattern sticks into the warp. Warp 97 cm, weft 45 cm. Collection of Mrs Lelita Klainatorn, Bangkok.

35. Tausug *kambut* silk sash from Jolo, which is patterned with geometric designs in a tapestry weave. Warp 200 cm, weft 50 cm. Collection of Ricardo Baylosis Antiques, Manila.

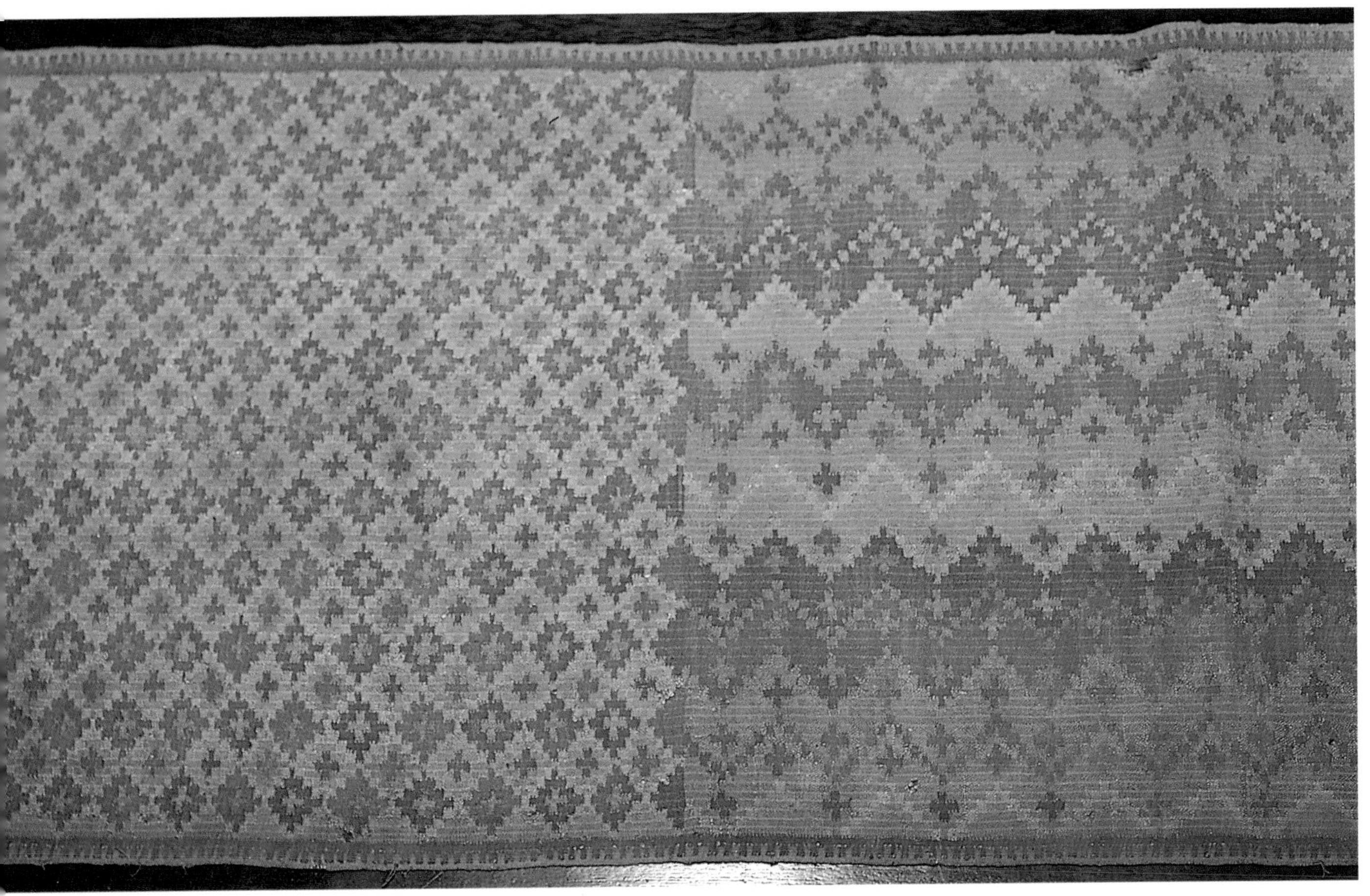

36. A silk *kain lepus limar* sarong from Palembang, Sumatra. The centre field possibly depicts the wings and tail of the mythical *garuda* bird in weft *ikat*. The sarong terminates in a *kepala* area embellished with gold yarns in a supplementary weft. Warp 210 cm, weft 93 cm. Collection of Atjs Kuta, Bali.

37. A young Minangkabau girl from Silungkung, West Sumatra, in traditional dress. She is wearing a gold *songket*-patterned *baju kurung*, sarong, *selendang* shawl, and horned *tanduk* head-dress.

38. Balinese ceremonial silk *kain songket* sarong patterned with *wayang* figures in gold and silk supplementary yarns. Klungkung. Arts of Asia Gallery, Denpasar. Warp 150 cm, weft 135 cm.

39. *Geringsing* double *ikat*-patterned cloth woven by the Bali Aga of Tenganan Pageringsingan village, East Bali. This textile is patterned with *cemplong* motifs. Because of its complicated method of manufacture, which is associated with many taboos, *geringsing* cloth is credited with magic and protective powers. Warp 150 cm, weft 39 cm.

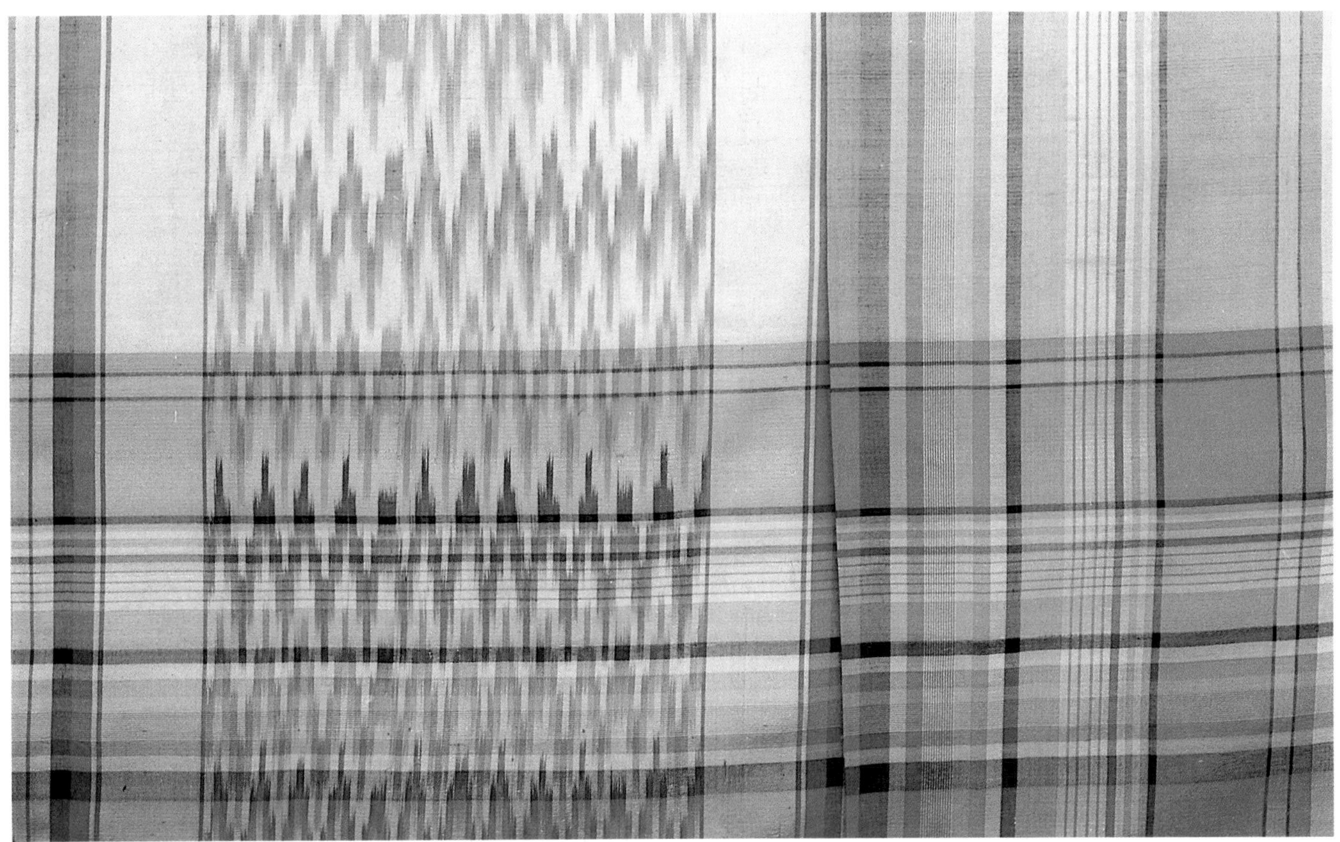

40. Buginese silk sarong from Sengkang, South Sulawesi. It is woven on a body-tension loom in two identical parts and joined along the warp edges to make a sarong length. The large plaid pattern is called *balo lobang*, and the *kepala* features a weft *ikat bombang* wave design in different colours. Warp 180 cm, weft 112 cm.

41. Toradja *sekomundi* funeral shroud from Galampang in Central Sulawesi. It is made from cotton and patterned with *sekon*-like motifs called *sekokandauri*. Crosses and small human figures lie between the curling appendages of the *sekon* motifs. The warp ends are finished with borders which end in a row of small triangular *tumpul* motifs. Warp 180 cm, weft 120 cm.

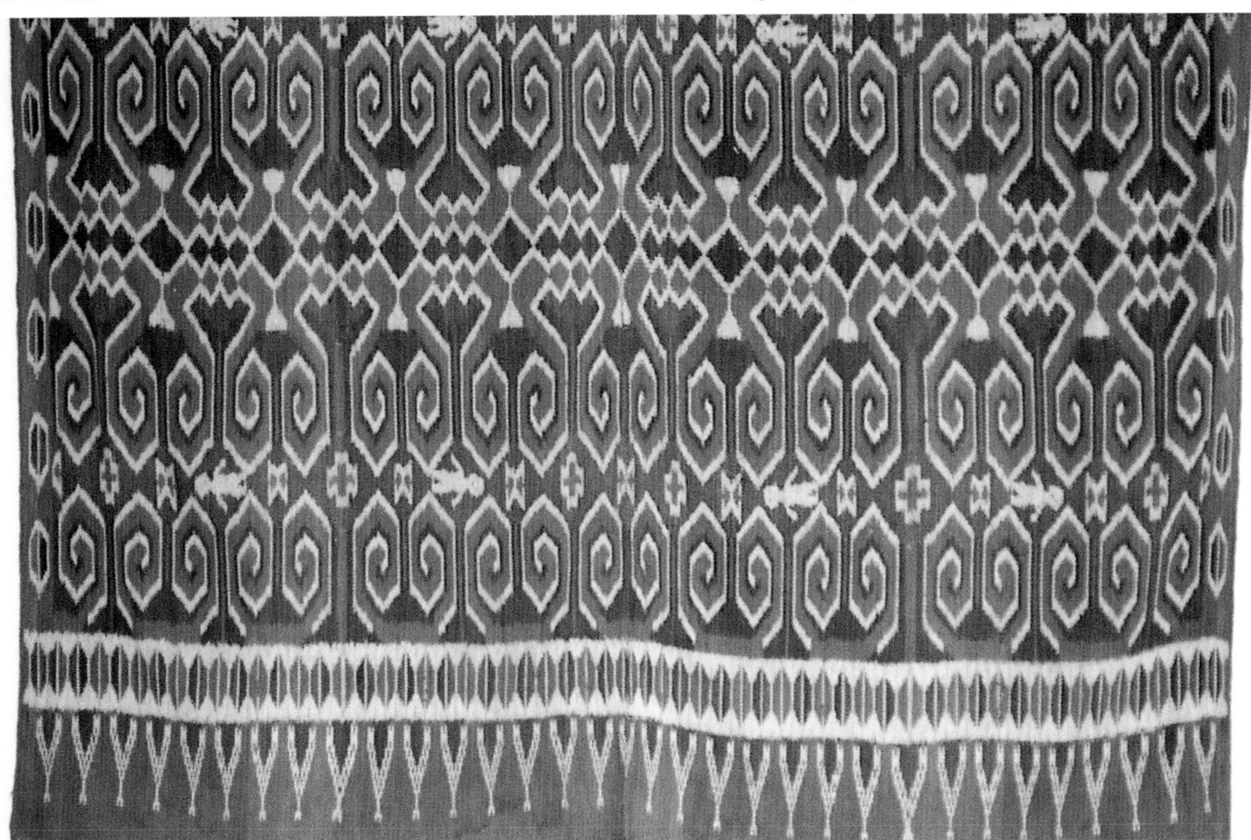

42. A colourful silk supplementary weft-patterned textile from Lombok. It is decorated with fierce animal masks artfully connected to attenuated vegetal motifs to form an extremely well-integrated composition. Butterflies alternate with a row of *tumpul* motifs along the edges of the cloth. This textile is very similar to Balinese *songket*. Museum Negeri Nusa Tenggara Barat, Mataram, Lombok.

43. A new *hinggi kombu* warp *ikat*-patterned blanket from East Sumba. Pairs of twisting dragons inspired by Chinese ceramics dominate the lower field. The centre of the cloth is patterned with a band of *habak* royal symbols followed by a row of *mamuli*, a Sumba ear ornament. Small heraldic figures and local fauna, such as squirrels, sea horses, snakes, and insects, fill the background between the main motifs. Warp *ikat* patterns end in a row of splendidly antlered deer above a striped *kabakil* band which terminates in twisted fringes. Warp 256 cm, weft 127 cm. Ikat Art Collection, Waingapu, Sumba.

44. Long tubular cotton sarong from Central Timor. It is decorated with a wide band of different coloured warp stripes, followed by a strip warp *ikat* in hook and rhomb designs on an indigo ground. The lower portion of the sarong is filled with small squares of fine geometric motifs worked in the *sotis* supplementary weft technique. Warp 130 cm, weft 131 cm. Collection of Mrs Marta Ostwald, Jakarta.

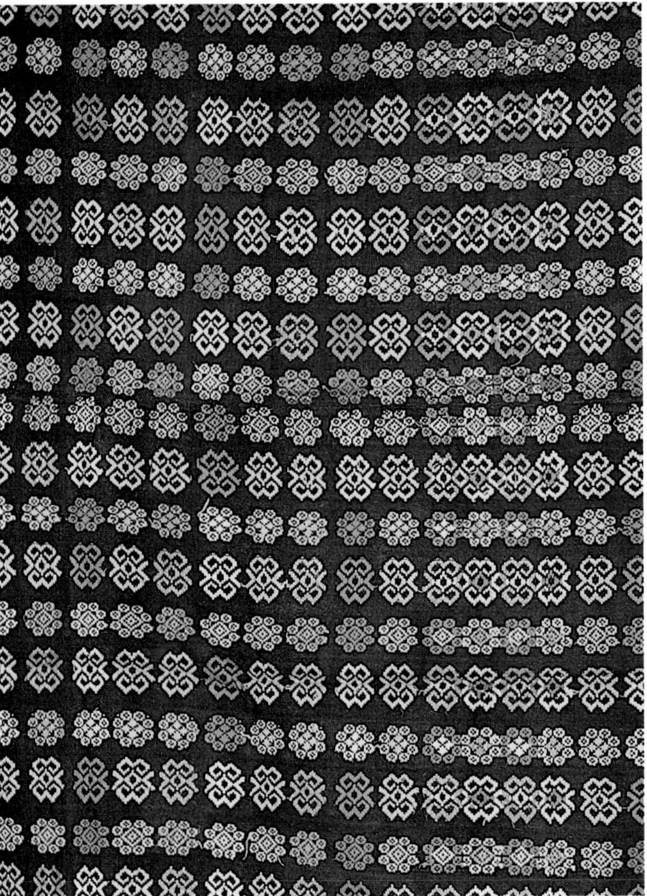

45. Supplementary weft-patterned sarong from Manggrai, West Flores. Coloured silk yarns in floral motifs on a black cotton ground. Warp 180 cm, weft 115 cm. Ledalero Catholic Seminary Museum, East Flores.

(Following page)

46. Cotton *selendang* shawl from the island of Alor. Bands of geometric motifs between warp stripes have been created by a warp-faced alternating float weave technique. The ends are finished with brightly coloured star motifs in a kelim tapestry weave. Collection of Dharma Bhakti, Kupang, Timor.

CHAPTER TEN

Indonesia: Main Islands

THE REPUBLIC OF INDONESIA, consisting of a vast archipelago strung some 4 800 km between the Indian and Pacific Oceans, is a country of great cultural diversity. Lying at the base of South-East Asia, all the influences which have shaped textile production in the region have filtered down into Indonesia. No other country in South-East Asia can match Indonesia for the sheer diversity of its textiles. All the major techniques of patterning—warp and weft *ikat*, supplementary warp and weft, tapestry weave, float weaves, tablet weaving, and twining—are known to the people of Indonesia. Cotton, silk, and bast fibres are used for yarns. Some of the most exquisite textiles continue to be produced on simple body-tension looms. Floor looms are also widely used in cottage industries. Despite modern innovations, the weaving of cloth in many areas continues as a skill which is handed down from mother to daughter. Textiles in Indonesia assume an even greater ritual importance than elsewhere.

Sumatra

Sumatra, the largest island of Indonesia, lies along the western side of the Straits of Malacca which, historically, has been an important trade route in South-East Asia. This location has brought coastal Sumatra into contact with many foreign traders. Chinese, Indian, Arab, Portuguese, and Dutch traders are known to have brought luxury items, such as ceramics, precious metals, exotic silks, dyes, and metallic threads, in exchange for gold, tin, wood, resins, pepper, ivory, and other jungle produce gathered from a vast equatorial hinterland peopled by a number of distinct and isolated ethnic groups. This combination of an outward-looking merchant élite, coastal towns dependent upon foreign imports, and a number of fiercely independent, self-sufficient tribal communities in the interior, have led to the production of a remarkable variety of textiles.[1]

Coastal Sumatra

The textiles that are most likely to first catch the eye of the visitor to Sumatra are the sumptuous light-to-medium-weight silk sarongs, head-cloths, and shoulder wraps in vibrant red, gold, purple, blue, and green. Finely patterned in either weft *ikat* or in the *kain songket* technique with supplementary gold and silver yarns, these silks have evolved from a history of foreign contacts. The yarns and dyes for these silks traditionally have been imported, while the motifs and their spatial arrangement on cloth appear to have been greatly influenced by Indian *patola* textiles. These textiles are important for formal wear on special occasions, such as at weddings and on religious holidays, but they do not seem to be imbued with as much ritual significance as some textiles in other parts of Indonesia.

Some of the finest Sumatran silk textiles seen today come from the city of Palembang, the second largest town in Sumatra. From the seventh to the eleventh centuries AD, this city was considered the capital of the Indianized Mahayana Buddhist Kingdom of Sri Vijaya, which at one time dominated much of South-East Asia. Today, it is the chief town for Sumatra's oil industry.

A few villages on the outskirts of Palembang continue to weave weft *ikat* with both body-tension and modern floor looms. Traditional weft *ikat* designs include geometric stars and circles, floral arabesques, rosettes, vines, fruits, paisley patterns, and stylized fanciful birds. These motifs occupy the centre field, while the end borders have rows of triangular *tumpul* patterns and narrow bands of small geometric and floral elements. Modern weft *ikat* designs are usually simpler and the colours brighter than their predecessors because of the increasingly widespread use of chemical dyes. One popular pattern for men consists of a series of arrowhead-type motifs between stripes, while for women, simple floral designs predominate.

There are about 250 women weaving *kain songket* on sturdy body-tension looms in their homes in and around Palembang. A rich red is the predominant ground colour although other colours, such as green, blue, and purple, are also used. To provide a firmer base for the supplementary gold threads, Palembang *songket* is often woven with a cotton warp and a silk weft. Silk thread is imported from Singapore while metallic threads come from India. For special textiles, the weavers unravel and reuse the gold threads from old sarongs. Prior to World War II, a high quality, 14 carat gold thread was used on special textiles. Due to import restrictions, such thread is no longer available today.[2]

Floral motifs predominate on Palembang *kain songket*. In addition to their decorative properties, some designs are imbued with special connotations, such as *bunga melati* or jasmine—a symbol of purity and courtesy, *bunga mawar* or rose—for good luck, and the *tanjung* flower, which is synonymous with welcome (Fig. 212). Other popular motifs include *kembang tabul*, *kaya sari* (the star fruit slice), and *kembang manggis* (the mangosteen flower).[3] Some floral *songket* patterns may have their centres embroidered in different-coloured silk yarns which provides a pleasing contrast to the gold threads. Birds and animals, such as winged lions and snakes, sometimes surreptitiously appear in the border patterns of old Palembang *kain songket*. The *tumpul* motif in the *pucuk rebung* (bamboo shoot) pattern is also found at the ends of Palembang *selendang* shawls and in the *kepala* area of some sarongs.

Fig. 212. A new *kain songket*-patterned *selendang* shawl from Palembang. The centre field is patterned with *bintang* star and *melati* jasmine flower motifs. Warp 178 cm, weft 44 cm.

Kain songket textiles which are densely patterned with gold are called *songket lepus*, while those with scattered designs are referred to as *songket tawur*, or *songket berbatur*. In *kain lepus limar*, the weft *ikat* and *songket* techniques are combined to produce a textile which has a centre field of *ikat* often decorated in the form of stylized *garuda* forms. These forms are beautifully integrated with delicate floral tracery patterns finished along the edges with gold *songket* motifs (Plate 36).

Palembang *songket* traditionally served as family heirlooms to be handed down through the generations. They were considered a source of wealth, and the ostentatious wearing of *kain lepus* was regarded as a sign of affluence. *Kain songket* was a required part of the bride price. When presenting the marriage proposal, the groom's family was expected to present the bride's family with three sets of handwoven sarongs and *selendang*, one for everyday wear, one for the wedding ceremony, and one for formal occasions. With the birth of a child, the paternal side of the family gave a *selendang* to serve as a baby carrier and a *singep* (a square silk handkerchief) to cover the head of the baby during the hair cutting ceremony.[4]

To the north of Palembang lies the island of Bangka, famous for its tin deposits. Muntok, a small Chinese town on the east coast, produces outstanding weft *ikat* in red, highlighted by blue and yellow tones. Further north, Batu Bara and Lingga Islands in the Riaou Archipelago weave two-tone textiles called *siang malam*, or day and night cloth, with different-coloured warps and wefts which give a marked 'shot' effect to the fabric according to the light. The coastal town of Siak, which faces the Riaou Archipelago, also makes some very finely patterned *kain songket* with geometric and floral designs in small squares. The Palembang region is also noted for decorating cloth with resist patterning techniques, such as *batik*, *tritik*, and *pelangi*, and for fine embroidery and appliqué work.[5]

Aceh in the far north, which was the first area in

Indonesia to be converted to the Muslim faith, has a long tradition of weaving silk textiles. During the seventeenth century, when Aceh was at its height politically, it was reported as being an exporter of raw silk. Aceh also has a tradition of producing warp *ikat* patterns on silk rather than cotton.[6]

Traditional dress for the Acehnese consists of a shirt or blouse worn with Chinese-style trousers, over which is worn a knee-length sarong for men and a longer sarong for women. Head-cloths are worn by both sexes. The Acehnese palette is very subdued with black, dark blue, purple, and maroon shades in silk predominating both in solid colours and in plaids. The latter are sometimes edged with *kain songket* borders in gold yarns enlivened with the odd touches of colour on selected motifs (Fig. 213). Acehnese *songket* designs consist of stylized flowers, leaves, and bird forms constructed from diamonds, hooks, and meanders.[7]

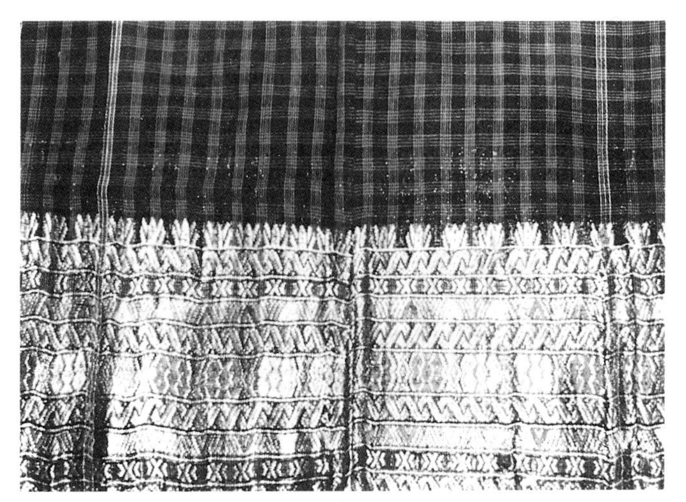

Fig. 213. An Acehnese red and black plaid silk sarong with a *kain songket* border composed of geometric and wavy designs highlighted with touches of coloured threads in addition to gold yarns. Warp 209 cm, weft 106 cm. Arts of Asia Gallery, Denpasar.

The Minangkabau of West Sumatra, a group of people organized on matrilineal lines, comprise about one-quarter of Sumatra's population. They weave an array of traditional, or *adat*, textiles which include distinctive horn-like head-cloths called *tanduk*, or *tilakuang*, shoulder cloths, sarongs, and pieces of material for tailored blouses, trousers, waistbands, banners, and receptacle covers (Plate 37). These are worn or displayed on important occasions, such as weddings, funerals, or at the inauguration ceremony of a chief, or *penghulu*. In the Bukit Tinggi area, silk weaving has been organized into a cottage industry. Many weavers work for a co-operative which supplies them with yarn and Malay floor looms. *Kain songket* (called *kain balapak* by the Minangkabau) is the most popular patterning technique used to decorate textiles. On some cloth, the gold threads are often so densely applied that the silk foundation is scarcely visible.

The best-known area for Minangkabau *kain songket* is Pandai Sikat, some 12 km out of Bukit Tinggi. Here, about 1,000 women, working either at home or in small workshops, produce richly woven cloth (Fig. 214). The ground cloth may be entirely of cotton or silk. In some cases, however, the warp may be of cotton or rayon, while the weft is of silk. Most silk is imported from Japan; however, a little silk is produced locally at Paya Kumbuh, some 36 km west of Bukit Tinggi, and may be used for the weft. Metallic yarns are imported from Singapore. Quality varies from sturdy two-ply yarns to light tinsel. The Pandai Sikat area also makes bright, multicoloured *songket* for everyday wear.

Fig. 214. *Kain balapak* (*songket*) *selendang* shawl from Pandai Sikat. Well-known Minangkabau textile patterns depicted on this shawl include *kaluak* (fern tendrils), *batang padi* (rice stalk), *sajamba makan* (ceremonial plate), *basisiak batang pinang* (areca nut tree), *balah kachang gadang*, and *selapa-selapa*, and border patterns, such as *ayam tadir hilalang* and *cintadu bapatah*. Warp 188 cm, weft 48 cm. Pusako Weaving House.

Nyonya H. Sannar Ramli, who operates the

Fig. 215. Some common textile patterns seen on Minangkabau *kain songket*. (After *Pakaian Adat Wanita Daerah Paya Kumbu.*)

ayam tadir hilalang

cintadu bapatah

Pusako Weaving House, has a number of weavers who use traditional Malay floor looms. They specialize in producing heavy, fine quality, closely woven *songket* on a red or black ground, the typical colours for traditional Minangkabau cloth. A fine sarong and *selendang* shawl may take up to three months to weave.[8] Designs are arranged in horizontal bands across the weft of the cloth. There are small borders along the selvages; there may also be *tumpul* borders at the warp ends of the cloth. Motifs are strongly geometric and owe their inspiration to local flora and fauna, such as bamboo shoots, fern tendrils, betel leaves, snakes, ducklings, and such cultural phenomena as rice cakes, wicker patterns, and ceremonial receptacles (Fig. 215). Names of motifs are sometimes based on Minangkabau proverbs and sayings of philosophical significance.[9] Zigzag designs include *biku biku* (saw-tooth), *tali burung* (bird tracks), *itek pulang patang* (duckling), and *barayan baaka*. Some of the more well-known diamond patterns include *basisiak batang pinang* (areca nut tree), *bari rendang* and *batang padi* (rice stalks), *bugis barantai* (interlocking Bugis design), *bunga tanjung* (star-shaped *tanjung* flower), *telur burung* (bird's eggs), *saik kalamai* and *sajamba makan* (ceremonial plate). Hook motifs include *kaluak*, *pinang baaka cina* and the border patterns *ayam tadir hilalang* (chicken design), and *cintadu bapatah* (insect design).[10]

Silungkang, some 70 km south-west of Bukit Tinggi, has a few workshops which continue to produce *kain songket* on a cotton ground with traditional Malay looms. Traditional Silungkang motifs include geometric designs, peacocks, and flowers (Fig. 216). It takes about ten to fifteen days to make a sarong. Silungkang *songket* workmanship today is generally not as fine as that of Pandai Sikat. Silungkang is also an important centre for weaving cotton sarongs in conservative plaids and stripes. Many of the weavers are men who work hand-operated looms with flying shuttles, often referred to as the ATBM loom in Indonesia.[11]

The Paya Kumbuh area produces both traditional Minangkabau *songket* and *songket* in new designs. Traditional weaving is done in several of the villages near the city of Paya Kumbuh, while the more commercialized weaving is done at Kubang, some 12 km from the city centre.[12] Besides floral and geometric motifs, Minangkabau items such as houses (*rumah adat*) with large sweeping roofs, and buffalo, appear in some modern designs.

The Minangkabau are also noted for their richly patterned, gold-couched embroidery filled with minute knots in different gradations of colour to

Fig. 216. A sample of traditional Silungkang *kain songket* depicting a peacock motif in white on a black cotton ground. A. Munir Sito, Silungkang.

form flower and bird motifs on velvet, satin, and muslin cloth. The *selayu* head-cloth, the *baju kurung* blouse, the *selendang* shawl, pillows, and household banners are the principal items to be decorated in this way. Embroidery is done at the craft centre of Kota Gadang near Pandai Sikat, at Ampek Angkek in the Agam district, and in Padang city. In recent years, bobbin lace has become a speciality of some Kota Gadang women. It is sometimes used as trim for handwoven *selendang*.[13]

The Pasemah highlands near Lahat in south-west Sumatra, an area noted for its megalithic sculptures, at one time produced distinctive weft *ikat* in combination with supplementary weft and tapestry weave decoration. Some of this material features horizontal bands of weft *ikat* in simple geometric shapes, such as diamonds and diagonal crosses in beige on a brownish to purplish wine-coloured ground. The *ikat* stripes may be separated by bands of supplementary weft featuring small *tumpul*, crosses, and hook designs.[14]

Central Sumatra

The Batak, a fiercely independent people numbering approximately three million, farm the fertile valleys of the interior uplands adjacent to Lake Toba in north-central Sumatra, and produce noteworthy textiles. In comparison with coastal silks, Batak cloths are very subdued and understated from an aesthetic point of view. They are of great ritual and social importance in rites of passage and in reaffirming social relationships. Batak cloths, called *ulos*, which may be worn as sarongs, shoulder cloths, or headcloths (Fig. 217), are usually woven in a plain weave in sombre blue-black, rust, or maroon shades, sometimes enlivened by pale stripes and/or bands of simple warp *ikat* consisting of small dashes, simple rhombs, and arrowheads. Finely woven supplementary weft patterns often adorn the ends and parts of the centre field. Beads also decorate some Batak textiles. The ends of some *ulos* are finished with finely twined borders in geometric designs. Formerly, hand-spun yarns and natural dyes were used. These have now been largely replaced by commercial thread and chemical dyes. In the more remote villages on Samosir Island in Lake Toba, it is still possible to observe women weaving on body-tension looms anchored to the front of the house.

Fig. 217. Batak woman from Tomok, Samosir Island, in traditional dress. She is wearing an *ulos ragidup* as a sarong, a beaded *ulos sadum* as a shawl, and an *ulos mangiring* as a headcloth.

Over two dozen types of Batak textiles have been documented. One of the best known is the *ragidup*, which is woven by the Toba Batak who live on the southern shores of the lake. The *ragidup* is an approximately 200 cm by 100 cm length of cloth, composed of a warp *ikat*-patterned or striped central panel, linked at each end to an interlocking length of white cloth (*pinar halak*) embellished by horizontal bands of very finely woven geometric patterns in a supplementary weft. Motifs depicted on the *pinar halak* include stylized representations of the water beetle, centipede, teeth, and human figures. Strips of black cloth sewn along the selvages of the central panel and *pinar halak* complete the textile. This cloth is regarded as protective and may be given to a woman by her family during the last stages of her pregnancy (Fig. 218). The supplementary patterns which vary slightly on the *pinar halak*, have a 'male' and 'female' end. These patterns are thought to contain information pertaining to the future and may be interpreted by a knowledgeable elder or shaman, called a *dukun*.[15]

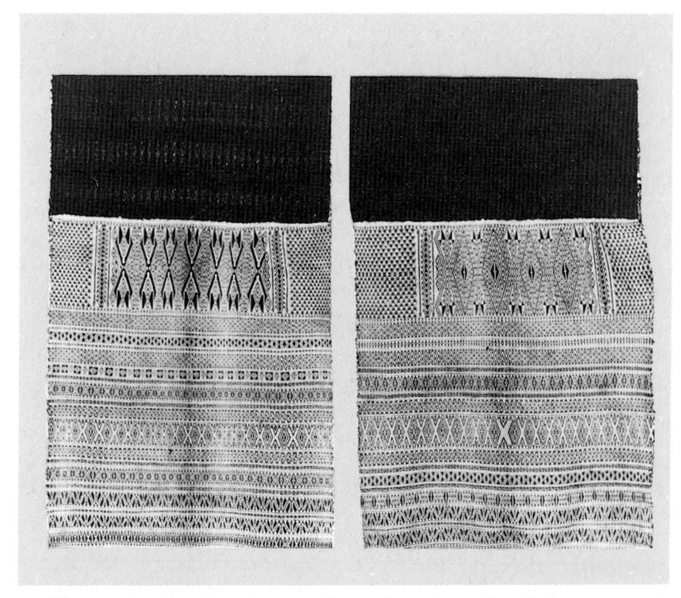

Fig. 218. Batak *ulos ragidup*, showing detailed geometric supplementary weft patterns at the *pinar halak*. 'Male' designs are depicted in the centre panel of the *pinar halak* (on the left) while 'female' designs are depicted on the right.

The Toba Batak also weave the *ragi hotang*, which consists of a single strip of cotton in a warp-faced weave. A blue-black central zone, contained within a reddish brown band along the lateral edges, is

Fig. 219. The *sibolang*, an important Batak ritual textile, which is blue or black in colour and patterned with bands of whitish-blue *ikat*. It is finished with twining. Warp 250 cm, weft 116 cm. Collection of Mrs James Simandjuntak, Jakarta.

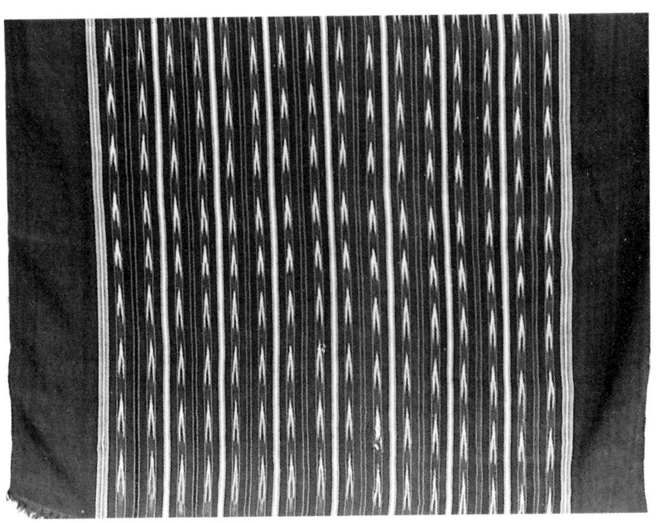

Fig. 220. The *ulos mangiring*, a maroon rectangle of cloth decorated with pin stripes and arrow heads in warp *ikat*. Warp 165 cm, weft 77 cm.

patterned with numerous narrow white warp stripes and light flecks of warp *ikat* which are thought to imitate the rattan plant, a symbol of longevity to the Batak. The warp ends of the cloth are finished with small bands in geometric supplementary weft, followed by an elaborate twined border. The *ragi hotang* is presented by the bride's father to the bridal couple at the wedding ceremony.[16]

A sombre textile of pale blue or black patterned with up to five bands of whitish blue *ikat*, called *sibolang*, is very important in gift exchanges during rites of passage (Fig. 219). On rare occasions it is given in lieu of a *ragidup* or *ragi hotang* in certain ceremonies. The *sibolang* may be presented by the bride's parents to the groom's father and to their daughter when she marries. It is an important cloth presented to in-laws when there is a death in the family. It is also the main cloth that a widow receives from her mother at the time of bereavement. During the mourning period, this cloth is worn on the head rather than around the waist.

The *mangiring* is a maroon rectangle of cloth measuring approximately 150 cm by 70 cm. The selvages are white followed by a wide plain maroon band on each side. The central area has small pinstripes in different colours. Some of these are decorated with small arrowheads in white warp *ikat* (Fig. 220). The *mangiring* serves as the *ulos parompa* (carrying cloth) presented by the maternal grandparents on the birth of their daughter's first child. It is also used as a head-cloth on festive occasions.[17]

The Angkola Batak weave a distinctive festive looking textile called the *sadum*. If it has two panels,

Fig. 221. A Batak *parompa sadum* cotton textile which is decorated with beaded designs in many colours. Collection of Mrs James Simandjuntak, Jakarta.

the *sadum* may serve as a garment or *ulos godang* in a gift exchange. A smaller version, called the *parompa sadum*, is used as a ceremonial baby carrier. The *sadum* is patterned with horizontal rows of rhomb-shaped, brightly coloured supplementary weft patterns in a plain warp faced or twill weave on a maroon or black foundation. There may also be a little tapestry weave in wool yarns. Small bead patterns sometimes appear throughout the textile and may be heavily concentrated along the warp edges at the ends of the cloth. Salutations and greetings may be also worked into the textile.[18] With the current penchant for brighter colours amongst the Batak, the *sadum* is being imitated by other groups. Influenced by their coastal neighbours, supplementary metallic threads have also crept into some Batak weaving (Fig. 221).

Southern Sumatra

Some of the most remarkable weaving ever done in Indonesia is seen in the famous ship cloths which were once made by the Paminggir peoples of Lampung Province. Woven on a body-tension loom, these cloths are patterned by coloured supplementary wefts in silk or cotton on a natural, unbleached, plain weave cotton foundation.[19]

The main motif is usually a ship or pair of ships with dramatically curving prows. On board, a veritable Noah's ark of intriguing structures and figures are depicted as subsidiary motifs. These include stylized 'trees of life', houses, shrines, rows of human figures, animals, such as the horse, buffalo, and elephant, birds, fish, and heraldic items, such as flags, umbrellas, and banners. All motifs are portrayed in geometric form against a background crowded with smaller birds, boats, and floral and star motifs. Great use is made of Dongson-style spiral, rhomb, hook, key, and meander patterns to give an abstract dimension to the cloths, which are enhanced by the subtle use of natural dyes in red-brown, indigo, and yellow. Metallic threads and small pieces of glass occasionally highlight a special feature or enhance a border.[20]

These cloths were once widely used by a society divided into rigid social classes and founded on a culture richly steeped in animistic beliefs. Ship cloths may be classified into two special ceremonial textiles, the *palepai* and the *tampan*. These differ in size, usage, and some design elements.

The *palepai*, measuring about 350 cm in length and about 100 cm in width, appears only in the Kroe area around Lake Ranau on the south-west coast. The cloth is usually multicoloured and is worked in a discontinuous supplementary weft. The ship and its contents may appear as a huge panoramic picture often arranged as a mirror image, or it may be repeated a number of times across the cloth.[21]

Its use was formerly limited to the aristocrats of the *penyimbang* class who were leaders of the clan, or *suku*, and at one time the cloth was inherited by the eldest son. During important rites of passage ceremonies, such as birth, circumcision, tooth filing, marriage, obtaining a rank, and death, the *palepai* was hung on the right wall in the inner room on the woman's side of the house. The ship motif, formerly associated with death and the afterlife, in a broader context symbolized the transition from one stage to another in the journey through life. The tree of life motif was once a symbol of fertility and the life force. It also served as a link between the upper and lower worlds. Given these various connotations, the *palepai* provided a fitting backdrop to the dramatis personae in the observation of rites of passage. The seating of a bride before her husband's *palepai* signified her change in status from a single to a married

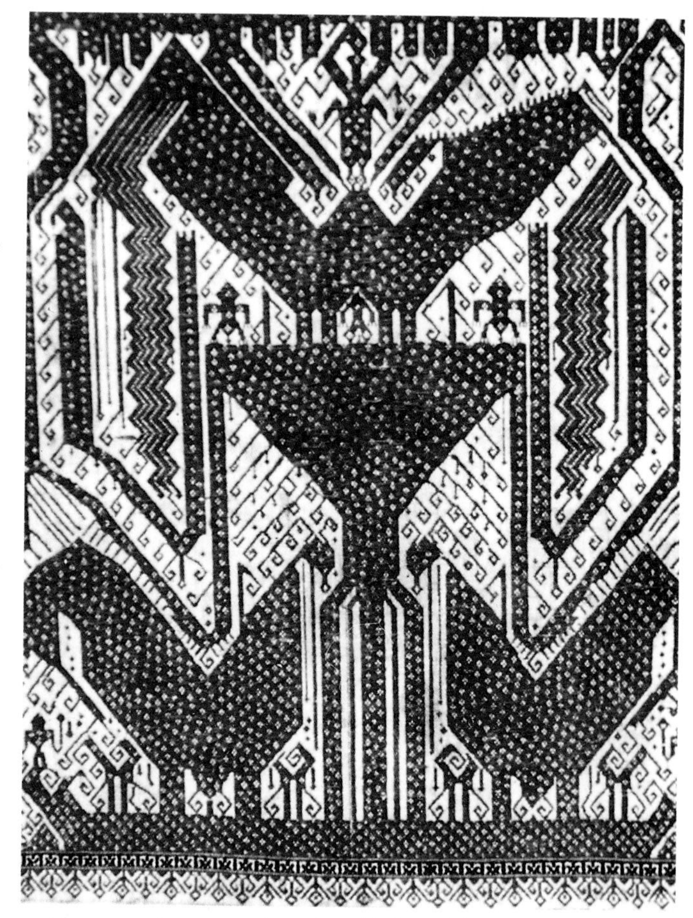

Fig. 222. A *tampan* depicting birds and humanoid figures in pink on a white ground. Warp 50 cm, weft 62 cm. Collection of Mr and Mrs Rodger Peren, Tokyo.

woman and her acceptance into her husband's clan. The first-born was welcomed into the family fold by receiving a name before the *palepai.*

The *tampan* is a smaller, almost square cloth ranging in size from about 50–100 cm in width (Fig. 222). The designs are more divergent than those expressed on the *palepai.* Motifs include large birds, or a number of smaller ones, trees, ships carrying animals and people, and repetitive geometric motifs. Worked in one or more colours in a continuous supplementary weft, the *tampan* had a wider geographic distribution than the *palepai.* It was at one time widely seen along the south and south-west coastal regions of Sumatra. Unlike the *palepai,* its use was not confined to the upper echelons of the aristocracy. Through a very comprehensive system of gift exchanges designed to reaffirm social relationships, the *tampan* circulated through the whole of society. During marriage negotiations and at other important events, ceremonial food wrapped in a *tampan* was exchanged between lineages. The *tampan* also covered the seat of honour at these ceremonies. It also served as a food cover and a table-cloth for elders on these important occasions. Prior to a funeral, the head of the deceased might rest on a *tampan.* At house consecration ceremonies, *tampan* were fixed to the ridge pole and remained there for the duration of the house.[22]

Due to changing economic and social conditions, the *palepai* and *tampan* ceased to be made around the turn of the century. Most that survive are either in museums or in the possession of Lampung families who still use them on special occasions. There are now very few for sale in Indonesia and those available are very expensive. Prospective buyers should be careful, for many reproductions are being made expressly for the tourist trade.

Another unusual and important cloth formerly made in the Lampung area is a cotton sarong, or *tapis,* decorated with warp *ikat* and embroidery. This cloth once served as ceremonial dress for local women in the mountainous interior of Liwa, Kenali, and Talar Padang. Bands of subdued brown *ikat* in angular scrolls, hooks, and rhombs, alternate with plain bands filled with strange horn-like creatures or 'ships of the dead' embroidered in cream satin-stitch and outlined in a darker colour (Fig. 223). These two design areas present a sharp contrast on a single piece of cloth. To explain this, it has been suggested that the embroidered motifs might have come from the woodcarving tradition as practised by men who have been credited with designing the embroidery motifs for their fiancées. The motifs expressed in the warp *ikat,* on the other hand, were created by women and have evolved from a long-standing weaving tradition. Like the ship cloths, this particular type of *tapis* is no longer woven today.[23]

Along the south-west coast, the Kauer people in the Kroe–Bengkulu area used to decorate their distinctive handwoven *tapis* by embroidering hundreds of small *cermuk* mirror pieces in rows of scrolling and geometric designs alternating with panels of satin-stitch, metallic thread, and patches of appliquéd cloth. A *cermuk tapis* could take up to a year to make and could weigh up to 5 kg. This elaborate *tapis,* and a complementary long-sleeved jacket, were made by a young woman prior to her marriage (Fig. 224).

Handwoven cotton *tapis* in black and rust, which depend mainly on embroidery for their design features, are still being woven today in south-east Lampung by the Abung people. Motifs include geometric forms, wavy designs, stars, flowers, foliage,

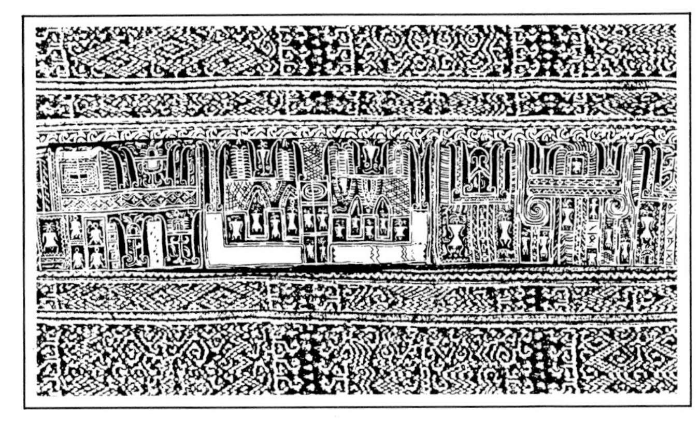

Fig. 223. A *tapis* decorated with bands of warp *ikat* and ship motif embroidery. Collection of Mrs Eiko Kusuma.

Fig. 224. A Kauer woman's jacket made from striped handwoven cotton fabric and decorated with cotton appliqué, sequins, and shells. Length of jacket 40 cm. Collection of Mrs Marta Ostwald, Jakarta.

Fig. 225. Detail of a *tapis* from South Sumatra, handwoven in black and brown and embellished with gold couched yarns. Warp 124 cm, weft 112 cm.

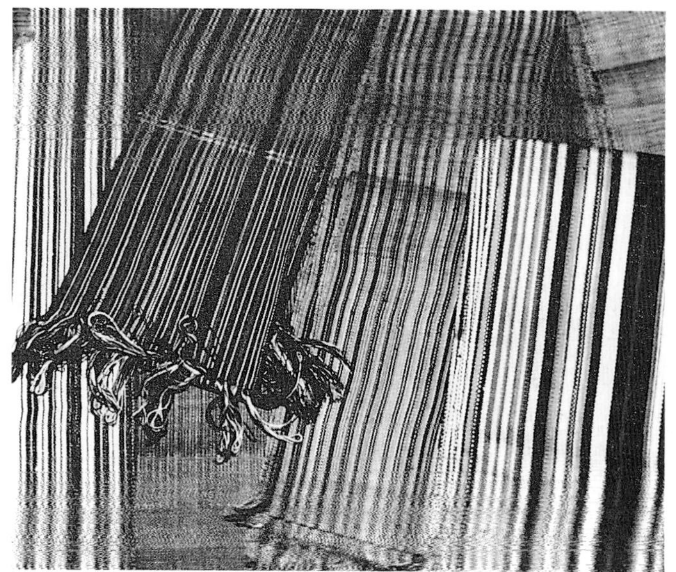

Fig. 226. Some examples of handwoven *kain lurik*. Logro, Jogjakarta.

and silhouettes of ships with people on board, trees, birds, and composite creatures. These are applied to the clothing in the form of couched metallic yarns (Fig. 225). A new sarong takes about three weeks to weave and up to three months to embroider.[24]

Java

At one time, handweaving was very widespread in Java. Over the last century, however, weaving appears to have declined in favour of batik. Today, weaving is largely confined to a few pockets of activity in Central and East Java.

Central Java

Central Java has long been famous for *kain lurik*, a sturdy, hand-spun, plain weave cotton cloth patterned with stripes and sometimes small plaids in conservative hues (Fig. 226). This all-purpose cloth is used for everyday clothing, household 'linen', and for carrying loads. With age and use, it acquires a pleasing softness. Its strength and durability are quite remarkable. Although widely regarded as 'peasant' cloth for the common folk, there is evidence to suggest that *kain lurik* used to play an important role in palace rituals.[25]

Kain lurik woven on a body-tension loom comes in a variety of widths. The widest width is about 60 cm and can be used for making sarongs or *kain gendong*, a carrying cloth. The earliest *kain lurik* was a black and white plaid or stripe.[26] Today, the colour combinations for *kain lurik* are endless. Each type of stripe combination has a special name to weavers. For example, a plain cloth in a natural home-spun cotton with a weft stripe at the ends is called *kain pankung*. A black and white or green and white combination may be referred to as *pulowatu*. *Kain pribul* consists of a two- to three-stripe combination, while *kain nanmanya* refers to stripes in many shade gradations within the same colour range.

A narrow, tightly woven, approximately 15 cm wide cloth is often referred to as *kain bendo*, and this is woven on a small loom. In a plain black or white colour, this cloth may be used as a *stagen*, the long cloth wound around a woman's waist to secure the sarong.

While much *kain lurik* today is produced in factories by machine-powered looms, there are at least 100 women in the outlying areas of the Jogjakarta–Surakarta (Solo) region who weave special orders of *kain lurik* on traditional body-tension looms. Natural-coloured, hand-spun, locally grown cotton yarns are occasionally used for weaving special cloth where a certain coarseness and durability are required. In most cases, however, the weavers prefer to use pre-dyed, store-purchased yarns which have been processed by factories in Surakarta. Much of the handwoven *kain lurik* is made into shirts, dresses, and *selendang*, or is used for trim and reinforcing. A few retail outlets in Jogjakarta (such as Logro) specialize in selling and promoting handwoven *kain lurik*. They find it difficult, however, to compete

Fig. 227. A cotton weft *ikat*-patterned sarong from Gresik. This example in soft mauve and white was acquired by the author in the Philippines. Warp 174 cm, weft 108 cm.

with the numerous factories in the area which churn out vast yardages of cheap *kain lurik*. While colourful machine-produced *kain lurik* is attractive to those of lesser means, it cannot compare with the handwoven product for durability. There are a number of Central Javanese who are particularly fond of handwoven *kain lurik* and will proudly show the interested visitor some 20–30 year old pieces which are still in constant use.[27]

Gresik

Formerly, centres such as Priangan, Tjidjulang, Pekalongan, Gresik, and Banjumas were known to produce *ikat*. Gresik, a Muslim centre and trading port in north-east Java, a few kilometres out of Surabaya, is one of two centres in Java still producing weft *ikat* today. Since the sixteenth century, Gresik has been an important port for handling Gujerat trade textiles brought to eastern Indonesia by Muslim traders. Early in the seventeenth century, the Dutch established a trading post there to control this lucrative trade.[28]

Today, the industry is centred on the old whitewashed Arab quarters in the Jalan K. H. Agus Salim and Jalan Malik Ibrahim area. Here, Indonesians of Arab descent have organized the making of weft *ikat* into a flourishing backyard cottage industry. Some concerns employ as many as 200–300 workers. Cotton sarongs, made in muted shades of rust, olive green, light mustard yellow, and navy blue, are patterned with realistic floral and geometric motifs. In many cases, the floral motifs fill the *badan*, or body of the cloth, while geometric patterns cover the *kepala*, or ends (Fig. 227). On silk *ikat*, the motifs are similar but the colours are generally much brighter. Some *ikat* are patterned in a lighter colour against a striped or checked ground, while the *badan* on some other cloths contains, in addition to a little *ikat*, small geometric supplementary weft designs set in pattern heddles during the course of the weaving. Gresik *ikat* is characterized by narrow, brightly coloured bands of splashed *ikat* along the selvages. Dyeing is mainly by the *cetak* process and the ground colour, in most cases, is applied last. All weaving is done on floor looms with hand-operated flying shuttles. Women are able to weave at least one, if not two, sarongs a day on this loom. Gresik *ikat* finds a ready market throughout Indonesia, in neighbouring ASEAN countries, such as Malaysia and the Philippines, and in the Middle East.[29]

Troso

The second important centre for weaving *ikat* on Java is at Troso, a village 15 km out of Jepara on the north coast of Central Java. Here, virtually every family is connected to the weaving industry in some way. As at Gresik, weaving is organized as a cottage industry on the premises of the owner. In front of each establishment women may be seen spinning the yarn. In a side room there will be two or three men reeling cotton onto *ikat* frames. Designers using templates and stencils block out *ikat* patterns onto the threads held taut in the frames. The yarns are then handed over to women who use the *cetak* method to tie and colour the yarns by painting on chemical dyes according to the dictates of the design. Once all the minor colours have been applied, the bound threads are turned over to the men who are responsible for dip-dyeing the ground colour. At Troso, the actual weaving is very much a male preserve; over 90 per cent of the weavers are young men from the village. The owners of the various concerns, when interviewed, expressed their satisfaction with male weavers. Although they had to pay them a little more than women, they felt that they were faster and stronger in operating the floor looms. They also took less time off for domestic duties.

Troso specializes in floral weft *ikat* designs in cotton, which are virtually indistinguishable from those of Gresik. Repetitive geometric designs, such as diamonds and zig-zags, similar to those seen on cloth from Sengkang in Sulawesi, are locally referred to as *kain kreasi*. White floral designs on a coloured ground are referred to as *kain bidang*, while a two-coloured cloth with small *ikat* designs may be called *buntal kursi*.[30] In recent years, Troso has been copying Nusa Tenggara warp *ikat* designs, particularly those of Sumba, in a big way (Fig. 228). Taking some of the simpler repetitive motifs arranged in horizontal bands, Troso weaving concerns have been able to produce warp *ikat* in fairly long yardages. 'Sumba' cloth woven in a heavy cotton is a popular furnishing fabric and is currently being used in many Indonesian hotels. Table-cloths, napkins, and cushion covers produced for the tourist market, are also patterned with Nusa Tenggara designs.

The Troso approach to weaving is strictly commercial. There is no altruistic desire to preserve and use local motifs. Whatever designs have proved popular elsewhere soon find their way into the Troso *ikat* repertoire. Troso cloth is produced largely for the local markets in the heavily populated areas of Central Java. Because of a lack of a local style, coupled with Troso's practice of imitating designs, few people in Indonesia are aware of this very commercialized weaving centre.[31]

Fig. 228. Blue and white banded warp *ikat*-patterned cloth from Troso, decorated with Sumba motifs. This cloth is made expressly for the tourist trade. Warp 120 cm, weft 60 cm.

Bali

The Hindu island of Bali at one time displayed an extraordinary diversity of textiles. Rich, brightly coloured cottons and silks, batiks, gold-patterned *prada* cloths, intricately patterned sarongs in weft *ikat*, and *kain songket* were all worn on important occasions. Textiles also played an important role in rites of passage, and at ceremonies honouring ancestors and invoking favours from the gods. Cloth banners were suspended from altars or from cremation towers, and the lintels and stone guardians flanking the entrances to temples were draped with cloth. Textiles also formed part of the ritual offerings made to a temple. Certain textiles and design motifs, in addition to their aesthetic value, were thought to be imbued with magical powers to ward off evil influences, promote healing, and confer good luck on the wearer.[32]

Over recent years in Bali, there has been a remarkable revival of interest in making weft *ikat*, especially in a cotton sarong called an *enduk*. Weft *ikat* now rivals batik in popularity for everyday wear in Bali. The industry is concentrated in and around Gianyar in south-east Bali. There are at least six weaving firms employing over 100 workers each. The cotton used comes from Indonesia, while the silk yarns are imported. A cotton/rayon mix is used

Fig. 229. Weft *ikat* shoulder or breast cloth from Bali depicting *wayang* puppet figures. Arts of Asia Gallery, Denpasar. Warp 300 cm, weft 55 cm.

on cheaper cloth. At most factories, it is the men who do the tying and dyeing of threads by the *cetak* method, while women do the reeling of thread for the warp and the weft and the weaving on hand-operated floor looms with flying shuttles.

Bright colours, such as magenta, purple, green, and royal blue, are generally preferred for the ground colour. Motifs include geometric patterns, Chinese-inspired designs, Balinese floral sprays, *patola* designs, and human and animal figures drawn from Hindu mythology and the *wayang* puppet repertoire (Fig. 229). In addition to weft *ikat*, some of the larger concerns, like those of Troso, make imitations of 'Sumba' cloth for table linen. A little *pelangi*, batik, and simple *songket* may also be made in these factories. These weaving concerns sometimes commission women in outlying villages to make finer pieces of weft *ikat* from yarns and dyes supplied by the factory.[33]

In addition to weft *ikat*, the nearby village of Batuan also produces *sabuk*, or Balinese sashes, patterned with traditional motifs in a supplementary weft. Women, seated at benches, weave on a body-tension loom. The warp beam is slotted into a 40 cm high, elaborately carved wooden frame. The supplementary weft patterns, consisting mainly of vegetal designs placed inside various grids, are set into the warp with fine sticks. Different-coloured supplementary yarns are passed through the weft by small flat shuttles. Over 100 women in the area weave such *kain sabuk*.[34]

Some 40 km east of Denpasar is Klungkung, a former royal cultural centre, which is noted for its *songket* textiles of silk and synthetics overlain with gold threads (Plate 38). *Songket* in Klungkung was regarded as an aristocratic art and was formerly made by women associated with the royal courts. Today, Klungkung *songket* are worn as ceremonial garments for dances and temple rituals and for weddings. The silk ground is usually plain. Motifs include *tumpul*, *ceplokan*, *jelamprang*, and zig-zag designs. Since there is no Balinese prohibition against portraying living things, Hindu figures, fearsome masks, *wayang* scenes, and royal animals such as the lion, the *garuda* bird (the mount of the Hindu God Vishnu), and even farm animals such as chickens and pigs, may appear on Balinese *songket* (Fig. 230). On traditional work, various coloured threads, as well as metallic yarns, were used to create some interesting processional-type scenes. Today, some twenty-five women weave *songket* in and around Klungkung. It takes about two months to weave a *kain songket* sarong on a traditional body-tension loom.[35]

Fig. 230. Balinese ceremonial *kain songket* sarong with stylized floral patterns in silver and different-coloured silk supplementary weft threads. Klungkung. Warp 150 cm, weft 138 cm. Collection of Mr and Mrs John Reagan, Tokyo.

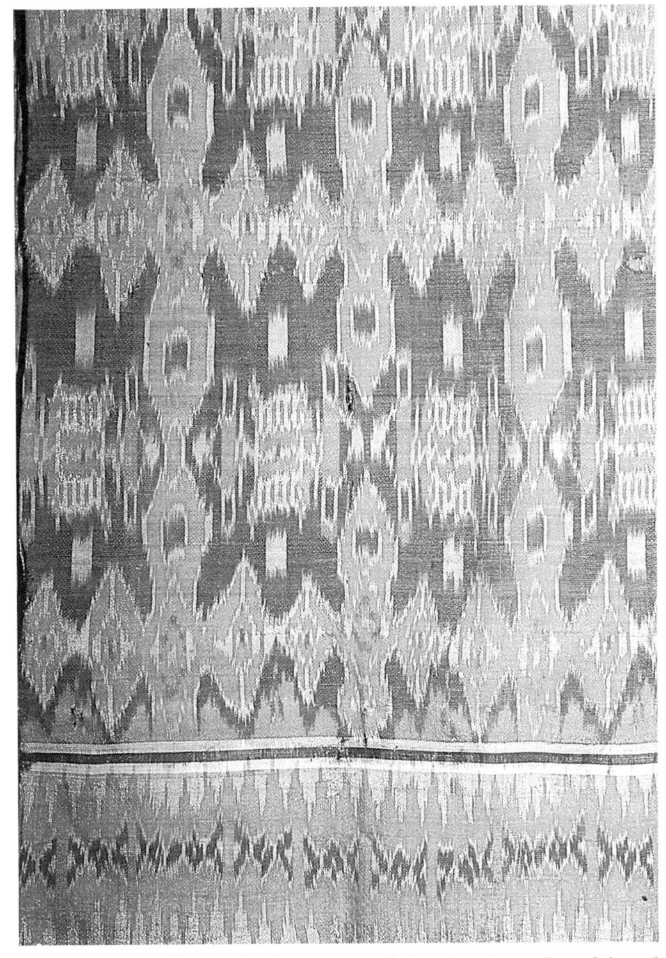

Fig. 231. Silk weft *ikat*-patterned *kamben* in colourful red tones from Singaraja, northern Bali. This textile was made in the 1920s. Warp 220 cm, weft 79 cm. Collection of Isteri Komang Alit, Berdikari, Singaraja.

In northern Bali is the town of Singaraja, the former Dutch capital of Nusa Tenggara. This town, at one time, was famous for its finely patterned cotton and silk weft *ikat* in complicated geometric designs in glowing red, touched with brown and yellow hues (Fig. 231). Many were finished with *songket* borders. Today, there are two factories which specialize in making fine quality weft *ikat*. The larger Berdikari, established in 1966, has a staff of over twenty. Fifteen women do the weaving, while six men attend to the tying and dyeing. Designs include adaptations of well-known Balinese patterns such as those found on the *geringsing* cloths of Tenganan, as well as Hindu and animal figures. This company has a thriving export business to Australia, Japan, and the USA. Puri Agung, which was established some sixty-five years ago, specializes in simple floral and geometric designs, as well as traditional motifs. The sarongs produced are mainly for the local Balinese market.[36] At Baratan, a small brass working village on the outskirts of Singaraja, a few women weave *kain songket* for the local market.

It is still possible to visit the village of Tenganan Pageringsingan in East Bali to see women at work weaving the famous rust red-purplish black *geringsing* double *ikat* cloth on a small body-tension loom in a continuous warp (Plate 39). On the newer cloths, the dyeing time appears to be greatly reduced so that the colours appear very pale and washed out compared with older cloths.[37] *Geringsing* cloth, which averages 140 cm by 35 cm, shares some affinities with *patola* textiles in terms of pattern layout and motifs. A white 3 cm stripe forms the selvage. The centre field may consist of one repetitive design, or it may consist of wide bands of up to four or five different patterns on one cloth. The warp edges are finished with two or three small bands in a repetitive *ceplokan*-type design followed by three stripes in red, blue, and white, before ending in warp fringes.

Fig. 232. Some popular designs seen on double *ikat geringsing* cloth from Tenganan Pageringsingan, East Bali. (After Urs Ramseyer.)

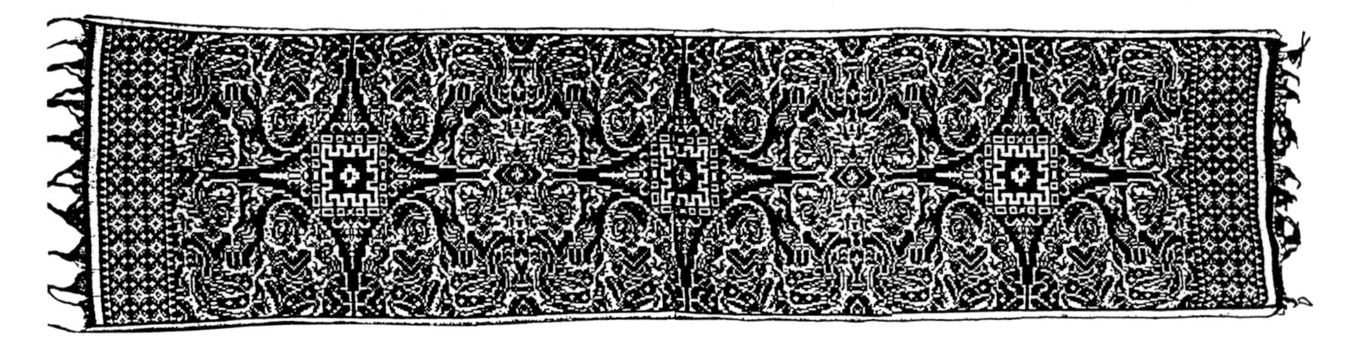

Fig. 233. Double *ikat* cloth from Tenganan Pageringsingan, East Bali. This cloth depicts pairs of *wayang* figures arranged around three large star-shaped motifs which dominate the centre field. This design is called *wayang puteri* and was used for ceremonial events.

Occasionally, some *geringsing* motifs may be overlain with gold, as in Balinese *prada* cloth.

Geringsing has a repertoire of up to twenty basic designs which do not change (Fig. 232). Floral and geometric motifs predominate. Some of the best-known designs include the *si gading* and *dingdinggai*, with motifs set within checked squares. Diamond forms predominate in the *patelikur talidandan*, the *cemplong*, and *cicempaka petang dasa* designs. Other repetitive patterns include *pepare*, a circular motif with a large cross-like form in the centre, and *enjekan siap* and *batun tuhung*, which are aligned horizontally across the body of the cloth. Some of the most striking of the larger cloths have *wayang* figures patterned in beige against a purplish black ground. Referred to as *wayang kebo* or *wayang puteri*, these figures are usually arranged around a central four-point star in a way somewhat reminiscent of a Javanese temple scene (Fig. 233).[38]

Nusa Penida

The island of Nusa Penida off the east coast of Bali, which was once the penitentiary for undesirables from the Kingdom of Klungkung, also specializes in producing fine weft *ikat* in the form of a red *kamben*, a 75 cm by 120 or 220 cm length of cloth which serves as a short sarong or a shoulder cloth (Fig. 234). Made usually from cotton rather than silk, the Nusa Penida *kamben* strongly resembles *patola* cloth in layout. A series of white lines, occasionally embellished by small *tumpul*-like triangular motifs, delineate the border areas and the lateral edges of the cloth. A trellis-work of intricate diamond patterns in beige to white, highlighted by traces of yellow and blue predominate. These cloths were at one time very popular with the Balinese for temple decorations and garments for royal court ceremonies.[39]

Fig. 234. Purple silk and cotton *kamben* from Nusa Penida patterned with geometric designs in weft *ikat*. Warp 124 cm, weft 78 cm.

1. Mattiebelle Gittinger, *Splendid Symbols, Textiles and Tradition in Indonesia*, Textile Museum, Washington, DC, 1979, p. 79.
2. Dra Suwati Kartiwa, *Kain Songket Weaving in Indonesia*, Penerbit Djambatan, Jakarta, 1986, p. 34.
3. Ibid., p. 35.
4. Ibid., pp. 34 and 36.
5. Gittinger, *Splendid Symbols*, pp. 104–5.
6. Mattiebelle Gittinger, 'Sumatra', in Mary Hunt Kahlenburg (ed.), *Textile Traditions of Indonesia*, Los Angeles County Museum of Art, Los Angeles, 1977, p. 28.
7. Kartiwa, op. cit., pp. 19–20.
8. Visit to Pusako Weaving House and interview with Nyonya H. Sannar Ramli at Pandai Sikat, 3 July 1986.
9. Peggy R. Sanday and Suwati Kartiwa, 'Cloth and Custom in Western Sumatra', *Exhibition*, University Museum of Pennsylvania University, Philadelphia, Vol. 26, No. 4, 1984, pp. 19–23.
10. Anon., *Ragam Hias Songket Minangkabau* (Types of Decorative Minangkabau Songket) (in Bahasa Indonesia), Proyek Pengembangan Permuseuman, Sumatera Barat, 1982, pp. 28–32; and Anon., *Pakaian Adat Wanita Daerah Paya Kumbuh* (Traditional Women's Clothing of Paya Kumbuh District) (in Bahasa Indonesia), Proyek Pengembangan Permuseuman, Sumatera Barat, 1980, pp. 40–8.
11. Visit to workshop of A. Munir Sito, Silungkang, 3 July 1986.
12. Sanday and Kartiwa, op. cit., p. 23.
13. Visit to Kerajinan Amai Setia, Kota Gadang, 3 July 1986.
14. Gittinger, *Splendid Symbols*, p. 109.
15. Mattiebelle Gittinger, 'Selected Batak Textiles: Technique and Function', *Textile Museum Journal*, Vol. 4, No. 2, 1975, pp. 13–15 and 19–21.
16. Ibid., pp. 15–16.
17. Ibid., p. 27.
18. Ibid., p. 28.
19. Mattiebelle Gittinger, 'The Ship Textiles of South Sumatra: Functions and Design System', *Bijdragen tot de Taal-Land- en Volkenkunde*, Vol. 132, 1976, p. 208.
20. Toos van de Dijk and Nicole de Jonge, *Ship Cloths of the Lampung South Sumatera*, Gallerie Mabuhay, Amsterdam, 1980, pp. 34–42.
21. Gittinger, 'The Ship Textiles', pp. 210–11.
22. Ibid., pp. 211–20.
23. Gittinger, *Splendid Symbols*, pp. 82–4.
24. Ross Elizabeth Dalrymple, 'Gold Embroidered Sarong from Southern Sumatra', *Arts of Asia*, January–February 1984, pp. 90–9.
25. Olga Yogi, 'Lurik, A Traditional Textile in Central Java', in Mattiebelle Gittinger (ed.), *Indonesian Textiles: Irene Emery Roundtable on Museum Textiles, 1979 Proceedings*, Textile Museum, Washington, DC, 1979, p. 283.
26. Ibid., p. 282.
27. Visit to Logro, Jogjakarta, 27 June 1986.
28. Wanda Warming and Michael Gaworski, *The World of Indonesian Textiles*, Kodansha, Tokyo, 1981, p. 120.
29. Visits to the factories of Catur and Hadra, and Alchotib, Gresik, 24 June 1986.
30. *Kain bidang* in the Indonesian language refers to a sarong for women.
31. Visit to Srikandi Ratu and other weaving centres in Troso, 26 June 1986.
32. Bronwen Solyom and Garrett Solyom, 'Bali', in Kahlenburg (ed.), *Textile Traditions of Indonesia*, p. 73.
33. Visit to 'Cili' workshop, Gianyar, 5 June 1986.
34. Visit to 'Dewata' at Batuan, 5 June 1986.
35. Interview with Isteri Oka, weaver, Klungkung, 5 June 1986.
36. Visits to Puri Agung and Berdikari workshops, Singaraja, 6 June 1986.
37. Visit to village of Tenganan, 5 June 1986.
38. Alfred Buhler, Urs Ramseyer, and Nicole Ramseyer-Gygi, *Patola und Geringsing*, Museum fur Volkerkund, Basel, 1975, photographs, pp. 31–57.
39. Gittinger, *Splendid Symbols*, p. 204.

CHAPTER ELEVEN

Indonesia: Outer Islands

Sulawesi

SULAWESI, also known as the Celebes, is located directly east of the island of Borneo. Up until the early 1900s, this island produced a wide array of textiles from many different fibres in a variety of patterning techniques. Today, textile production is largely limited to the southern province of Sulawesi.

The Bugis

The Makassarese and Buginese, hardy seafaring peoples once feared for their piracy and respected for their trading skills, have a history of making finely woven iridescent silk cloth. As early as 1701, Nicolas Gervais, a foreigner residing in Makassar (now called Ujung Pandang), in his writing remarked on the wide variety of silk and cotton goods available in the godowns of that city.

The weaving industry today is centred on Sengkang in the Wajo district on the shores of Lake Tempe, 150 km north of Ujung Pandang. Some 4,700 women weave nearly half a million sarongs a year. Many women weave at home in their spare time on body-tension looms, while others work in larger concerns organized as cottage industries. Here, electrically powered spinning and reeling devices, the use of the *cetak* dyeing technique, and hand-operated floor looms with flying shuttles help speed up production.[1]

Apart from a tiny project in Sumatra, Sulawesi is the only place in Indonesia which produces its own silk. At Tajuncu, on the road to Toradja Land, the Government, in 1976, with Japanese assistance, embarked on a sericulture project to produce indigenous silk for the Sengkang silk weaving industry. Local silk is now used for the weft of most woven cloth. Imported silk continues to be used for the warp threads. At one time, natural dyes were used, but since 1936 they have been supplanted by aniline dyes.

Fig. 235. Buginese woman from Sengkang, Sulawesi, in *sarong kebaya* and *selayu* head-cloth. Her brightly coloured silk sarong is patterned with a large plaid called *balo lobang*. The *kepala* carries the *bombang* wave pattern in weft *ikat*.

Bright crimson, rich purples, yellows, pinks, turquoise, and lime green are the preferred modern colours for Buginese sarongs. Traditional patterns consist of stripes, checks, and plaids. Small plaids are referred to as *balo renni*, while larger tartan-like squares bounded by lines of varying thickness, are called *balo lobang* (Plate 40). Threads of gold and silver may highlight the boundaries of a plaid. Many Bugis sarongs are patterned with a zig-zag design of different colours produced by the weft *ikat* technique, which is called *bombang* or wave pattern (Fig. 235).

This motif may be an all-over design or confined to the *kepala*, or centre portion of the cloth. It may also be integrated into a plaid background. Colour gradations in the *bombang* pattern are sometimes subtle, sometimes very bold. Sarongs patterned with large flowers in weft *ikat* are called *sarong Samarinda* after the capital of East Kalimantan, which is noted for its colourful durable silks.[2]

Instead of weft *ikat*, simple floral patterns are sometimes worked in a supplementary weft of fine metallic-coloured silk threads. Patterns are very simple and are inlaid on the right side of the cloth by carefully counting the warp threads. A small pick, or *subik*, is used to help insert the supplementary threads. Pattern sticks are not used to pick out the designs in the warp. These silks are very tightly woven so that the outlines of the supplementary weft are scarcely visible on the reverse side of the cloth. South Sulawesi was at one time noted for weaving a silk or cotton textile called *lipa garusu* with a fine plaid or gold stripes. *Lipa garusu* were normally starched with sago flour, pressed, and polished with a sea shell.[3]

Buginese women wear the sarong pleated in front with a small flap, which is borne gracefully over the left arm. It is worn with a sheer, loose-fitting overblouse, the *baju bodo*, which is made from the finest muslin, fine raw silk yarns, or pineapple leaf fibres. The *baju bodo* may be either woven by villagers in their homes on body-tension looms, or be fashioned from store-purchased cloth (see Fig. 85). Choice of colour varies with age. Young girls and unmarried maidens are permitted to wear bright festive colours. After marriage, more subdued colours are considered appropriate. A black *baju bodo* is worn at the onset of old age.[4]

The Mandar district on the north-west coast of South Sulawesi also produces finely woven silk sarongs. The colours are generally more subdued than those of Sengkang. Black, maroon, and dark green hues in plaid patterns are generally preferred.[5]

Donggala, north of Mandar on the west coast of Sulawesi, has a tradition of weaving cloth called *buya subi sabbe*, with geometric and scattered floral motifs in coloured silk and metallic threads in a supplementary weft. Some Donggala sarongs are patterned in weft *ikat* embellished with supplementary weft decoration to highlight key elements in floral and bird motifs.[6]

Toradja

Further to the north are the Toradja people, who inhabit the mountainous central portion of Sulawesi. These people, having been isolated from outside contact, managed to retain their Bronze Age culture and animistic beliefs up until the early twentieth century. In Toradja society, the ancestors are revered and propitiated, so funeral rites are very important. These involve constructing a ceremonial complex for animal sacrifices and communal feasts. Both primary and secondary burial are practised. Elaborately carved sacophagi and wooden-articulated effigies of the deceased, called *tau tau*, are made. Although the majority of the population today is Christian, elaborate funeral ceremonies continue to be an integral part of Toradja life.[7]

Large, magnificent warp *ikat* blankets are very important in funeral rites (Plate 41). Ceremonial pavilions for receiving guests are decorated with these blankets, and the body of the deceased may be wrapped in warp *ikat* blankets while lying in state. The best of these blankets are made in the remote villages of Rongkong and Galampang (To Maki) about five or six days trek from the main Toradja town of Rantepao. Formerly woven from homegrown, hand-spun cotton, these blankets took many months to weave because of the time needed to produce the required colour from natural dyes. Red came from morinda, to which was added chili and dammar resin. Blue was made from the indigo plant, and from the leaf of *torae* grass mixed with *lalardo* (ink fruit). Black was produced by combining black earth with charcoal and leaf mould.

Blankets are boldly patterned with angular geometric designs, such as hooks, arrows, rhombs, zigzags, and triangular forms in red or blue outlined in white against a blue or purplish-black ground. Some scholars have interpreted a diamond-shaped motif with four inward curving appendages, called *sekon*, as a stylized anthropomorphic figure.[8] Toradja blankets are warp faced and woven with a blue weft thread. The edges of blankets may have stripes and small zig-zag borders. These blankets were acquired through barter by other Toradja groups for burial ceremonies. There are a few women in the To Maki area who continue to make blankets in the traditional way, but store-purchased yarns have largely replaced hand-spun cotton.

Due to tourist demand, copies of the traditional blankets are now being made in the Rantepao area. These imitations use chemical dyes and are fairly easy to spot. Over recent years, a small weaving centre has developed at Sa'dan, a few kilometres north of Rantepao. Here, women weave cloths patterned with typical Toradja motifs, such as buffalo and *tongkonan* houses with upturned roofs, in a

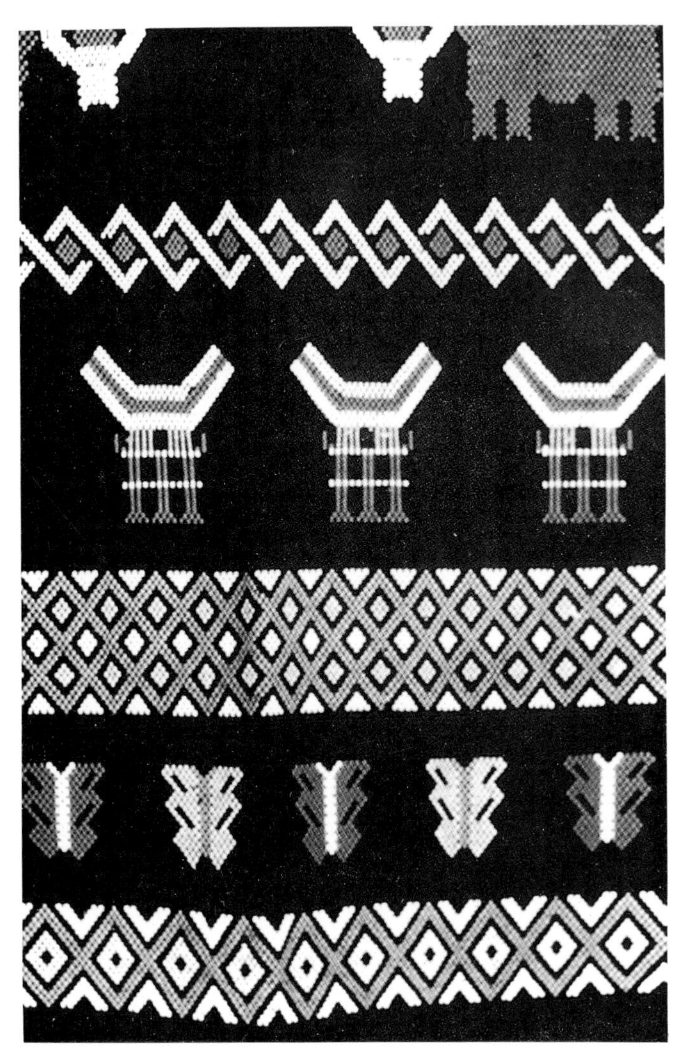

Fig. 236. Some modern Toradja weaving from Sa'dan. Supplementary weft-patterned cloth with buffalo, *tongkonan* houses, butterflies, human figures, and zig-zag and trellis designs are now being made for the tourist trade.

supplementary weft. Like the blankets, these textiles are mainly for the tourist trade (Fig. 236).

In addition to warp *ikat*, the Toradja at one time wove patterned belts, sashes, headbands, and strips for sewing into purses. These items were woven with a tortoise shell tablet loom. Up to 182 tablets were used for some particularly complicated motifs. The Toradja used to also make batik and *pelangi* resist-decorated cloths. Weaving was not limited to cotton fibres; pineapple fibres were once used for weaving ritual clothing and screens. The fibres of the *Corrypha* palm, bamboo, and rushes have all been used to create textiles. Beaten bark cloth, called *fuya*, patterned with beautiful painted designs was also used for clothing at one time.[9]

The Toradja have always held textiles in high esteem. They have a history of trading with the Dutch for *patola* cloths and the Dutch imitations. Along with their own batik *sarita* cloths, these imported textiles were thought to be imbued with magical powers. They were often displayed in designated areas in the form of bunting and banners for special events. When not in use, they were carefully stored away. Textiles play an important role in funerals, marriage gift exchanges, initiation rites, and in the payment of fines.[10]

Northern Sulawesi

People living in the Minahassa Peninsula in north-east Sulawesi at one time made garments from a wide variety of fibres. Cotton weavings, collectively called *kain bentenan*, were particularly well known and were patterned by a number of techniques such as warp *ikat*, woven stripes, and repetitive patterns in warp floats. Unfortunately, Western influence in the late nineteenth century led to the demise of weaving in North Sulawesi.[11]

Sanggir and Talaud Islands, to the east of Minahassa, weave an abaca cloth called *koffo*, embellished with geometric motifs such as diamonds, stars, angular scrolls, and rosettes in a supplementary weft. Lengths of *koffo* are sewn together to create 10.0 m long by 1.5 m wide lengths of cloth for curtains, room dividers, screens, and banners, both for domestic use and for ceremonial events. *Koffo* was at one time used for clothing. Nobles at ceremonial events could be distinguished by the extra length of their tunics and head-cloths.[12]

Kalimantan

Kalimantan, comprising approximately two-thirds of the equatorial island of Borneo, has two distinct textile traditions: that of the Malay coastal inhabitants, and that of the Dayak and related peoples of the interior river basins.

On arrival, the Malay peoples who migrated to Borneo over the last few centuries set up small trading kingdoms along the coastal areas. One such example is the former Kingdom of Sambas located north of Pontianak on the west coast of Kalimantan. Sambas was once an important weaving centre noted for its fine cotton and silk *ikat* (called *kain cual*) and its richly coloured *kain songket* made from imported silk and metallic threads. *Kain cual* is no longer made today although *kain songket* continues to be woven on a traditional Malay loom.

Kain songket fabrics are generally classified according to their designs. Motifs are derived from both geometric and floral sources. Striped patterns are referred to as *kain lurik*, while the term *kain petak* refers to checks. *Kain tabur* is the term for scattered motifs, and a diagonal alignment of motifs is called *kain serong*. The most sumptuous textile is the *kain padang terbakar*, which is completely covered with closely patterned *kain songket*. With the exception of yellow, a colour reserved for royalty, Sambas silks have long used chemical dyes. Since World War II, the production of fine *songket* has been hampered by the unavailability of good quality metallic thread.[13]

Samarinda, the capital of East Kalimantan, was once famous throughout South-East Asia for its fine silks. Its weaving tradition originated in Sengkang, in South Sulawesi. A war between Bone and Wajo, in 1665, caused the royal family of Wajo to seek a new life in the Kutai area of Kalimantan. Intermarriage between the local royal family and Buginese nobles led to the establishment of a silk industry at Samarinda.[14] Fine silk continues to be woven in the Serberang area across the river from the town of Samarinda.[15] At one time, Samarinda specialized in producing floral-patterned weft *ikat*; throughout Indonesia, floral designs created by this technique are referred to as *kain Samarinda*.

The Dayak tribes of the interior of Borneo which have a weaving tradition include the Ot Danum, the Apo Kayan, the Bahau, and the Iban. The Bahau people of the Mahakam River basin are noted for weaving a striking warp-patterned textile from cotton and bast fibres commonly called *kutai ikat* (Fig. 237). The *ikat* is applied to wide warp bands separated by red stripes. The spotted, beige-coloured designs, representing stylized birds and crocodiles, are sharply angular and attenuated in form. They are enlivened by arrowheads and discontinuous zig-zags in red and green outlined in deep purplish red.[16]

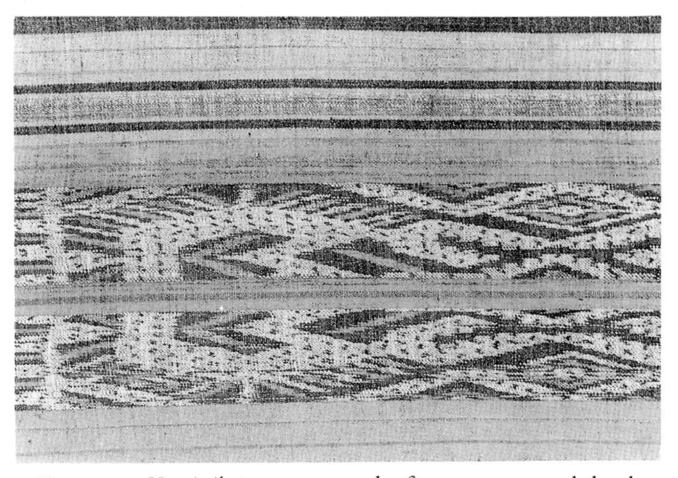

Fig. 237. *Kutai ikat* sarong made from cotton and lemba fibres (*Curculigo latifolia*). From the Mahakam River basin area of Kalimantan. Designs represent stylized birds and crocodiles.

The Iban of Kalimantan, like members of their tribe in Sarawak, are notable weavers. They have largely kept aloof from Western contacts and continue to lead their traditional longhouse way of life. For clothing, they continue to weave traditional garments using time-hallowed Iban decorative techniques. *Pua* ceremonial textiles continue to play an important role in Iban ritual life.[17] The Maloh, a Dayak people of the upper Kapuas River valley, as successful farmers, instead of weaving appear to acquire most of their textiles from the neighbouring Iban. They spend their time instead embellishing these cloths with elaborate, colourful beadwork.[18]

Nusa Tenggara

The islands of Nusa Tenggara, also known as the Lesser Sundas, extend some 1 500 km towards Australia. On these islands, textiles are the chief means of artistic expression and the diversity of the inhabitants is mirrored in the range of textiles produced. Because of their relative isolation, textiles continue to permeate the socio-religious life of the people. Textiles in Nusa Tenggara at one time not only indicated a particular location, but defined the status of the individual, both in terms of costume and of wealth. Textiles continue to play an important role in observing rites of passage. They are also important items of trade both within and outside Nusa Tenggara.

Lombok

Since the seventeenth century, Lombok, a small island lying directly east of Bali, has been noted for growing cotton. Lombok continues to produce an array of weft *ikat*, supplementary weft, tapestry weave, and plain weave fabrics for both clothing and ceremonial use.

Weaving has long been a major activity amongst the Muslim Sasak people, who are considered to be the original inhabitants of Lombok, and a young Sasak woman was expected to weave up to forty pieces of cloth for her trousseau. The Sasak weave some unassuming coarse weave textiles, such as *kain usap* and *kain umbak*, which are regarded as being imbued with magic and protective powers. Various taboos are observed throughout their manufacture.[19] The *kain umbak* is a simple, warp-striped cloth, while the *kain usap* has more variation (Fig. 238). In addition to plain stripes, this textile may have geometric designs and loosely patterned floral motifs placed in bands between plain narrow stripes. These cloths are very important in observing rites of passage. The *kain umbak* is especially important in festivities associated with a child's first haircut, or *ngurisan*. The child is carried in a *lempot umbak* cloth and relatives may present other *umbak* cloths to the child as protection against illness.[20] *Umbak* textiles are also used in other childhood rites, such as in naming ceremonies, tooth filing, and circumcision. A ritual textile gift presentation also traditionally concluded dowry negotiations. A *kain umbak bergeringsingan* was presented by the groom to the bride's family at the *upacara sarong serah* ceremony. During the wedding ceremony, a special striped cloth, called *kain ragi kembang komak*, was twined together with white cloth and thread. At the time of death, the corpse was wrapped in a special cloth (*leyang putik*), and a small square of *kain usap* was placed over the face of the deceased.[21] *Kain usap*, in fact, may symbolize death, for it is the cloth which is taken to the house of the *imam*, or religious leader, to inform him that a member of the bearer's family has died.[22]

Traditional dress for the Sasak woman consists of a black sarong and matching, loose-fitting blouse, called *lambung*, which often has narrow borders of embroidered diamond patterns. It is worn with a colourful sash, called a *sabuk anteng*, which is held in place by a silver belt. A colourful *selendang* is also worn. The Sasak male wears a long sarong, called a *kereng*, which may be in a plaid or supplementary weft pattern. It is worn with a neat-fitting black jacket (*kalambi*), a long sash, and a colourful folded head-cloth (the *sapuk*) (Fig. 239).[23]

Sukarara, some 45 km south of Mataram, is the centre for traditional Lombok *kain songket*, which is woven in coloured supplementary wefts as well as in gold and silver yarns. Some 300 women, working in their homes on body-tension looms, weave sarongs, sashes, and *selendang* in traditional patterns against a brightly coloured cotton or silk ground. Lombok motifs are largely floral and geometric. The simplest motifs consist of regularly placed rosettes, zig-zags, and triangular motifs on a plain or checked ground. Balinese influences are also evident in the very popular traditional *subhanale* design, which consists of flowers, buds, and sprigs of foliage enclosed by a hexagonal grid repeated over the surface of the cloth (Fig. 240). The grid sometimes encloses a fierce mask figure cleverly integrated with flange-like tendrils of foliage springing from the head and mouth (Plate 42). Snake-like *naga* figures and birds appear in combination with long-stemmed leaves and hook-shaped tendrils. A pair of *wayang* figures

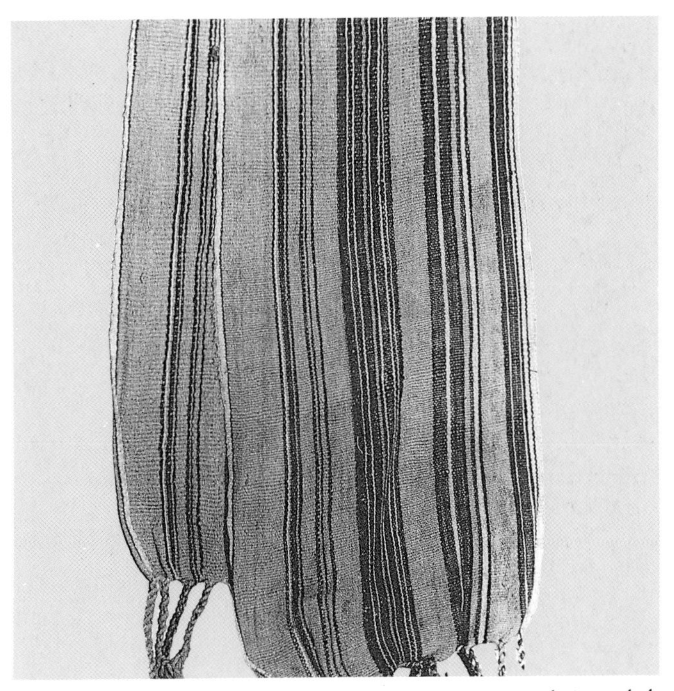

Fig. 238. This unassuming striped, coarse-weave *kain umbak* textile is very important in childhood rites of passage, such as naming ceremonies, tooth filing, and circumcision. Various taboos are associated with its manufacture and it is credited with being effective against illness. Warp 85 cm, weft 23 cm. Museum Negeri Nusa Tenggara Barat, Mataram, Lombok.

Fig. 239. Sasak couple from Sukarara, Lombok, in traditional dress. The woman is wearing a *selendang* shoulder cloth, a black sarong, and loose-fitting blouse called *lambung*, which has a little embroidery around the edges. The man is wearing a *kereng* sarong, a neat-fitting black *kalambi* jacket, and *sapuk* head-cloth.

Fig. 240. A supplementary weft-patterned sarong from Sukarara, Lombok, which is patterned in the hexagonal *subhanale* floral design. Warp 164 cm, weft 108 cm. Collection of H. Li. M. Ichsan Widasih, Sukarara.

under an umbrella is also a popular Lombok motif (Fig. 241). It takes a woman about two weeks to complete a colourful *selendang* patterned with scattered geometric motifs and tapestry weave borders. An intricately patterned *kain songket* sarong may take up to two months to complete. Sukarara cloth is made largely for local use. *Kain songket* is *de rigueur* for weddings and circumcision ceremonies.[24]

At Cakranegara, there are two large textile concerns which produce weft *ikat* textiles as a cottage industry. Rinjani, which was founded in 1948 by the present owner's grandfather, specializes in producing cotton cloth for the Balinese market in subdued greens, blues, and browns. Motifs are based on eastern Indonesian designs from Lombok, Sulawesi, Bali, and Nusa Tenggara. The Slamet Riady Company specializes in Balinese floral designs, simple repetitive patterns, and floral motifs against a checked or striped background. A few gold and silver supplementary weft threads may be added to highlight certain *ikat* patterns. Both factories make extensive use of mechanical spinning and reeling devices to speed up the processing of yarn and the setting up of warp threads. The *cetak* method is widely used to quicken the dyeing process. Material is woven in yardages rather than in sarong lengths on hand-operated looms with flying shuttles. Although each factory has over fifty looms in each of its establishments, much of the weaving is still done as piecework by women in the nearby villages.[25]

The village of Pringgasela in East Lombok spe-

Fig. 241. Some *kain songket* patterns from Lombok.

floral motif

subhanale

A ceremonial sarong featuring birds and floral designs.

wayang figures under an umbrella

cializes in making blankets (*selimut*) in bright, bold, primary colours. They are woven at home on a body-tension loom from a cotton warp and a weft of acrylic wool. They have a nubbly texture and consist of two lengths of cloth joined together. Stripes are the most popular form of decoration. Some may have a wide band of *tumpul* decoration in a tapestry weave. In addition to serving as covers, these blankets are commonly used as baby carriers. The women of Pringgasela also keep their menfolk supplied with *kain lurik* and plaid sarongs. Pringgasela textiles are largely for domestic use: some, however, are bartered or sold in Cakranegara for thread.[26]

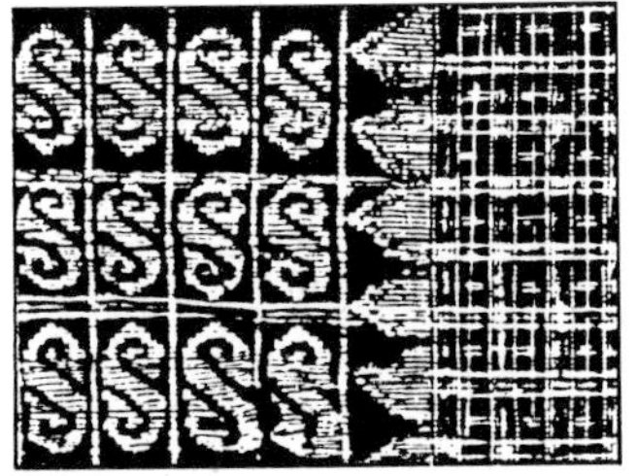
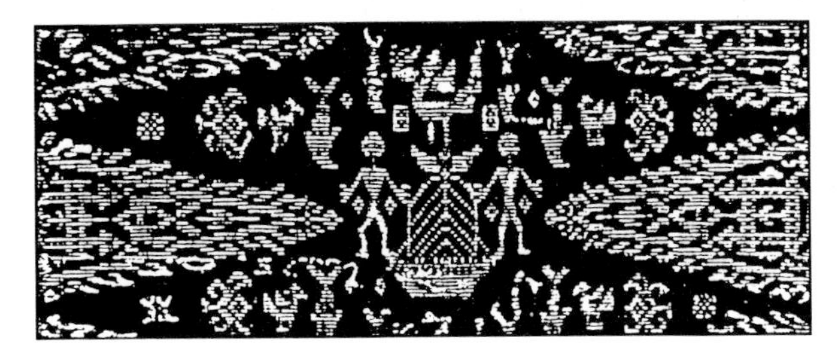

Fig. 242. Some *kain songket* patterns from Sumbawa.

Sumbawa

Sumbawa, one of the larger islands of Nusa Tenggara and lying immediately east of Lombok, produces an interesting array of textiles which have not received a great deal of attention from scholars. Historically, Sumbawa was at one time divided into a number of small kingdoms, such as Sumbawa Besar, Dompu, and Bima. It had links with East Java during the Majapahit period in the fourteenth century, and with the Buginese Kingdom of Makassar in South Sulawesi in the seventeenth century. It was the Bugis, as conquerors of Bima in East Sumbawa, who brought the Muslim faith to the island and, in all probability, it was they who introduced a number of Buginese weaving techniques to the Sumbawanese.

Traditional dress for the Sumbawa woman consists of a sarong, called a *tembe songke* or *kre alang*, which is worn with a short, loose blouse (the *baju boro*) and a shoulder cloth (the *selampe* or *perbasa*). A short veil (the *kerudung*) is also worn. Sumbawa women cover the head and part of the face with a second sarong, worn like a hood. Men wear a long-sleeved jacket, sarong, a sash, and a *sambolo* head-cloth.[27]

For everyday wear, checked sarongs are most popular. Sarongs with large square plaids are virtually identical to those of South Sulawesi, and are referred to as *tembe lompa*. Those with smaller checks are called *bali mpida*. Sumbawa is well known for its beautiful *kain songket* in gold and silver and cotton supplementary weft weavings, called *selungka*.[28]

The *sarong tembe songke* for ceremonial events consists of gold and silver supplementary wefts worked on a plaid ground. Red and black plaids appear to be the most prevalent. The *badan*, or body of the cloth, is sometimes closely patterned with floral or geometric designs. More often, it is lightly sprinkled with small rosettes placed inside plaid squares. The plaid stripes on the *badan* are sometimes outlined with gold supplementary yarns. The *songket* decoration on the *kepala* is usually quite dense. It consists of regular rows of closely placed rosettes or 'S'-shaped hooks framed by vertical *tumpul* borders (Fig. 242). More elaborately decorated *kepala* may depict beautifully crafted ship designs complete with human figures surrounded by birds and fish. The Sumbawanese also excel in depicting birds amidst flowers, leaves, and branches.

The ceremonial *selampe* shoulder cloth is a textile of great beauty (Fig. 243). It has an elongated centre field edged by small tumpul patterns usually worked

Fig. 243. Sumbawa ceremonial *selampe* shoulder cloth patterned with supplementary gold threads. Museum Negeri Nusa Tenggara Barat, Mataram, Lombok.

in a tapestry weave. This central area may be plain or have small scattered geometric and floral motifs. The darker border areas on all sides are patterned with a series of parallel bands containing small, repetitive *songket* patterns. Sumbawa weavers have traditionally used a fairly soft gimp-like metallic thread for much of their *songket*. Unfortunately, with age it tends to wear away causing some design areas to be lost. On some cloths, gold patterns are applied by the embroidery process rather than by weaving.[29]

Sumba

South of Sumbawa lies Sumba, a dry, barren island famous for its horses and for its distinctive warp *ikat* and supplementary warp textiles made for ritual and household use, local barter, and inter-island export. It is interesting to note that textiles played a role in the assumption of Dutch control of the island in 1901. A dispute had arisen between a local ruler and a Dutch captain over the return of some slaves who had run away and sought the captain's protection. Apparently, while weaving, the slaves had damaged a textile by tearing some threads, the penalty for which was most likely death. The refusal of the captain to accede to the ruler's demand to return the slaves led to a successful military action which brought Sumba under Dutch control.[30]

Formerly, the creation of fine textiles on Sumba was the exclusive right of noblewomen. Free from agricultural chores, they had the time and inclination to weave superior textiles which might take up to three years to complete. Today, weaving is no longer the preserve of the female aristocracy. Many village women in the eastern coastal district of Sumba continue to weave boldly patterned warp *ikat* textiles from imported cotton yarns on a traditional body-tension loom with a continuous warp. Natural indigo and red morinda dyes, as well as chemical substitutes, are used to colour yarns.

The most important traditional textiles seen in Sumba today are the man's *hinggi* mantle and the woman's *lau* sarong. Occasionally one sees a narrow, 7–9 m long warp *ikat* cloth, the *rohubanggi*, which was wrapped around the body of a warrior as extra protection under his leather armour.[31] *Hinggi* are large, warp *ikat*-decorated blankets worn by men. For everyday wear, a blue and white *hinggi*, called *hinggi kaworu*, is worn. For ceremonial occasions, a predominantly red *hinggi*, the *hinggi kombu*, is generally preferred (Plate 43).

The main design elements on a traditional *hinggi*

Royal symbols which often appear in the centre band of a warp *ikat hinggi*.

habak

Karihu, based on an open shell.

Animal Motifs

cockatoos

seagulls

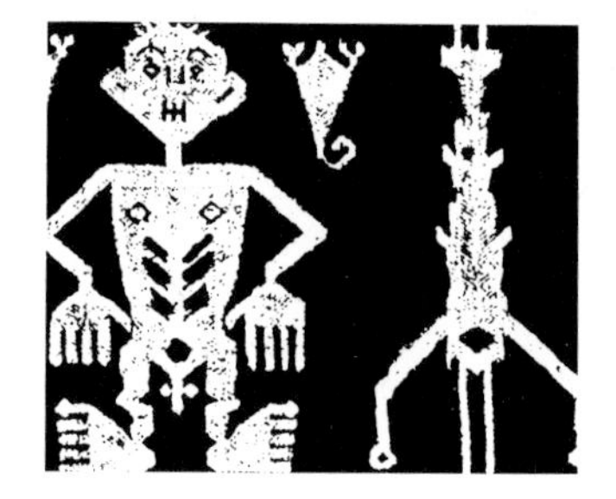

man and shrimp

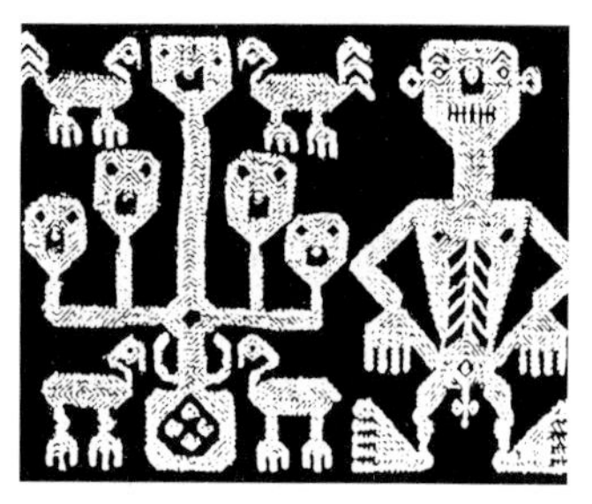

man with *andung* skull tree, and roosters

Heraldic motifs from Dutch coins

lions with sceptre

Fig. 244. Some popular motifs seen on warp *ikat* and supplementary warp textiles from Sumba.

Kanduhu design based on diamond and star shapes.

A design based on the gold ear ornament, the *mamuli*, a traditional marriage gift.

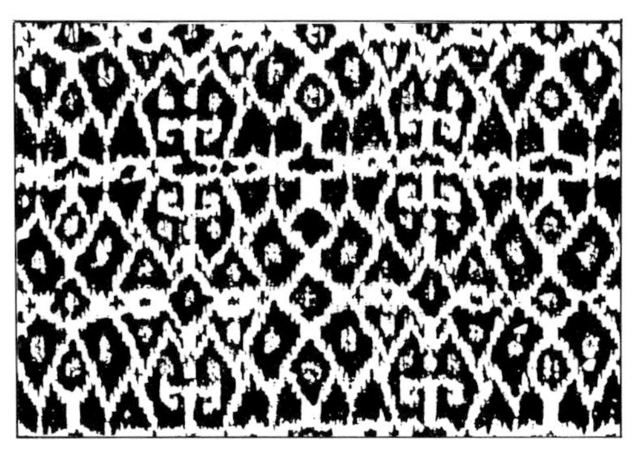

patola ratu

roosters

horses

horse, crocodile, snake, and birds

dragons, inspired by Chinese ceramics

monkey

Queen Juliana

coat-of-arms

kombu are usually arranged in three to five bands of varying widths. Major motifs appear as light-coloured figures against a background of alternating rust and bluish tones, obtained by overdyeing red on indigo (Fig. 244). On most cloths there is a centre band which spans the middle of the cloth. This band contains a series of repeated schematic motifs, such as an eight-point star and a diamond shape called *a kanduhu*, the four-crescent *habak*, the *karihu* (shell fish), *patola ratu*, and designs based on Sumba jewellery, such as the *mamuli* ear-ring. These motifs were formerly associated with Sumbanese royalty for whom the cloths were originally made.[32]

Above and below this centre band, the sequence of coloured bands and motifs is usually identical, for most *hinggi* are composed of mirror-image halves. These outer bands are patterned with figures, such as confronting pairs of animals, frontal human figures, and pairs of alternating motifs (see Fig. 244). Smaller figures fill the spaces in and around the major images. Most design elements are realistically portrayed and are immediately recognizable.

The major figures depicted on *hinggi* (horses, deer, dogs, monkeys, crocodiles, snakes, lizards, turtles, fish, shrimps, sea horses, roosters, and cockatoos) have been drawn directly from the local environment.[33] Horses are the chief means of transport on the island of Sumba. A horse with a rider is thought to represent a slave riding the deceased king's horse to the graveside where the king, through the slave, issues his last orders before both the slave and horse are sacrificed.[34] Dogs are associated with warriors, and deer—the focus of hunting, which was formerly limited to royalty—are a kingly symbol, as is the cockatoo. Snakes symbolize rebirth and long life, while the crocodile, the most dangerous animal found on Sumba, is associated with the afterlife. In Sumba mythology, the last obstacle facing the soul in its journey to the underworld is a crocodile-infested river which has to be crossed. Because the shrimp is able to discard its shell for a new one, it has come to be associated with longevity. Roosters are associated with masculinity, while a tree represents fertility.[35]

Other designs, such as the *andung* (skull tree), hark back to former cultural practices where it was the custom to suspend the heads of vanquished foes on a tree at the centre of the village to frighten away enemies and ensure prosperity (see Fig. 244). A pair of rampant lions, often seen on Sumba cloths, is derived from the Dutch coat-of-arms and is a symbol of prestige. The dragon motif, in all probability, comes from Chinese ceramics which have been ex-

Fig. 245. Woman from East Sumba, using a small body-tension band loom, weaves the *kabakil* border of the *hinggi kombu* blanket by incorporating the warp ends into the weft of the band.

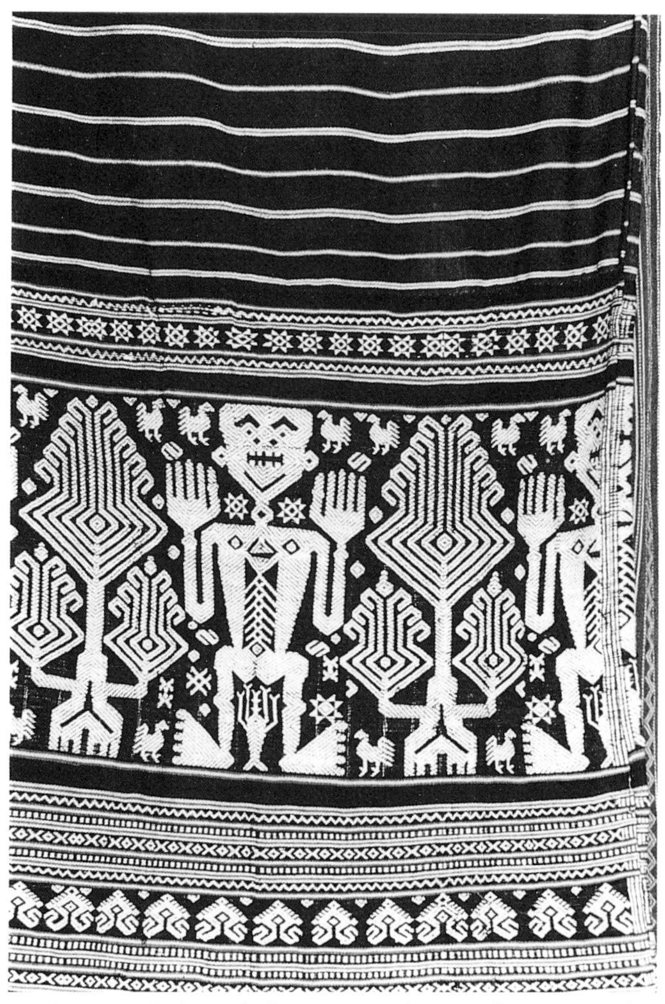

Fig. 246. A *lau pahudu*, a woman's striped sarong with a broad band of trees, human figures, birds, and shrimps formed by a complicated supplementary warp technique. Below is a smaller band containing the *mamuli* ear-ring motif. Rende, East Sumba. Warp 132 cm, weft 162 cm. Ikat Art Collection, Waingapu, Sumba.

cavated from many grave sites in Sumba. The royal symbols seen in the centre owe their origin to *patola* cloths, royal paraphernalia, and traditional jewellery.[36]

According to the Sumbanese, a good *hinggi* should be made from fine hand-spun cotton yarns and should be evenly woven with about 112 warp and 22 weft threads to the inch. The dye colours should be in deep, saturated vari-colour tones. The *ikat* outlines should be sharp and there should be no bleeding around the edges. It should be finished with a *kabakil*, a separately woven striped or patterned band which incorporates the warp ends to form the weft (Fig. 245). Corded fringes complete the cloth. Key motifs may be lightly touched by subtle tan overstaining in a vegetable dye at the conclusion of the weaving.[37]

The woman's *lau* is the ceremonial equivalent of the *hinggi*. For special occasions, such as a burial or a ritual dance, a Sumba woman might wear the *lau pahudu*, a plain, coloured, or striped sarong patterned with a broad horizontal band of patterns in heavy white yarn (Fig. 246). The motifs are similar to those found on the *hinggi*. The *lau pahudu* is woven in a twill weave in the Pau and Rende districts by a difficult supplementary warp technique. Some *lau*, called *lau pahudu padua*, may include bands of warp *ikat* as well as supplementary warp decoration. The *lau hada* is a sarong patterned with human figures embroidered with shell, beads, and tufted yarns.[38]

In the Kodi district of West Sumba, a rather different style of *ikat* is made in the form of a man's blanket called *kodi hinggi*. This cloth, in blue and white occasionally enlivened by patches of rust, has its lateral and end borders decorated in the style of a *patola* cloth. The designs are geometric and abstract and tend to be arranged in parallel bands. Some *hinggi* feature a semicircular design based on the *mamuli* ear-ring, once widely worn in Sumba.[39]

Textiles in Sumba formerly served a multitude of social and symbolic functions. Social rank was clearly indicated in the designs and motifs of various cloths. The number of cloths owned by an individual was an important criterion of wealth and status. A noble had to lay aside a vast store so that his extended family and retainers could be suitably clad at important ceremonial events. Textiles were major items in the gift exchanges between families of the bride and groom. To ensure a fitting send-off and good standing in the afterlife, huge numbers of textiles were required by the family for the burial of a noble.[40]

Hinggi blankets continue to be made today. Natural dyes are still widely used and cotton threads are purchased or hand-spun. Some weavers of East Sumba pride themselves in making exact replicas of the old cloths, complete with vari-coloured overdyeing and a myriad of small motifs in between the larger figures. Other weavers, more interested in a quick sale to tourists, have reduced their warp *ikat* patterns to the bare essentials of simple bands of major motifs against an alternating red and blue ground. White as a ground colour has become more popular in some of these cloths. Secondary motifs, if present, are often carelessly rendered. There is little evidence of overdyeing and cloths are not well finished along the warp edges.

Over recent years, there have been new developments in Sumba *ikat* design. In the 1960s, the Mangili area in south-east Sumba began producing textiles depicting traditional subjects, such as horsemen and roosters, in simpler compositions and in a more realistic style. The colours on these cloths are deeper and more saturated than in the traditional Sumba palette.[41] Weavers in other districts have followed suit by producing panoramic pictures with a few main motifs, such as horses and riders, warriors, and rows of human figures, dominating the centre field. Whimsical renditions of local fauna, such as snakes, squirrels, cockatoos (sometimes with two heads), and turtles, float around in the background. On some cloths, the traditional centre field filled with repetitive royal symbols has been retained, while on others it has been replaced with a band of supplementary warp weaving or beadwork. On some cloths, there is a blurring of the motifs during the dyeing process in an attempt to give a shimmering shot effect to the textile. A few of the modern cloths reflect influences from Timor in the naïve portrayal of human motifs, and in the spotted texturing effect applied to some design areas. On the best of the newer cloths, the colours remain strong and the standard of workmanship in both the tying and dyeing of the various motifs remains high. A few men in the Prailiu area have also been involved in designing modern warp *ikat*. They sketch out the motifs on the yarns and do much of the tying and some of the dyeing before giving the yarns to their womenfolk to weave.

Savu

Savu, a small island which lies to the south-east of Sumba, makes some very striking warp *ikat* in floral and geometric designs arranged in distinctive bands against a dark indigo and rust ground. Some motifs

Fig. 247. Some popular motifs from Savu.

ledo

hebe

makaba

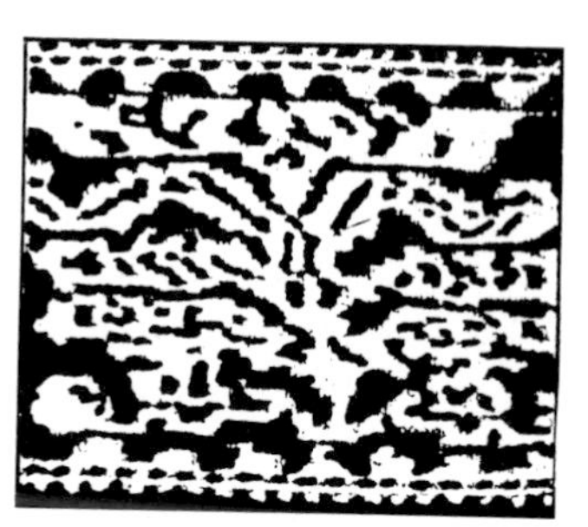

dula

tutu

coat-of-arms

European-inspired floral motifs

boda and *kekamahaba*

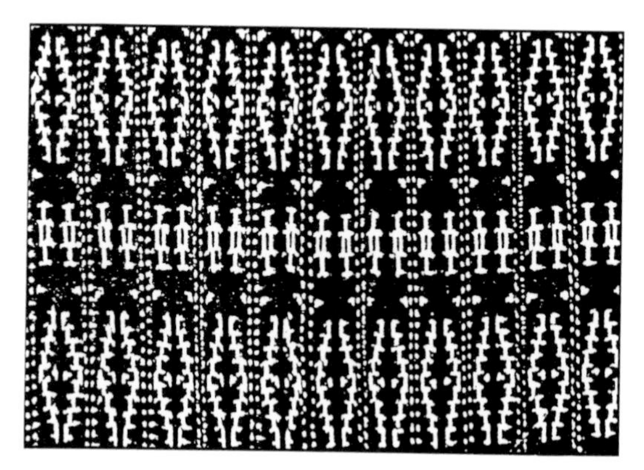

wohapi

on ceremonial cloth are considered to belong exclusively to certain groups of people. Over 300 years ago, the people of Savu were divided into two clans called the Greater Blossoms (*Hubi'Ae*) and Lesser Blossoms, or (*Hubi'Ike*), with about six subgroups called Seeds (*Wini*), which were based on female blood lines. Although the Savunese tend to live patrilocally in a male-dominated society, membership in the mother's clan traditionally determined what motifs could be woven on an individual's ceremonial clothes at important life-cycle activities, such as mortuary rites. Although not generally apparent to the outsider, clan motifs, in combination with special nuances of colour, are instantly recognizable to most Savunese.[42]

Designs for men's hip and shoulder cloths tend to be geometric and include traditional motifs like the diamond-shaped *wohapi*, a zig-zag *kekamahaba*, the *moto* (star), the *boda* (Savu flower), and the *dula* motif (Fig. 247). These motifs are evenly spaced in bands separated by narrow ribs of two-colour warp stripe decoration produced by a warp float technique. The woman's sarong is dominated by a fairly wide *ikat* band towards the top and bottom of the cloth which may have a rust ground. Next to this band is a plain undecorated area. The remainder of the cloth has numerous secondary bands containing smaller repetitive *ikat* designs. Traditional main motifs include the *hebe* and *makaba* floral medallions, possibly inspired by the *patola* cloth, the *tutu* (a stylized bird and flower design), and the *ledo* (snake motif).[43] Over the last century, Western-inspired motifs, such as vases of flowers, floral bouquets, roses, floral arabesques, grape vines, birds, and rampant lions, have been added to the design register and are very prevalent on women's apparel. Motifs on men's clothing have remained more traditional.

Roti and Ndao

Roti, a small crowded island 100 km south-east of Savu and close to Timor, is noted for its distinctive, well-crafted warp *ikat*. Textiles are simple rectangular lengths of cloth joined lengthwise to serve as hip cloths and shoulder cloths for the men, and as long tubular sarongs and *selendang* for the women (Fig. 248). The Rotinese at one time also made simple working clothes from the fibres of the lontar palm. They also make blankets, which serve as coffin covers and shrouds for the dead.

Rotinese textiles have been greatly influenced, both in layout and motif, by the *patola* cloths first brought to the island by the Dutch East India Company (Fig. 249). On the dissolution of the company in 1801, *patola* cloth became scarce and the nobility reserved *patola* motifs, such as the *jelamprang* (called black motif or *dula nggeo*) and the *dula penis* (a diagonal cross-shaped motif), for their own exclusive use. Commoners were permitted to pattern their cloths with simpler floral motifs. These floral patterns, called *dula bunga*, fill the centre field of the traditional Rotinese textile, which is divided into two identical parts, usually separated by a narrow undecorated band across the centre. A border pattern extends along the selvage edges. *Tumpul* triangles, followed by fringing, finish the warp ends of the cloth. The traditional Rotinese palette is fairly subdued and consists of natural-coloured *ikat* designs touched with red or pink and a little yellow against a blue-black indigo ground.[44] With the introduction of chemical dyes, the colours on newer cloths are much brighter and lack the subtle charm of the older examples.

Fig. 248. This cotton warp *ikat* textile made from two identical panels is patterned with *dula bunga* motifs. The border is finished with a row of triangular *tumpul* designs. Warp 146 cm, weft 82 cm.

Over recent years, there has been a gradual decline in weaving on the island of Roti. With the development of nearby Kupang as the capital of Nusa Tenggara, and the opening up of Timor to greater economic opportunities, many Rotinese have found a new market for their agricultural skills and business acumen. The population of Kupang, in fact, is one-third Rotinese.[45]

Ndao, a small island to the west of Roti, original-

Fig. 249. Some *patola*-inspired motifs which appear on textiles from Roti.

dua nggeo, black motif

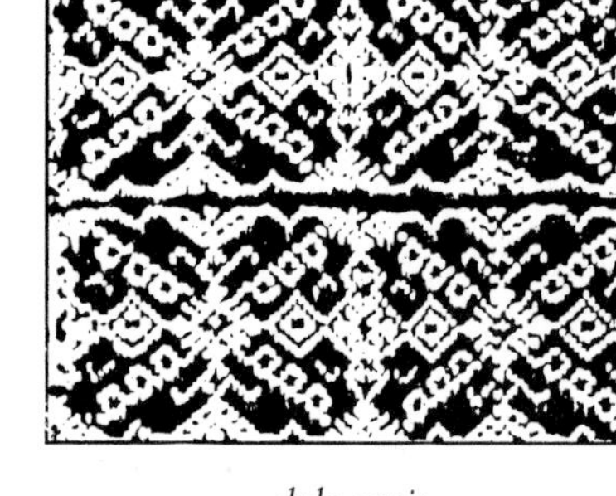

dula penis

The above four floral motifs are called *dula bunga*.

ly produced textiles which showed strong affinities with those of Savu. However, with the gradual decline of weaving on Roti, the women of Ndao are now producing textiles for sale with Rotinese designs using imported threads and chemical dyes. Their menfolk are the mendicant silversmiths of Nusa Tenggara. With Rotinese textiles draped over the shoulder and a sample of silver in hand, they are a familiar sight at the weekly markets held in the remote villages of Timor.[46]

Timor

Timor, a long, narrow, rugged, hilly island less than 500 km from Australia, produces some brightly coloured, boldly patterned textiles in the form of *selimut*, or blankets, approximately 200 cm long by 120 cm wide (Fig. 250). These are worn as a wrap-around sarong by men and as a long tubular sarong by women (Fig. 251). Sashes, head-cloths, belts, and bags are also woven.

A wide variety of patterning techniques, such as warp stripes, warp *ikat*, warp-faced alternating float weave, continuous and discontinuous supplementary weft, an embroidery supplementary weft technique called *sotis*, tapestry weave, and twining, are used by the peoples of Timor. The indigenous people of Timor are the Melanesian Atoni, who at one time were divided into ten princedoms. They produce some very distinctive textiles. Formerly, one glance

Fig. 250. An Atoni *selimut* with warp *ikat*-patterned humanoid figures set between parallel bands of red, yellow, and orange warp stripes. Warp 175 cm, weft 116 cm.

Fig. 251. Woman from Oelolok wearing a warp *ikat* sarong patterned in a traditional hook and romb design bounded by colourful warp stripes.

at a person's clothing would be enough to pinpoint the wearer's village and his station in life. With improved communications and the introduction of processed cotton yarns and chemical dyes, regional variations are less apparent today.[47] Other groups of people living on Timor, such as the Helonese, Belunese, and Amarasi, also do a little weaving.

Two basic variations of the *selimut* may be readily seen amongst the Atoni of Timor. One is composed of a wide central panel of striking warp *ikat* designs in white punctuated by small blue (or occasionally red) dots on an indigo ground. Composite rhomb and hook motifs tend to predominate. Humanoid shapes with out-turned and upraised extremities, which merge into animal and geometric forms, are also popular. Stylized birds, lizards, and snakes are featured as major motifs. This central panel is surrounded by parallel bands of narrow red, yellow, orange, and white stripes. There are often narrow bands of *ikat* on each side of the stripes. On some textiles, three narrow bands of *ikat* alternate with the stripes in lieu of a large central *ikat* band. The second type of blanket has a central panel and, occasionally, two lateral bands containing delicate renditions of stylized animal and human motifs worked in the *sotis* supplementary weft technique, a process akin to embroidery (Fig. 252). These are in direct contrast to the wide bands and colourful stripes which flank the main design area.[48]

Weavers in the Kefamananu–Miomafu–Oelolok areas can be seen from the road seated at their body-tension looms inside a *pondok*, a small, circular, open-sided pavilion with a rounded roof. As wives of tapioca and cattle farmers, they have less farm work to do and spend up to seven hours a day

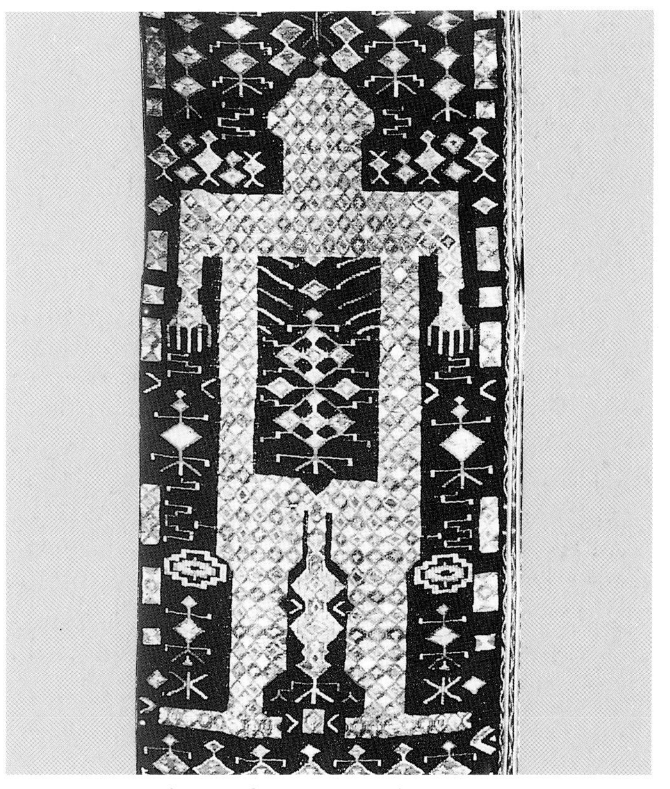

Fig. 252. A human figure patterned in many colours by the *sotis* supplementary weft technique. Arts of Asia Gallery, Denpasar.

weaving. In addition to warp *ikat*, the women use a warp-faced alternating float weave, *sotis* supplementary weft, and kelim tapestry weave in various combinations to produce some stunning textiles (Plate 44). Ceremonial sarongs for women consist of a wide central band of warp *ikat* in detailed hook and rhomb designs flanked by bands of dark-coloured indigo embellished with repetitive diamond patterns in the *sotis* technique. These patterns are either widely spaced, so that the ground cloth is clearly visible, or they are so densely patterned that they resemble needlepoint. It takes a woman working full-time about two months to weave such a sarong. Women in this part of Timor continue to weave most of their family's textile requirements. With the introduction of chemical dyes, colours are no longer traditional. Classical Timor blue and white *ikat* patterns are now rendered in yellow, orange, and green on a blue ground. Pink, magenta, green, black, and white stripes are now the preferred colours for border stripes. Motifs at this point in time continue to be largely traditional, but floral patterns are beginning to make their appearance.[49]

The Kapan area, in the hills south of Niki Niki, is noted for its textiles patterned in the warp-faced alternating float weave technique. Weavers may be seen with two parallel bands of different-coloured warp threads spread before them on their small body-tension looms. The design is picked out from the warp with a small pattern stick as the weaver proceeds to weave in the weft. Because the motifs are simple, repetitive geometric patterns well known to the weavers, the design is not usually set with numerous pattern sticks prior to weaving. The

Fig. 253. Some warp *ikat* designs from Timor.

A lizard design from Niki Niki.

Anthropomorphs and birds.

A modern warp *ikat* pattern.

Natam koroh bird motif from Amarasi.

Hook and rhomb pattern.

design appears in a reverse colour on the opposite side of the cloth. Oranges, reds, yellows, greens, and white seem to be the preferred modern colours. Recently, some weavers have begun introducing floral and peacock motifs into the design inventory.

Traditional warp *ikat* in the Niki Niki area of Central Timor is often patterned with stylized lizards, with arms outstretched, in a natural colour on a blue ground (Fig. 253). Blue and red spots may dot the body and surround the motif with small unobtrusive curvilinear forms. In the town of Niki Niki, there is the beginning of a small, home-based cottage industry for Timor textiles. The workshop of Feronika Bantamuan employs a number of workers to produce *selimut* and sarongs in both traditional and new patterns. New patterns on modern *ikat* include motifs such as peacocks, peonies, cherubs, and European dancing ladies.[50]

Amarasi, south of Kupang, formerly specialized in weaving *selimut* with complex hook and rhomb patterns, such as *natam koroh* (a bird), surrounded by a complicated hook and rhomb design in brownish red hues (see Fig. 253). Because of its close proximity to the boom town of Kupang, with its burgeoning polyglot population drawn from all over Indonesia, the weavers of Amarasi have recently turned to weaving floral and vegetal patterns favoured by the inhabitants of Kupang.

Despite ominous modern trends, handwoven textiles continue to be important in rural Timor where store-purchased clothing is still not readily available. In the more remote areas, textiles still play an important role in rites of passage and ceremonies associated with rice harvesting and house roofing. Timorese textiles, along with those from other islands, continue to be important items of trade with Irian Jaya, which does not have a textile weaving tradition.[51]

Flores

Flores, one of the most beautiful islands of Nusa Tenggara, is inhabited by a number of different people, many of whom weave interesting textiles. Warp *ikat* is the major means of textile expression. Supplementary weft patterning is also practised by some groups of people. Women wear long tubular sarongs in a variety of ways depending on local custom. Men seem to favour dark colours for their sarongs. Because of the difficulties of communication, regional differences in textiles are still quite evident. Now that the majority of the population are Roman Catholic, textiles have lost much of their former ritual importance.

In the markets of Ende, the capital of Flores, it is possible to see a wide variety of handwoven textiles from various parts of the island. The Ende district itself produces textiles. Due to its position as the leading port for the island, textiles reflect a variety of influences, ranging from *patola* to geometric, animal, and floral forms, Sumbanese compositional arrangements, and European designs. A number of weavers in Ende are Muslims and in keeping with the tenets of their faith, they have modified some of the former zoomorphic motifs of Flores. Today, floral and geometric motifs predominate on cloth from Ende.

Muslim influence is also evident in the textiles of West Flores. The Manggrai people, who were linked to the Islamic Kingdom of Bima on Sumbawa in the seventeenth century, weave a dark blue cotton sarong and decorate it with bands of bright, multicoloured, geometric and floral patterns in a supplementary weft (Plate 45). Some of the finer examples have *tumpul* patterns woven in a tapestry weave along the selvages. Some bright-coloured red, blue, and green plaid sarongs, which show signs of Bugis influence, are also seen. Women of the Manggrai area wear the sarong tied above the breast over a cotton blouse.[52]

The Ngada people, who are centred around the Bajawa–Baowai area in the more isolated areas of Central Flores, make traditional ceremonial women's sarongs and men's shawls from coarse home-spun cotton, which is patterned with naturally dyed blue or red-coloured bands containing very rudimentary *ikat* designs such as dots, dashes, arrows, zig-zags, and stick-like horses (Fig. 254). The most highly

Fig. 254. Warp *ikat*-patterned tubular sarong made from three panels. Simple bands of white *ikat* against a black ground are interspersed with red stripes. Central Flores. Warp 200 cm, weft 76 cm.

prized cloths are the *kain kudu*, which feature horse motifs in bands of warp *ikat*. Spidery threads of beadwork are sometimes added for further decoration on ceremonial cloth. These sarongs, worn knotted over one shoulder, are part of a young woman's dowry.[53]

East of Ende is the Lio district. For women in certain villages, such as Ngela and Joppu, weaving is virtually a full-time occupation. They are responsible for providing the other Lio groups with textiles, which are bartered or sold at small weekly markets. Lio textiles have been greatly influenced by *patola* cloths. Some sarongs and *selendang* in warp *ikat* are virtual copies of the original cloths produced in a more subdued dye range. Typical Lio *ikat* consists of finely patterned yellowish buff-coloured motifs on a blue-black to reddish-black ground, which on the best cloth resembles delicate tracery (Fig. 255). Motifs are primarily floral and the *jelamprang* design is quite prevalent. Trees, birds, and animals, such as snakes, lizards, and dogs, human stick figures, and traditional jewellery shapes are also seen. On some of the newer cloths, modern images, such as aeroplanes, motorcars, ships, and tea pots, are depicted. Motifs on Lio *ikat* are arranged in a large central field framed by parallel bands containing smaller repetitive elements. Some sarongs consist of horizontal bands of blue and white *ikat* of varying widths alternating with stripes of different weaving in a contrasting colour. *Ikat* motifs consist of a variety of zig-zag geometric patterns which are thought to represent a snake, which was an important cult figure in the former animistic religion of the Lio. A snake was at one time regarded as a harbinger of good news and may appear in a variety of stylized forms on ceremonial sarongs. Because of the high demand for Lio textiles, weavers are increasingly turning to synthetic dyes and rayon threads.[54]

Further east of Joppu and Ngela is the Sikka district, which has a tradition of weaving strong serviceable textiles from thick hand-spun cotton thread patterned with warp *ikat*. Some traditional Sikka textiles are similar to Savu cloth in terms of layout. Such cloth is patterned with two identical horizontal bands at the top and bottom of the fabric. These bands contain a major motif, such as the *jelamprang*, or a medallion-type form. Narrower bands, filled with smaller, less complex *ikat* patterns, lie on either side of the main band. The colours used are a little more reddish than those of the Lio and the *ikat* patterns are less delicate and bolder in conception. In addition to *patola*, geometric, and small vegetal patterns, realistic-looking deer, lions,

Fig. 255. A warp *ikat selendang* shawl made by the Lio people of Flores. *Patola*-inspired *jelamprang*-type motifs in beige against a reddish-brown ground. The warp ends are finished with *tumpul* patterns.

Fig. 256. A sarong length of warp *ikat* from Sikka patterned with a wide band of bird motifs on a blue ground, followed by narrower bands of floral designs separated by speckled bands of red, black, and white. European influences are very evident in this sarong. Warp 155 cm, weft 120 cm.

dragons, birds, and huge floral motifs may also appear in wide bands on cloth from Sikka (Fig. 256).

Another type of sarong produced in the Sikka region is composed of a series of blue and brownish-red bands containing geometric motifs, which have been tied in such a way as to give a pleasing speckled effect to the cloth (Fig. 257). The island of Palue off Sikka produces a similar type of warp *ikat* on a black indigo ground broken by bands of bright red warp stripes. The *ikat* patterns are a little less complex and more angular than those of Sikka. Touches of yellow and bits of beading may be added to this very arresting cloth.

At Maumere, it is possible to see a weaving demonstration at the Roman Catholic mission run by Pater Bollen. Here, one has the opportunity to see indigo plants and other substances which make traditional natural dyes. The women at the mission are very happy to show visitors how they gin the cotton with the *keho gung* and fluff up the fibres with the *wetir* (bow) before spinning them on the *jata* spinning wheel. Visitors may assist with preparing

Fig. 257. Some warp *ikat* patterns from East Flores.

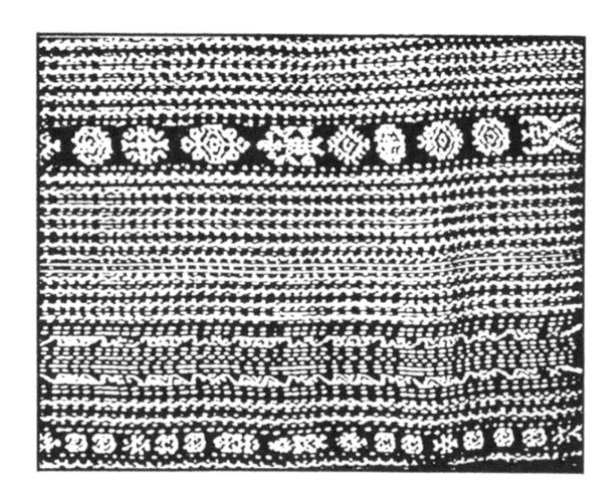

Some warp *ikat* patterns from Maumere in the Sikka area. Note the *patola* influences in some of the central bands.

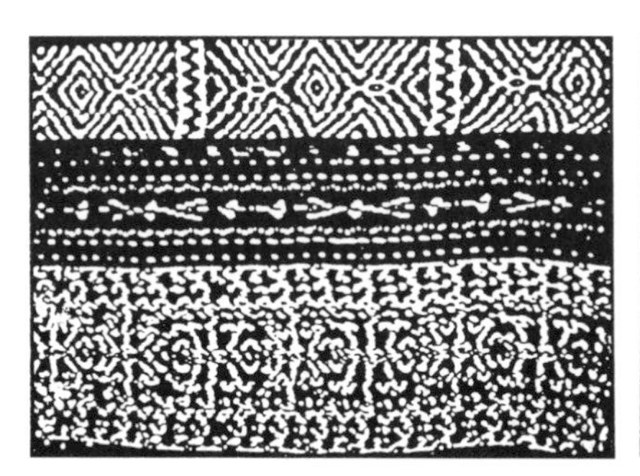

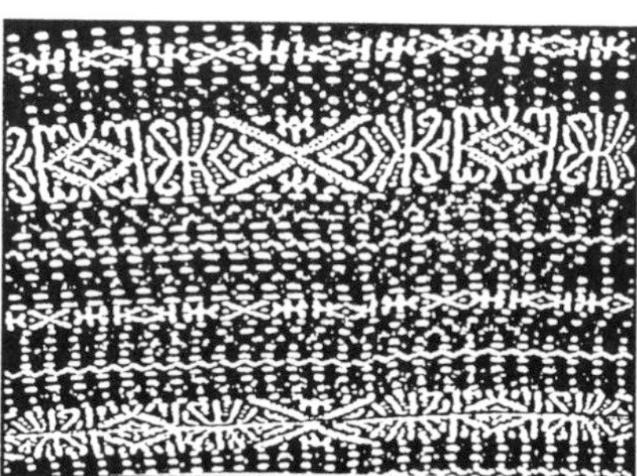

Two very early textile patterns which pre-date *patola* influences.

A modern Lio design.

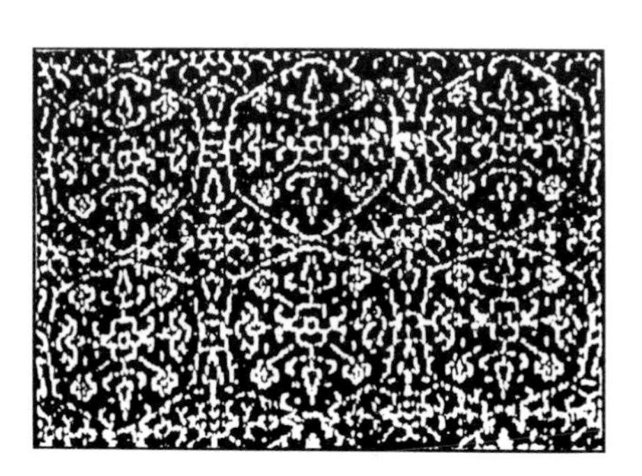

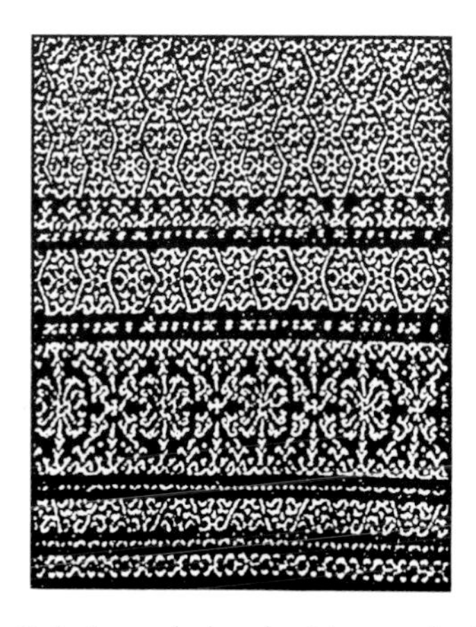

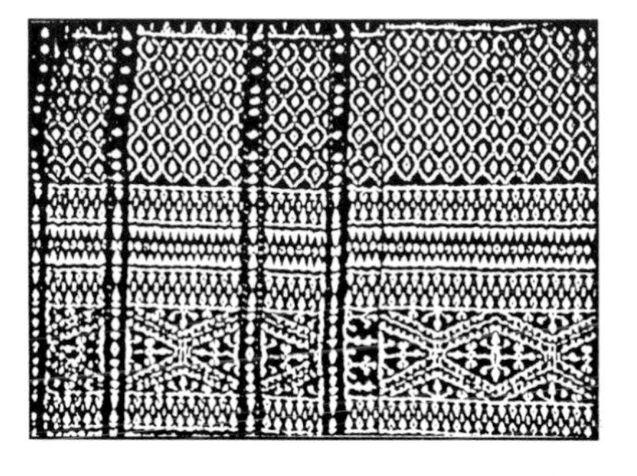

These three examples of cloth made by the Lio people show *patola* influences.

(These examples are from the Catholic Seminary Museum at Ledalero, Flores.)

dyes and learn how to tie in the warp designs using fibres from the coconut palm. They also see how the warp is set up and observe the women weaving their continuous warp *ikat* on body-tension looms (Fig. 258).

Not far from Maumere is a small museum of anthropology in the Catholic seminary at Ledalero. Here, one may view a representative collection of traditional textiles from all over Flores assembled as a labour of love by Father Piet Petu, the Curator, who is the foremost authority on the history and ethnography of Flores. The museum also has some outstanding examples of Indian *patola* cloth, and black and white *moko* cloth patterned in warp *ikat*, which was used to accompany the small megalithic hour-glass bronze drum, or *moko*, found on the island of Alor. The drum and textile were used in former fertility rites. There is also an excellent collection of photographs portraying the costumes and customs of the various peoples of Flores.

Fig. 258. Woman from Maumere, East Flores, wearing a long tubular warp *ikat* sarong patterned with European-inspired floral designs in white on a blue ground separated by narrow bands of subdued red and white speckled *ikat*.

Solor–Alor Archipelagos

East of Flores lie the Solor and Alor Archipelagos, which have weaving traditions greatly influenced by both Flores and Timor.

The women of the western half of Lomblen, or Lembata, in the Lama Lerup and Ille Api areas, produce distinctive warp *ikat* from heavy hand-spun cotton in a soft blue and brown dye range. The most important textile is an *adat* sarong patterned with simple geometric shapes, such as the eight-pointed star, diamond, *tumpul*, hearts, squares, and zig-zags, arranged in bands to make a most sophisticated and tasteful composition. Local phenomena, such as volcanic mountains, stick-like figures with upraised hands, and manta rays, also appear on *ikat* (Fig. 259). Two to three lengths of cloth are required to make the tubular sarong, which is rarely worn but serves as part of the marriage gift exchange. This textile traditionally became the property of the husband and might even be buried with him.[55]

Fig. 259. An *adat* sarong from Lomblen depicting volcanoes, stick-like figures, and manta rays. Arts of Asia Gallery, Denpasar.

The nearby island of Solor produces sarongs which are similar to those of Lomblen. There are four basic design arrangements, the use of which was formerly regarded as the exclusive property of a particular clan. Motifs are those of *patola* cloths and geometric designs which appear in white on a blue and reddish-brown ground. Motifs include interlocking heart shapes, 'S'-shaped hooks, and linear and crenellated zig-zag designs set between warp stripes.

Very fine work in a warp-faced alternating float

weave is done at Kolana, on the island of Alor (Plate 46). Motifs are in white and subdued colours on a black cotton ground. On the very finest work the ends of the cloth may be finished with a band of eight-sided stars or a similar motif woven in the kelim tapestry weave technique. The small island of Pantar, west of Alor, produces lengths of cloth which are patterned with bands of small geometric designs worked in a supplementary weft technique. Motifs show strong affinities with those of Timor. These are very popular as *selendang* in both Timor and Flores.

The Moluccas

The northern Moluccas have a weaving tradition which has its roots in Central Sulawesi, the original homeland of the forebears of the present-day Muslim population. *Songket*-patterned textiles, primarily on a red or checked ground, are made in Ternate and Tidor. Motifs are largely geometric, consisting of *tumpul* and lozenge shapes. A supplementary warp technique is also known and is used to pattern textiles with bands of geometric designs.[56]

The island of Ceram, in the central Moluccas, at one time produced some very striking textiles in a tapestry weave patterned with geometric designs in bright colours, reminiscent of Middle Eastern kelim. Supplementary weft, warp *ikat*, and twining are also known on Ceram.[57]

The islands of Kisar and Tanimbar, in the southern Moluccas, pattern loin cloths and sarongs with warp *ikat* on an indigo ground. Traditional sarongs in Tanimbar, which may be made from a combination of cotton and lontar fibres, feature elongated hook and lozenge motifs within narrow bands. The women also weave a distinguished loin cloth patterned with end stripes containing geometric designs in a supplementary weft. Appliquéd bands of brightly coloured cloth and small clusters of shells are added for further decorative effect.[58]

Sarongs from Kisar are similar to those of Tanimbar in that warp stripes predominate. Bands of blue and brown home-spun yarns are mixed with different-coloured narrow stripes formed from both commercially produced and home-processed threads.[59] Kisar sarongs are characterized by a wide band towards the selvage edges, which features a series of delightfully naïve pictorial motifs in the form of human figures with huge outspread hands, birds and riders, fish, and stars. Narrower stripes are filled with hook, lozenge, and lizard motifs (Fig. 260).[60]

In the southern Moluccas, textiles are important in gift exchanges and those of Tanimbar, in particular, were traded with neighbouring islands and with Timor.[61]

Fig. 260. Detail of a warp *ikat* pattern on a sarong from Kisar in the southern Moluccas which shows the 8 cm wide end border area depicting a human figure with upraised hands, a bird and rider, a star, and a fish. Warp 138 cm, weft 150 cm. Collection of Mr and Mrs Rodger Peren, Tokyo.

1. Charles Zerner, 'Silks from Southern Sulawesi', *Orientations*, February 1982, p. 48.

2. Ibid., pp. 46, 47 and 50; and interview with H. A. Muddarijah (Petta Ballasari) of Sengkang, 11 June 1986.

3. Suwati Dra Kartiwa, *Kain Songket Weaving in Indonesia*, Penerbit Djambatan, Jakarta, 1986, p. 63.

4. Petta Ballasari.

5. Eric Crystal, 'Mountain Ikats and Coastal Silks: Traditional Textiles in Sulawesi', in Joseph Fischer (ed.), *Threads of Tradition, Textiles of Indonesia and Sarawak*, exhibition catalogue, Lowie Museum of Anthropology and the University Art Museum, Berkeley, California, 1979, p. 54.

6. Kartiwa, op. cit., p. 63.

7. Anita Spertus and Jeff Holmgren, 'Celebes', in Mary Hunt Kahlenburg (ed.), *Textile Traditions of Indonesia*, exhibition catalogue, Los Angeles County Museum of Art, Los Angeles, 1977, p. 53.

8. Ibid., p. 54.

9. Mattiebelle Gittinger, *Splendid Symbols, Textiles and Tradition in Indonesia*, Textile Museum, Washington, DC, 1979, pp. 206–9.

10. Crystal, op. cit., p. 58. It is possible to see some of these textiles at a small museum in a Toradja house at Kete Kesu, 4 kilometres south of Rantepao.

11. Rita Bolland, 'Weaving the Pinatikan, a Warp Patterned Kain Bentenan From North Celebes', in Veronika Gervers (ed.), *Studies in Textile History*, Royal Ontario Museum, Toronto, 1971, p. 1.

12. Kartiwa, op. cit., p. 59; and Gittinger, *Splendid Symbols*, p. 211.

13. John Maxwell, 'Textiles of the Kapuas Basin—With Special Reference to Maloh Beadwork', in Mattiebelle Gittinger (ed.), *Indonesian Textiles: Irene Emery Roundtable on Museum Textiles, 1979 Proceedings*, Textile Museum, Washington, DC, 1979, pp. 127–8.

14. Kartiwa, op. cit., p. 52.

15. Bill Dalton, *Indonesia Handbook*, 2nd edn., Moon Publications, Hong Kong, 1980, pp. 343–4.

16. Mary Hunt Kahlenburg (ed.), *Textile Traditions of Indonesia*, exhibition catalogue, Los Angeles County Museum of Art, Los Angeles, p. 47; and Bronwen Solyom and Garrett Solyom, *Fabric Traditions of Indonesia*, exhibition catalogue, Museum of Art, Washington State University and Washington State University Press, Pullman, 1984, p. 2.

17. Maxwell, op. cit., p. 129.

18. Ibid., pp. 131–3.

19. See Rita Bolland and A. Polak, 'Manufacture and Use of Some Sacred Woven Fabrics in a North Lombok Community', *Tropical Man*, Vol. 4, 1971, pp. 149–79.

20. Kartiwa, op. cit., p. 77.

21. Anon., *Museum Negeri, Nusa Tenggara Barat* (in Bahasa Indonesia and English), Proyek Pengembangan Permuseuman, Nusa Tenggara Barat, Mataram, 1985–6, pp. 76 and 79.

22. Kartiwa, op. cit.

23. *Museum Negeri, Nusa Tenggara Barat*, op. cit., pp. 79–80.

24. Interview with H. Li. M. Ichsan Widasih, Manager of Kios Taufik Weaving Company, Sukarara, Lombok, 23 June 1986.

25. Visit and interviews with the managers of Rinjani and Slamet Riadi, Cakranegara, 23 June 1986.

26. Visit to Pringgasela, 24 June 1986.

27. *Museum Negeri, Nusa Tenggara Barat*, op. cit., pp. 80–1. Due to the independent histories of the major kingdoms of Sumbawa there are a few differences in textile terminology between East and West Sumbawa.

28. Kartiwa, op. cit., pp. 78–9.

29. Much of this information has been obtained by studying the excellent collection of Sumbawa textiles in the Museum Negeri Nusa Tenggara Barat, Mataram, Lombok.

30. Charles F. Ikle, 'The Ikat Technique and Dutch East Indian Ikats', *Bulletin of the Needle and Bobbin Club*, Vol. 15, Nos. 1 and 2, 1931, reprinted 1934, p. 16.

31. Wanda Warming and Michael Gaworski, *The World of Indonesian Textiles*, Kodansha, Tokyo, 1981, p. 79.

32. Ibid., pp. 82–3.

33. Marie Jeanne (Monni) Adams, 'Classic and Eccentric Elements in East Sumba Textiles: A Field Report', *Bulletin of the Needle and Bobbin Club*, Vol. 55, Nos. 1 and 2, 1972, p. 4.

34. Gittinger, *Splendid Symbols*, p. 159.

35. Marie Jeanne (Monni) Adams, *System and Meaning in East Sumba Textile Design: A Study in Traditional Indonesian Art*, Southeast Asia Studies Cultural Report Series, No. 16, Yale University Press, New Haven, 1969, pp. 129 ff.

36. Adams, 'Classic and Eccentric Elements in East Sumba Textiles', p. 5.

37. Jack Lenor Larsen *et al.*, *The Dyer's Art: Ikat, Batik, Plangi*, Van Nostrand Reinhold, New York, 1976, p. 150.

38. Steven G. Alpert, 'Sumba', in Mary Hunt Kahlenburg (ed.), *Textile Traditions of Indonesia*, p. 79.

39. Gittinger, *Splendid Symbols*, p. 161.

40. Alpert, op. cit., pp. 79–80.

41. Adams, 'Classic and Eccentric Elements in East Sumba Textiles', p. 16.

42. James J. Fox, 'Roti, Ndao and Savu', in Kahlenburg (ed.), *Textile Traditions of Indonesia*, p. 98.

43. Warming and Gaworski, op. cit., pp. 83–7.

44. Gittinger, *Splendid Symbols*, pp. 185–91.

45. Dalton, op. cit., p. 233.

46. Fox, op. cit., p. 100.

47. Gittinger, *Splendid Symbols*, p. 175.

48. Mary Hunt Kahlenburg, 'Timor', in Kahlenburg (ed.), *Textile Traditions of Indonesia*, p. 105.

49. Visit to weavers in Kefamananu, Miomafu, and Oelolok, 18 June 1986.

50. Visit to Kapan and to the workshop of Feronika Bantamuan, Niki Niki, 19 June 1986.

51. Kahlenburg, 'Timor', p. 105.

52. Gittinger, *Splendid Symbols*, p. 168; and Kent Watters, 'Flores', in Kahlenburg (ed.), *Textile Traditions of Indonesia*, p. 87.

53. Gittinger, *Splendid Symbols*, p. 169.

54. Warming and Gaworski, op. cit., pp. 89–90.

55. Gittinger, *Splendid Symbols*, pp. 172–3; and Watters, op. cit., p. 95.

56. Kartiwa, op. cit., pp. 89–90.

57. Gittinger, *Splendid Symbols*, p. 197.

58. Ibid., p. 193.

59. Kahlenburg, 'Timor' p. 108.

60. Solyom and Solyom, *Fabric Traditions of Indonesia*, pp. 21–3.

61. Gittinger, *Splendid Symbols*, p. 195.

CHAPTER TWELVE

Conclusion

Problems and Prospects

SOUTH-EAST ASIA possesses one of the world's most dazzling textile traditions. Both foreign influences and indigenous cultural diversity are captured in an overwhelming variety of textile forms, which use a wide range of materials and display a myriad of weaving techniques, colours, and design motifs. In addition to furnishing important information about the region's historical development, South-East Asian textiles play an important role in the social, economic, and religious life of the people. Through a study of textiles, it is possible to gain meaningful insights into the customs of the cultural groups who inhabit the region.

Today, many of these impressive textile traditions are under great pressure to adapt or face extinction. There are only a few groups of people on the outer islands and in the uplands of South-East Asia who still spend several months making a single piece of clothing or a ritual textile using the time-hallowed methods of yarn and dye preparation and weaving it on a hand loom. Progress is relentlessly creeping into even the most remote villages. With the development of modern communications, a bazaar stocking colourful yardages of floral and plaid prints, machine-processed yarns, and packets of cheap chemical dyes is only a horse or boat ride away. While excellent work was done by scholars during the colonial era, more needs to be done to document thoroughly the textile traditions of the region before they are irretrievably lost.

To keep pace with the rest of the world, South-East Asia has tried to industrialize rapidly and many crafts, including weaving, have declined as skilled people have been lost to new industries and occupations. Industrialization has greatly altered traditional ways of life by undermining basic religious and social structures. With changing beliefs and ways of life, some textiles lost much of their *raison d'etre* and became extinct. The ship cloths of South Sumatra ceased to be made around the turn of the century. The warp *ikat*-embroidered *tapis* of South Sumatra and the weft *ikat* of Pasemah in Sumatra are no longer made. The *kain bentenan* of Minahassa in North Sulawesi has not been made for over 100 years. The Shans of Burma no longer create their beautiful striped sarongs using weft *ikat*, tapestry weave, and supplementary weft techniques. The Lao Neua of Laos and Thailand no longer weave their intricately patterned *pha beang* shawls. The Isinai in the northern Philippines stopped weaving warp *ikat* blankets early this century, while the Gaddang of northern Luzon no longer weave their finely patterned jackets and loin cloths. Malaysia no longer makes its famous *kain cindai* weft *ikat*.

Some South-East Asian textiles now only exist in museums and, in some cases, better collections exist in Western museums than in the country of origin. Where they still exist, antique textiles are systematically being purchased by dealers to be sold to an eager clientele abroad.[1] As old textiles are bought up, new ones of inferior quality are being made. Imitations of well-known cloths, such as those from Sumba and other outer islands of Indonesia, are being made in weft *ikat* and by the screen-printing process for sale to tourists.

The screen-printing of traditional designs on machine-produced cloth is one of the greatest challenges to the indigenous South-East Asian weaving industry. By this process, a factory can turn out vast yardages of patterned material in one day that would take skilled weavers weeks, or even months, to produce by traditional methods. Although these new prints cannot compare in craftsmanship with the genuine article, the price is attractive to those of lesser means.

South-East Asian weaving traditionally has had two parallel traditions, that of the royal courts in the cultural centres, and that of the common people in the countryside. While royal patronage has largely disappeared, textiles and other crafts survive in many old court centres. The descendants of former royal weavers eke out a livelihood by making cloth for administrators and tourists. Jogjakarta and Solo continue to produce traditional batik. Klungkung in Bali, and Bima on Sumbawa, still make *kain songket.* Weavers near Kota Bharu and Trengganu in north-east Malaysia, Chiangmai, the old northern capital of Thailand, and Mandalay, the seat of the former Burmese monarchy, continue to make a number of handwoven textiles. Despite a drift to the towns in search of factory employment, a number of women in the rural areas continue to weave. Although in terms of time, effort, and skill the returns on finished textiles are low, weaving fits in well with domestic chores and the agricultural cycle.

Both colonial and early post-independence governments have been ambivalent towards the plight of indigenous hand weaving industries. Using Western models for economic advancement, handicrafts have not generally been accorded much of a role in national development. Fortunately, this is changing. In recent years, the Western world has become increasingly disillusioned with large-scale technology and mass production. The attendant pollution, resource depletion, congestion, and mind-numbing uniformity, have caused grave concern. Surrounded by a world of synthetics, people are searching out older values to enrich modern life. In many places, crafts are undergoing a minor renaissance and are being offered as an antidote to the relentless progress of industrialization.

Some of this is beginning to affect Asia. Many South-East Asian leaders, who formerly regarded crafts and ancient traditions as irrelevant to modern economic growth, are beginning to realize their importance in fostering national pride and social and economic self-reliance. Today, in many countries of South-East Asia, it is the government and prominent individuals in the private sector who are the new patrons concerned with the preservation of traditional crafts.

The governments of Thailand, Burma, and Indonesia have made efforts to increase the acreage under cotton to reduce their country's dependence on raw cotton imports. In countries where sericulture is practised, governments have been seeking Japanese assistance to improve the quality of the silk produced. Some governments also assist in procuring metallic threads, chemical dyes, and other necessary supplies to sell to weavers at reasonable prices.

Governments are also examining the problem of endemic rural poverty and are coming up with rural community development schemes which include irrigation, improved crops, and the making of handicrafts as a supplementary occupation. While these schemes are not new, they have not previously been attempted on such a large and co-ordinated scale. Throughout Thailand and Malaysia, the development of rural women through income-generating activities is an important goal. Because there is a reservoir of skill, weaving is the most important craft being promoted by these schemes. In these two countries, weaving has been revived and extended. In Thailand, various missionary groups have been active amongst the hill tribes in sponsoring weaving and needlework projects as sources of supplementary income. Their promotion and marketing activities have made great use of unpaid volunteers, so that most of the money earned from the sale is returned directly to the craftswomen. While the income received by rural weavers is still far from princely, the extra money has helped families make ends meet by tiding them over until the harvest is in. The money also helps with school fees and provides a few simple household comforts.

Government-sponsored weaving competitions, craft fairs, and visits to the capital by outstanding weavers have helped rekindle a sense of pride in weaving. Governments and financial institutions have also made credit available on easy terms to approved entrepreneurs and village leaders to establish weaving projects. Weaving schools, craft councils, and textile and craft museums have been established in some countries. Many of these suffer from inadequate funding and are unable to undertake large-scale projects; however, from time to time, they do mount interesting textile exhibitions and publish very informative catalogues. The University of Chiangmai in Thailand has begun offering courses on traditional Thai textiles. As part of their course work, students are sent to investigate weaving in communities known for their unique textiles.[2] Local artists and textile connoisseurs concerned with maintaining excellence and purity of style, have been advocating the establishment of institutes where weavers can study weaving and dyeing techniques to upgrade the quality of their work. At such centres, weavers would be able to see superb examples of their craft and have access to information about methods and techniques of weaving.[3]

In certain areas, handwoven textiles continue to supply a strong local demand. In Burma, for example, everyone wears national dress, and locally produced textiles are widely worn. Textile production in Burma is hampered by a shortage of supplies rather than by the lack of a market. With its vast population, Indonesia has a strong market for local textiles. Batik and warp and weft *ikat* sarongs are still widely worn by much of the rural population. Ceremonial and ritual textiles continue to be important, and the domestic Indonesian producer is well adapted to produce textiles for this custom-made market. Burial blankets are important in northern Luzon in the Philippines. The closure of Indo-China's borders to the outside world in the mid-1970s and the non-availability of imported supplies has caused many weavers in Laos, in particular, to return to the time-hallowed methods of their forebears to supply the population with cloth.

In South-East Asian countries, where Western dress and Western-type fabrics are widely worn, governments have been exhorting the population to use more indigenous fabrics through 'buy local' campaigns. Government servants and diplomats are encouraged to wear national dress on important state occasions and on national holidays. When they appear in public at home or abroad, First Ladies often wear either national dress or beautifully coutured clothes in sumptuous handwoven local fabrics. HM Queen Sirikit of Thailand has been particularly active in promoting Thailand's handwoven products abroad. The example set by such leaders is often followed by the local populace.

The place of royal weavers in South-East Asian society has largely been assumed by a number of modern designers who produce exclusive textiles by combining old designs with new techniques. One such person is Iwan Tirta, the well-known batik designer of Jakarta, who is noted for using a variety of fabrics in wider widths for fashion garments, furnishings, and wall hangings. Over recent years, he has turned his attention to weft *ikat*. His factory in Ende, Flores, is now producing pleasing handwoven fabrics inspired by classic Nusa Tenggara motifs.[4] A number of *kain songket* weavers in Kelantan and Trengganu in north-east Malaysia, such as Tengku Ismail bin Tengku Su, have revived long-forgotten classical patterns and have trained young weavers to execute them in flawless craftsmanship. A number of talented designers in the Philippines make use of local handwoven *piña* and *jusi* cloth to create high-fashion apparel. The late Jim Thompson was largely responsible for the post-war revival of Thailand's silk industry. By working with local weavers, and through the imaginative use of German and Swiss dyes, he began producing high-quality handwoven cloth in longer yardages and in more marketable colours than the traditional *pha sin*. Selling first to local tourists and then abroad, he built up a very profitable silk enterprise. He was soon followed by others, and, today, Thailand has a flourishing silk industry. Throughout Thailand, other entrepreneurs are attempting to do the same for north-eastern weft *ikat* and hill tribe weaving and needlework.

Government agencies are attempting to promote handwoven products as an export item. Traditional cloth is now being woven in longer yardages and different weaves for Western clothing, bags, hats, table and bed linen, wall hangings, lampshades, and upholstery materials. In addition to fine cotton and silk, other fabrics, such as heavyweight cottons, canvas, wool, and synthetics, are being woven and patterned with traditional designs. Well-known Western designers have come to South-East Asia to work with local weavers on creating new designs in different weaves. With the recent heightened interest in handwoven fabrics, some products are beginning to find a market abroad. Western museums have mounted exhibitions featuring South-East Asian weaving, particularly that of Indonesia. This has alerted Western textile enthusiasts to a hitherto unknown wealth of beautiful (and still affordable) textiles. Unfortunately, many of the textiles featured in such exhibitions are no longer woven today.

Over recent years, there has been a remarkable revival of the art of weft *ikat* throughout Indonesia, Thailand, and Burma. Through the use of semi-mechanized spinning and winding devices, it is possible to draw off large numbers of threads quickly and efficiently for both the warp and weft. The use of templates to copy designs, as well as *cetak* spot-dyeing techniques, have considerably reduced the time spent on the tying and dyeing process. Frame looms with hand-operated flying shuttles quickly weave the *ikat*-patterned yarns into lengths of cloth. While connoisseurs generally deplore the mass production aspects and the accompanying crass commercialism of this revival, these new weft *ikat* sarongs are extremely popular with locals and are sold at prices the average wage-earner can afford. Weft *ikat* now offers strong competition to machine-printed cloth. Many factories produce weft *ikat* of varying grades. For discerning patrons, fine weft *ikat* continues to be made at small ateliers, or as piecework in the homes of skilled craftswomen who have the time and inclination for such work.

At this point, it is impossible to predict the future of handwoven textiles in South-East Asia. While their continued existence is precarious in some areas, there still exists, throughout the region, an impressive textile tradition supported by a reservoir of skilled and talented weavers. The current problems facing South-East Asian weavers are not new. Historically, the wealthy and the élite have always tended to favour imported fabrics over local cloth. To keep pace with changing demands, local weavers have obligingly incorporated imported design elements into their cloth at the behest of patrons. Throughout the history of weaving, the South-East Asian weaver has been exposed, not only to new motifs, but to new techniques, different ways of utilizing materials, and to new ideas. Over the course of time, weavers have taken these new elements and skilfully combined them with indigenous elements to produce textiles which are distinctly South-East Asian.

Despite dire predictions, fine-quality handwoven textiles can survive in South-East Asia, not by trying to compete with machine-produced cloth for the mass market, nor by looking longingly back to an era where the weaver wove to meet the family's textile needs, but by building on the one area where handwoven products are clearly superior: aesthetic quality, which cannot be replicated by machine. For connoisseurs of fine textiles, mass-produced cloth may keep the body warm, but it leaves the heart cold. Handwoven textiles, to survive, will need to appeal to this discerning slice of the consuming public, whose criteria for purchasing cloth is based on quality and aesthetic appeal rather than cost. To be economically viable, prices will have to rise, but that need not be the death knell of hand weaving. Once established as a luxury item, there is no reason why handwoven textiles cannot survive as well as other items, such as hand-knotted rugs and carved furniture, which find a ready market amongst people who value the finer things of life.[5]

In conclusion, some of the finest traditional textiles of South-East Asia will ever remain a living testimony to the abilities of the fairer sex to make use of simple materials within the environment, and with the aid of simple implements, combined with patience and painstaking effort, transform them into objects of great beauty, imbued with all manner of social and ritual significance. While many modern textiles being produced in South-East Asia today, divorced from their cultural traditions, cannot compare in craftsmanship and design with former cloths, there are still many fine weavers at work producing both high-quality traditional and new cloths. These cloths continue to be woven with fine attention to detail and meticulous craftsmanship. It is to be hoped that there will always be an appreciative and increasing clientele for such textiles. National, community, and artistic leaders throughout South-East Asia have a special responsibility to continue to encourage, promote, and give substantial patronage to weavers to enable them to continue their time-hallowed craft with dignity. Failure to do this could mean that their work eventually may only exist as museum exhibits rather than as a continuing treasured contribution enriching the cultural and aesthetic life of the nation.

1. This is changing. Over recent years, a number of affluent South-East Asians have been collecting and are most anxious to preserve their cultural heritage.
2. Interviews with Acharn Vitti Parnchapak, Chiangmai University, 7 September 1985 and 27 July 1986.
3. Carol V. Doran, 'New Style From Old Traditions', *Living in Thailand*, July 1986, p. 86.
4. Interview with Iwan Tirta, Jakarta, 2 June 1986.
5. Personal communiqué with Robert Retka, Bangkok.

Bibliography

General

Benda, Harry J., and Larkin, John A., *The World of Southeast Asia: Selected Historical Readings*, Harper & Row, New York, 1967.

Bolland, Rita, 'Three Looms for Tablet Weaving', *Tropical Man*, No. 3, 1970, pp. 160–89.

Brown, Rachel, *The Weaving and Spinning Book*, Alfred A. Knopf, New York, 1983.

Buhler, Alfred, 'The Essentials of Handicrafts and the Craft of Weaving among Primitive Peoples', *Ciba Review*, No. 30, 1940, p. 1078.

——, 'Turkey Red Dyeing in South and Southeast Asia', *Ciba Review*, No. 39, 1941, pp. 1423–6.

——, 'The Ikat Technique'; 'Dyes and Dyeing Methods for Ikat Threads'; 'Origin and Extent of the Ikat Technique', *Ciba Review*, No. 44, 1942, pp. 1586–1612.

——, 'Plangi, The Tie and Dye Work', *Ciba Review*, No. 104, 1954, pp. 3722–50.

——, 'Patola Influences in Southeast Asia', *Journal of Indian Textile History*, Vol. 4, 1959, pp. 1–46.

Buhler, Alfred; Ramseyer, Urs; and Ramseyer-Gygi, Nicole, *Patola und Geringsing*, Museum fur Volkerkunde, Basel, 1975.

Burnham, Dorothy K., *A Textile Terminology—Warp and Weft*, Routledge Kegan Paul, London, 1981.

Bullough, Nigel, *Woven Treasures of Insular Southeast Asia*, Auckland Institute and Museum, Auckland, 1981.

Callenfels, P. V. van Stein, 'The Age of Bronze Kettle Drums', *Bulletin Raffles Museum*, Series B, Vol. 1, No. 3, 1937, pp. 150–3.

Coedès, George, *The Indianized States of Southeast Asia*, 3rd edn., East–West Center Press, Honolulu, 1971 (first published in French, 1964).

Crockett, Candace, 'Card Weaving', in Irene Emery and Patricia Fiske (eds.), *Looms and their Products: Irene Emery Roundtable on Museum Textiles, 1977 Proceedings*, Textile Museum, Washington, DC, 1979, pp. 27–9.

Dobby, E. H. G., *Southeast Asia*, 11th edn., University of London Press, London, 1973.

Emery, Irene, *The Primary Structure of Fabrics*, Textile Museum, Washington, DC, 1964.

Fisher, Charles A., *South East Asia*, Methuen, London, 1964.

Geijer, Agnes, *The History of Textile Art*, Sotheby, Parke Bernet, London, 1979.

Gittinger, Mattiebelle, 'An Introduction to the Body-Tension Looms and Simple Frame Looms of Southeast Asia', in Irene Emery and Patricia Fiske (eds.), *Looms and their Products: Irene Emery Roundtable on Museum Textiles, 1977 Proceedings*, Textile Museum, Washington, DC, 1979, pp. 54–68.

——, 'Master Dyers to the World: Early Indian Dyed Cotton Textiles', *Orientations*, Vol. 14, No. 2, 1983, pp. 12–24.

Hall, D. G. E., *A History of South-East Asia*, 2nd edn., Macmillan, New York, 1964.

Heine-Geldern, Robert, 'Some Tribal Art Styles of Southeast Asia: An Experiment in Art History', in Douglas Fraser (ed.), *The Many Faces of Primitive Art*, Prentice-Hall, Englewood Cliffs, 1966, pp. 165–221.

International Association of Costume, *Proceedings of the Fifth Asian Costume Congress*, Tokyo, 1986.

Jenyns, R. Soame, *Chinese Art III*, rev. edn., Rizzoli, New York, 1985.

Kunstadter, Peter (ed.), *Southeast Asian Tribes, Minorities and Nations*, 2 vols., Princeton University Press, New Jersey, 1967.

Labarthe, Jules, *Textiles: Origins to Usage*, Macmillan, New York, 1968.

Larsen, Jack Lenor, with Buhler, Alfred; Solyom, Bronwen; and Solyom, Garrett, *The Dyer's Art: Ikat, Batik, Plangi*, Van Nostrand Reinhold, New York, 1976.

LeBar, Frank M.; Hickey, Gerald C.; and Musgrave, John K., *Ethnic Groups of Mainland Southeast Asia*, Human Relations Area Files Press, New Haven, 1964.

——, *Ethnic Groups of Insular Southeast Asia*, Human Relations Area Files Press, New Haven, 1972.

Le May, Reginald, *The Culture of South-East Asia*, George Allen and Unwin Ltd., London, 1954.

Ling Roth, Henry, *Studies in Primitive Looms*, 1918, 3rd edn., Bankfield Museum, Halifax, 1950.

Newman, Thelma R., *Contemporary Southeast Asian Arts and Crafts*, Crown Publishers, Inc., New York, 1977.

Rawson, Philip, *The Art of Southeast Asia*, Thames and Hudson, London, 1967.

Seiler-Baldinger, Annemarie, *Systematik der Textilen Techniken*, Pharos-Verlag Hansrudolf Schwabe AG, Basel, 1973.

Sheares, Constance, 'Southeast Asian Ceremonial Textiles in the National Museum [of Singapore]', *Arts of Asia*, May–June 1987, pp. 100–7.

Steinmann, Alfred, 'The Ship of the Dead in Textile Art', *Ciba Review*, No. 52, 1946, pp. 1870–96.

———, 'The Art of Batik', *Ciba Review*, No. 58, 1947, pp. 2090–109.

Taber, Barbara, and Anderson, Marilyn, *Backstrap Weaving*, Watson Guptill, New York, 1975.

Tidball, Harriet, *The Weaver's Book*, Collier Books, New York, 1976.

Trotman, E. R., *Dyeing and Chemical Technology of Textile Fibres*, 4th edn., Charles Griffin, High Wycombe, England, 1970.

Vollmer, John E., 'Archaeological and Ethnological Considerations of the Foot-braced Body-Tension Loom', in Veronika Gervers (ed.), *Studies in Textile History*, Royal Ontario Museum, Toronto, 1977, pp. 343–54.

———, 'Archaeological Evidence for Looms from Yunnan', in Irene Emery and Patricia Fiske (eds.), *Looms and their Products: Irene Emery Roundtable on Museum Textiles, 1977 Proceedings*, Textile Museum, Washington, DC, 1979, pp. 78–89.

Wen Yu, *Selected Ancient Bronze Drums Found in China and Southeast Asia* (in Chinese), International Bookstore, Peking, 1957.

Wilwerth, Ardis, 'Basics of Weaving', unpublished notes for members of the Southeast Asian Textile Group of the National Museum Volunteers, Bangkok, 26 April 1985.

Wingate, Dr Isobel, *Fairchild's Dictionary of Textiles*, Fairchild Publications, Inc., New York, 1974.

Wong, Grace, 'Tributary Trade between China and Southeast Asia in the Sung Dynasty', in *Chinese Celadons and Other Related Wares in Southeast Asia*, compiled by the Southeast Asia Ceramic Society, Ars Orientalis, Singapore, 1979.

Burma

Amarapura Cooperatives, 'Things We Should Know About Weaving', pamphlet (in Burmese), Mandalay, 1985.

Aung Thaw, 'Neolithic Culture of the Padalin Caves', *Journal of the Burma Research Society*, Vol. 52, No. 1, 1969, pp. 9–17.

———, *Historical Sites in Burma*, Ministry of Union Culture, Government of the Union of Burma, Rangoon, 1972.

Aye Aye Myint *et al.*, 'The Structure and Designs of Zinme Silk of the Shan States of Burma' (in Burmese), unpublished BE thesis, Rangoon Institute of Technology, 1971.

Bailey, Jane Terry, 'Burmese Textiles', *Burma Art Newsletter*, Denison University, Granville, Ohio, Vol. 1, No. 4, 1969.

Bernot, Lucien, *Les Cak: Contribution a l'Etude Ethnographique d'une Population de Langue Loi*, Editions du Centre National de la Recherche Scientifique, Paris, 1967, pp. 61–9.

Carrapiett, W. J. S., *The Kachin Tribes of Burma: For the Information of Officers of the Burma Frontier*, Superintendent of Government Printing, Rangoon, 1929.

Collis, Maurice, *Land of the Great Image*, Alfred A. Knopf, New York, 1943.

Di Crocco, James V., 'Burmese Textiles: A Summary of a Lecture by Virginia M. Di Crocco, 2 December 1986', *Newsletter, National Museum Volunteers*, Museum Volunteers, Bangkok, March 1987, pp. 12–14.

Enriquez, Maj. C. M., *Beautiful Burma*, London, 1924.

Fraser-Lu, Sylvia, 'Kalagas, Burmese Wall Hangings and Related Embroideries', *Arts of Asia*, July–August 1982, pp. 73–82.

Fytche, Lt. Gen. Albert, *Burma, Past and Present* (2 vols.), Kegan Paul, London, 1878.

Gilhodes, Revd C., *The Kachins*, Catholic Orphan Press, Calcutta, 1922.

Hansen, Henny Harald, 'Some Costumes of Highland Burma at the Ethnographical Museum of Gothenburg', *Etnologiska Studier*, Vol. 24, Goteborg, 1960.

Harvey, G. H., *History of Burma*, Longmans Green, London, 1925.

Hla Tun Byu, U, *The History of Spinning and Weaving Technology (Burma)* (in Burmese), Sarpay Beikman, Rangoon, 1970.

Htin Aung, Maung, *Folk Elements of Burmese Buddhism*, U Hla Maung, Rangoon, 1959.

———, *A History of Burma*, Columbia University Press, New York, 1967.

Htwe Khin, Daw, 'Textbook for Students at Saunder's Weaving Institute', cyclostyled (in Burmese), Amarapura, 1965.

Innes, R. A., *Costumes of Upper Burma and the Shan States in the Collections of Bankfield Museum*, Halifax, 1957.

Izikowitz, K. G., 'Quelques Notes Sur le Costume des Puli-Akha', *Ethnos*, Vol. 8, Goteborg, 1943, pp. 133–52.

Khin Myo Chit, *Burmese Scenes and Sketches*, Nilar Publications, Rangoon, 1977.

———, *A Wonderland of Burmese Legends*, Tamarind Press, Bangkok, 1984.

Lehman, F. K., *The Structure of Chin Society*, Illinois Studies in Anthropology, No. 3, University of Illinois Press, Urbana, 1963.

Lowis, C. C., *A Note on the Palaungs of Hsipaw and Tawpeng*, Ethnographical Survey of India, Burma, No. 1, Superintendent of Government Printing, Rangoon, 1906.

Lowry, John, *Burmese Art*, Her Majesty's Stationery Office, London, 1974.

Luce, Gordon H., 'The Ancient Pyu', *Journal of the Burma Research Society, Fiftieth Anniversary Publications*, No. 2, 1960, pp. 307–22.

———, 'The Economic Life of the Early Burman', *Journal of the Burma Research Society, Fiftieth Anniversary Publications*, No. 2, 1960, pp. 323–75.

Mandalay Division of Cooperatives, 'Luntaya Acheik', pamphlet (in Burmese), Mandalay, 1985.

Marshall, Harry Ignatius, *The Karen People of Burma: A*

Study in Anthropology and Ethnology, Ohio State University Bulletin, Vol. 26, No. 13, 1922, reprinted AMS Press, New York, 1980.

Mi Mi Khaing, *Burmese Family*, Indiana University Press, Bloomington, 1962.

Milne, Leslie, *Shans at Home*, John Murray, London, 1910, reprinted Paragon Book Reprint Corp., New York, 1970.

Moore, W. Robert, 'Strange Tribes in the Shan States of Burma', *National Geographic Magazine*, Vol. 58, 1930.

Myo Min, U, *Old Burma as Described by Early Foreign Travellers*, Hanthawaddy Press, Rangoon, 1947.

Ono Toru, and Inoue Takao, *Pagan Mural Paintings of the Buddhist Temples of Burma*, Kodansha, Tokyo, 1979.

Pe Kywe, Maung, 'Woven with 100 Shuttles', *Forward*, Vol. 3, No. 6, 1964, pp. 11–16.

San Win, 'Robes for the Buddha, Burma: Golden Country', *Horizons Magazine*, n.d. pp. 14–16.

Scott, Sir George, *Burma: A Handbook of Practical Information*, Moring, London, 1906.

Searle, H. F., *Burma Gazetter: The Mandalay District*, Vol. A, Superintendent of Government Printing, Rangoon, 1928.

Shakespear, Lt. Col. J., *The Lushei Kuki Clans*, Macmillan, London, 1912.

Shway Yoe (Sir George Scott), *The Burman, His Life and Notions*, Macmillan, London, 1896, reprinted Norton Simon, New York, 1963.

Smart, R. B., *Burma Gazetter: Akyab District*, Superintendent of Government Printing, Rangoon, 1917.

Spearman, H. R., *British Burma Gazetter*, Vol. 1, Government Press, Rangoon, 1880.

Start, Laura E., *Burmese Textiles from the Shan and Kachin Districts*, Bankfield Museum Notes, 2nd Series, No. 7, Halifax Museum, Halifax, 1917.

Stevenson, H. N. C., *The Hill Peoples of Burma*, Burma Pamphlets, No. 6, Longmans Green, London, 1944.

Superintendent of Cottage Industries, *Report of the Superintendent of Cottage Industries*, Superintendent of Government Printing, Rangoon, 1930 and 1936.

Symes, Michael, *An Account of an Embassy to the Kingdom of Ava, Sent by the Governor-General of India, 1795*, Nicol and Wright, London, 1800.

Taung Pauline, 'The Kachin Loom', *Forward*, Vol. 17, No. 6, 1979, pp. 20–2.

Theikpa, Maung, 'Textiles from Inle Lake', *Forward*, Vol. 6, No. 24, 1968, pp. 16–20.

Trager, Helen (ed.), *We the Burmese, Voices from Burma*, Praeger, New York, 1969.

Tydd, W. B., *Burma Gazetter: Sandoway District*, Superintendent of Government Printing, Rangoon, 1912, reprinted 1962.

Willis, Elizabeth Bayley, 'The Textile Arts of India's North-East Borderlands', *Arts of Asia*, January–February 1987, pp. 93–115.

Yule, Capt. Henry, *A Narrative of the Mission Sent by the Governor General of India to the Court of Ava in 1855*, Smith Elder & Co., London, 1858, reprinted Oxford University Press, Kuala Lumpur, 1968.

Thailand

Anon., 'Silk Weaving Centre Set Up For Villagers', *Bangkok Post*, 8 March 1985.

Aranyanak, Chiraporn, 'Ancient Fragments from Ban Chiang', *Muang Boran Journal*, Vol. 2, No. 1, 1985, pp. 83–4.

Archambault, Michelle; Dupaigne, Bernard; Drosson, Monique; and Pornchai, Suchitta, *Tissus Royaux Tissus Villageois de Thailande*, Musée de l'Impression sur Étoffes, Mulhouse, France, 1988.

Binks, Anne, 'An Unusual Pattern-Loom from Bangkok', *Bulletin of the Needle and Bobbin Club*, Vol. 44, Nos. 1–2, 1960, pp. 15–21.

Brown, Roxanna, 'Collecting Surin Silks', *Living in Thailand*, August 1985, pp. 60–5.

Browne, Stephen, 'The Origins of Thai Silk', *Arts of Asia*, September–October 1979, pp. 91–100.

Butler-Diaz, Jacqueline, *Yao Design of Northern Thailand*, rev. edn., Siam Society, Bangkok, 1981.

Campbell, Margaret, *From the Hands of the Hills*, Media Transasia, Bangkok, 1978, 2nd edn. 1981.

Charles, Nancy, 'Textiles of Laos and Thailand', unpublished notes from a lecture given to the Southeast Asian Textiles Group of the National Museum Volunteers, Bangkok, May 1985.

Charoenwongsa, Pisit, and Diskul, M. C. Subhadradis, *Archeologia Mvndi: Thailand*, Nagel, Switzerland, 1978.

Chongkol, Chira, 'Textiles and Costume in Thailand', *Arts of Asia*, November–December 1982, pp. 121–31.

Di Crocco, Virginia M., 'Highlights from Ikat Isarn: A Lecture given by Khun Nisa Sheanakul', *Newsletter, National Museum Volunteers*, Museum Volunteers, Bangkok, August 1983, pp. 20–1.

Diskul, M. C. Subhadradis, *Art in Thailand, A Brief History*, Amarin Press, Bangkok, 1969.

Doran, Carol V., 'New Style from Old Traditions', *Living in Thailand*, July 1986, pp. 62–3, 86.

Ellis, Mary Elana, 'Life Cycle Ceremonies: Courtship and Marriage', in *Sawaddi Special Edition. A Cultural Guide to Thailand*, American Women's Club of Thailand, Bangkok, 1978?, pp. 74–8.

French, Cherie, 'Thai Fashions Then and Now', in *Sawaddi Special Edition. A Cultural Guide to Thailand*, American Women's Club of Thailand, Bangkok, 1978?, pp. 127–31.

Greenwalt, Betty Lou, 'Thai Silk', *Sawaddi*, American Women's Club of Thailand, Bangkok, September–October 1985, pp. 43–5.

Griffin, Robert S., 'Thailand's Ban Chiang: The Birth Place of a Civilisation?', *Arts of Asia*, November–December 1973, pp. 32–4.

H & M Thai Silk, *Mud Mee* (in Thai), Bangkok, n.d.

Henrikson, Merete Aagaard, 'A Preliminary Note on Some Northern Thai Woven Patterns', in S. Egerod and P. Sorensen (eds.), *Lampang Reports*, Scandanavian Institute of Asian Studies, Special Publication No. 5, Bangkok, 1978, pp. 137–53.

Hinton, E. M., 'The Dress of the Pwo Karen of North Thailand', *Journal of the Siam Society*, Vol. 62, 1974, pp. 27–34.

Hoagland, Loretta, 'Chitralada Shops—Handicrafts by Thai Artisans', *Look East*, Oriental Plaza Magazine, Bangkok, September 1985, pp. 13–17.

Hutchinson, E. W., 'The Lawa in Northern Siam', *Journal of the Siam Society*, Vol. 27, Part 2, 1935, pp. 153–82.

Hvitfeldt, Christine, 'Yao History in Stichery', *Sawaddi*, American Women's Club of Thailand, Bangkok, January–February 1985, pp. 14–17.

Hyatt, Cynthia, 'A Dyeing Art, or an Introduction to Natural Dyes', unpublished typescript, Thailand, 1985.

Klausner, William J., *Reflections in a Log Pond*, 2nd edn., Suksit Siam, Bangkok, 1974.

Krug, Sonia, and Duboff, Shirley, *The Kamthieng House, Its History and Collections*, Siam Society, Bangkok 1982.

Larsen, Jack Lenor, 'Evolution of Thai Silk: Homage to Jim Thompson', *Arts of Asia*, May–June 1978, pp. 75–8.

Leesuan Viboon (ed.), *Thai Textiles: Industrial and Social Development*, Thai Art and Industry Project No. 2 (in Thai with English abstract), Industrial Finance Corporation of Thailand, Amarin Printing Group Co. Ltd., Bangkok.

Lefferts, Leedom H., 'A Collection of Northeast Thai Textiles: Content and Descriptions', typescript, Department of Anthropology, Smithsonian Institution, Washington, DC, 1980.

Lemoine, Jacques, *Yao Ceremonial Painting*, White Lotus, Bangkok, 1982.

Lewis, Paul, and Lewis, Elaine, *Peoples of the Golden Triangle*, Thames and Hudson, London, 1984.

MacAnnallen, Marian, *Lao Song Handicrafts*, catalogue, Church of Christ in Thailand, Bangkok, n.d.

McCauley, Susan 'Thai Mudmee', *Sawaddi*, American Women's Club of Thailand, Bangkok, November–December 1982, pp. 19–25.

Panyacheewin, Saowarop, 'Now Soldiers' Wives Benefit from SUPPORT', *Bangkok Post*, 25 August 1985.

Peetathawatchai, Vilmophan, *Esarn Cloth Design*, Faculty of Education, Khon Kaen University, Khon Kaen, Thailand, 1973.

——, *Folkcrafts of the South*, Housewives' Voluntary Foundation Committee, Bangkok, 1976.

Rainart, Joyce, 'Village That Came Back To Life', *Bangkok Post*, 22 July 1985.

Rajadhon, Phya Anuman, *Life and Ritual in Old Siam: Three Studies of Thai Life and Customs*, translated by William J. Gedney, Human Relations Area Files Press, New Haven, 1961.

——, *Essays on Thai Folklore*, Duang Kamol, Bangkok, 1968.

Rosenfield, Clare S., and Mabry, Mary Connelly, 'Discovering the Art of Teenjok', *Sawaddi*, American Women's Club of Thailand, Bangkok, September–October 1982, pp. 23–6.

Sananikone, Thao Peng, 'Further Notes on Dyes of Thailand', *Brooklyn Botanic Garden*, 1964, p. 46.

Sanasen, Uab, 'Notes on a Weaving Village', *Muang Boran Journal*, Vol. 16, No. 1, 1979, pp. 13–14.

Segaller, Denis, *Thai Ways*, Allied Newspapers Ltd., Bangkok, 1979.

——, *Traditional Thailand: Glimpses of a Nation's Culture*, Hong Kong Publishing Co., Hong Kong, 1982.

Seidenfaden, Eric, *The Thai Peoples*, Siam Society, Bangkok, 1958.

Shiratori, Yoshiro, *Ethnographic Survey of the Hill Tribes of Northern Thailand with Special Reference to the Yao*, Kodansha, Tokyo, 1978.

Sisters of the Good Shepherd, *Village Weaver Handicrafts*, catalogue, C. K. Press, Bangkok, 1985?

Suvatabandhu Kasin, 'Buddhist Rules Prescribe Dyes for Monks' Robes', *Brooklyn Botanic Garden*, 1964, pp. 45–6.

Takuji Takemura, 'Funeral Rites', in Yoshiro Shiratori, *Ethnographic Survey of the Hill Tribes of Northern Thailand with Special Reference to the Yao*, Kodansha, Japan, 1978.

Tambiah, Stanley J., *Buddhism and the Spirit Cults in North-East Thailand*, Cambridge University Press, Cambridge, 1970.

Tank, Marelyn, 'Life Cycle Ceremonies, Minus Nine to Plus Twelve', in *Sawaddi Special Edition. A Cultural Guide to Thailand*, American Women's Club of Thailand, Bangkok, 1978?, pp. 85–9.

Turner, Violet, 'The End of the Cycle', in *Sawaddi Special Edition. A Cultural Guide to Thailand*, American Women's Club of Thailand, Bangkok, 1978?, pp. 79–84.

Umemoto, Diane L., 'Prehistoric Archaeology: Impact of Ban Chiang', *Artistic Heritage of Thailand*, *Sawaddi* Magazine and National Museum Volunteers, Bangkok, 1979, pp. 49–56.

Vibhanand, Rangsit, 'Thai Silk, the Magic Material', *Bangkok Post*, 4 October 1981.

Walker, Anthony R., *Farmers in the Hills: Upland Peoples of Northern Thailand*, Chinese Association for Folklore, Taipei, 1981.

Wells, Kenneth E., *Thai Buddhism, Its Rites and Activities*, 1939, 3rd edn., Suriyabun Publishers, Church of Christ in Thailand, Bangkok, 1975.

Wilwerth, Ardis, 'Thai Textiles', *Newsletter, National Museum Volunteers*, Museum Volunteers, Bangkok, April 1985, pp. 18–21.

Wright, Michael, 'Where Was Sri Vijaya? Another Approach', *Siam Society Newsletter*, Vol. 1, No. 1, 1985, pp. 4–12.

Young, Gordon, *The Hill Tribes of Northern Thailand*, 5th edn., Siam Society, Bangkok, 1974.

Whitfield, Danny J., *Historical and Cultural Dictionary of Vietnam*, Scarecrow Press Inc., New Jersey, 1976.

Indo-China: Laos, Kampuchea, and Vietnam

Berval, Rene de (ed.), *France Asie*, Special Edition, No. 12, 1955, pp. 138–47.

Boulbert, Jean, 'Modes et Techniques du Pays Maa', *Bulletin de l'Ecole Francais d'Extreme Orient*, Vol. 52, No. 2, 1965, pp. 359–413.

Cheesman, Patricia, 'Antique Weavings from Northern' Laos', *Living in Thailand*, August 1981, pp. 35–9.

——, 'The Antique Weavings of the Lao Neua', *Arts of Asia*, July–August 1982, pp. 120–5.

——, 'Laos: Indigo Dyed Fabrics', *Craft Australia Supplement*, Autumn/1 1984, pp. 87–91.

——, *Lan Na Textiles: Yuan Lue Lao*, Center for the Promotion of Arts and Culture, Chiang Mai University, 1987.

——, *Lao Textiles: Ancient Symbols—Living Art*, White Lotus Co., Bangkok, 1988.

Cuisinier, Jeanne, *Les Muong, Geographie Humaine et Sociologie*, Institut d'Ethnologie, Paris, 1948.

Dournes, Jacques, 'Le Vetement Chez Les Jorai', *Objets et Mondes*, La Revue du Musee de l'Homme, Tome III, Fasc., 2, Ete, 1963, pp. 99–113.

Fitzsimmons, Thomas (ed.), *Cambodia*, Country Survey Series, Human Relations Area Files Press, New Haven, 1957.

Galotti, Jean, 'Les Samphots du Cambodge', *Art et Decoration*, Vol. 50, 1926.

Gough, Kathleen, *Ten Times More Beautiful: The Rebuilding of Vietnam*, Monthly Review Press, New York, 1978.

Hickey, Gerald C., *Sons of the Mountains, Ethnohistory of the Vietnamese Central Highlands to 1954*, Yale University Press, New Haven, 1982.

——, *Free in the Forest, Ethnohistory of the Central Highlands, 1954–1976*, Yale University Press, New Haven, 1982.

Huard, R., and Maurice A., 'Les Mnong du Plateau Central Indochinois', *Institut Indochinois pour l'Etude de l'Homme, Bulletin et Travaux*, Hanoi, Vol. 2, 1939, pp. 27–148.

Mizzi, Donna, 'In Rescue of Historic Heirlooms', *Nation Review*, Bangkok, 29 May 1985.

Mole, Robert, *The Montagnards of South Vietnam*, Tuttle, Tokyo, 1970.

Moore, Elizabeth H., 'Meaning in Khmer Ritual', *Arts of Asia*, May–June 1981, pp. 98–109.

Seitz, Paul L., *Men of Dignity, Montagnards of South Vietnam*, Bar-le Duc, France, 1975.

Stoeckel, Jean, 'Etude sur le Tissage au Camboge', *Art et Archaeologie Khmers*, Vol. 1, No. 4, 1921–3, pp. 387–402.

Ratnam, Perala, *Laos and Its Culture*, Tulsi Publishing House, New Delhi, 1982.

Malaysia and Brunei

Alman, John H., 'Bajau Weaving', *Sarawak Museum Journal*, Vol. 9, Nos. 15–16, 1960, pp. 603–18.

——, 'Dusun Weaving', *Sabah Society Journal*, No. 2, 1962.

Alman, John, and Alman, Elizabeth, *Handcrafts in Sabah*, Borneo Literature Bureau, Kuching, 1968, reprinted 1973.

Brunei Arts and Handicrafts Training Centre, 'Pusat Latihan Kesenian dan Pertukangan Tangan Brunei', pamphlet (in Bahasa Malaysia), n.d.

Chin, Lucas, *Cultural Heritage of Sarawak*, Sarawak Museum, Kuching, 1980.

Cole, Fay-Cooper, *The Peoples of Malaysia*, Van Nostrand Reinhold, New York, 1945.

Evans, I. H. N., *Religion, Folklore and Customs in North Borneo and Malaya*, Cambridge University Press, London, 1923.

Freeman, Derek, *Report on the Iban*, Athlone Press, New York, 1970.

Gullick, J. M., 'Survey of Malay Weavers and Silversmiths in Kelantan in 1951', *Journal of the Malayan Branch of the Royal Asiatic Society*, Vol. 25, Pt. 1, 1952, pp. 134–48.

Haddon, Alfred, and Start, Laura E., *Iban or Sea Dayak Fabrics and Their Patterns*, Cambridge University Press, London, 1936, reprinted Ruth Bean, Bedford, 1982.

Harrison, T. H., *The Peoples of Sarawak*, Kuching, 1959.

Hill, A. H., 'Weaving Industry in Trengganu', *Journal of the Malayan Branch of the Royal Asiatic Society*, Vol. 22, No. 3, 1949, pp. 75–84.

Hose, Charles, and McDougal, William, *The Pagan Tribes of Borneo*, Macmillan, London, 1913.

Howell, W., 'Sea Dayak Method of Making and Dyeing Thread from Their Homegrown Cotton', *Sarawak Museum Journal*, Vol. 1, No. 2, 1912, pp. 61–4.

Ling Roth, Henry, *The Natives of Sarawak and British North Borneo* (2 vols.), Truslove and Hanson, London, 1896.

Malaysian Handicraft Development Corporation, *Serian Songkit* (in Bahasa Malaysia), Perbadanan Kemajuan Kraftangan, Malaysia, Kuala Lumpur, n.d.

——, 'Seni Telepuk' (Gilded Cloth), pamphlet (in Bahasa Malaysia), Perbadanan Kemajuan Kraftangan, Malaysia, Kuala Lumpur, n.d.

——, 'The Handcrafted Textiles of Malaysia', pamphlet, Perbadanan Kemajuan Kraftangan, Malaysia, Kuala Lumpur, n.d.

——, 'Infokraf Malaysia', pamphlet, Perbadanan Kemajuan Kraftangan, Malaysia, Kuala Lumpur, n.d.

——, *National Dress of Peninsula Malaya* (in English and Bahasa Malaysia), Perbadanan Kemajuan Kraftangan, Malaysia, Kuala Lumpur, n.d.

———, 'Objective, Policy, Programme and Activities', typescript kindly given to the author by Raja Fuziah bte Raja Tun Uda, 1986.
Ministry of Environment and Tourism, *Pua Kumbu*, AGAS, Kuching, n.d.
Norwani Nawawi, 'Malaysian Songket', unpublished MA thesis, Textiles/Fashion, School of Art and Design, Manchester Polytechnic, England, 1985.
Ong, Edric, 'Pua, Iban Weavings of Sarawak', Society Atelier, Kuching, Sarawak, 1986.
Palmieri, Michael, and Ferentinos, Fatima, 'The Iban Textiles of Sarawak', in Joseph Fischer (ed.), *Threads of Tradition, Textiles of Indonesia and Sarawak*, exhibition catalogue, Lowie Museum of Anthropology and the University Art Museum, Berkeley, California, 1978, pp. 73–80.
Peacock, B. A. V., *Batek, Ikat, Pelangi and Other Traditional Textiles from Malaya*, exhibition catalogue, Urban Council, Hong Kong, 1977.
———, *Malaysian Traditional Crafts*, exhibition catalogue, Urban Council, Hong Kong, 1981.
Ryan, N. J., *The Cultural Heritage of Malaya*, Longmans, Kuala Lumpur, 1962, reprinted, 1971.
Sabah Tourist Association, *A Guide to Sabah*, Vol. 4, Kota Kinabalu, 1986–7.
Sandin, Benedict, *Iban Way of Life*, Borneo Literature Bureau, Kuching, 1976.
Sarawak Museum Staff, *Sarawak in the Museum*, Sarawak Museum, Kuching, 1983.
Sim, Katharine, *Costumes of Malaya*, Eastern Universities Press Ltd., Singapore, 1963.
Shell Oil, Brunei, *Preserving a Proud Heritage* (in English and Bahasa Malaysia), Brunei, n.d.
Sheppard, Mubin, *Living Crafts of Malaysia*, Times Books, Singapore, 1978.
Skeat, W. W., 'Silk and Cotton Dyeing by Malays', *Journal of the Straits Branch of the Royal Asiatic Society*, No. 38, 1902, pp. 123–7.
Tweedie, M. W. F., 'Prehistory in Malaya', *Journal of the Royal Asiatic Society*, Vol. 26, No. 2, 1942, pp. 9–63.
Wheatley, Paul, *The Golden Chersonese*, University of Malaya Press, Kuala Lumpur, 1961.
Winstedt, Richard O., *Arts and Crafts. Papers on Malay Subjects*, Part 1, FMS Government Press, Kuala Lumpur, 1925.
———, *The Malays, A Cultural History*, Routledge Kegan Paul, 1947, revised by Than Seong Chee, Graham Brash, Singapore, 1981.
Wray, L., 'Notes on the Dyeing and Weaving as Practised at Sitiawan in Perak', *Journal Anthropological Institute*, Vol. 32, 1902, pp. 153–5.
Zainie, Carla, *Handcrafts of Sarawak*, Borneo Literature Bureau, Kuching, 1969.

The Philippines

Baradas, David B., 'Art in Maranao Life', *Filipino Heritage*, No. 4, 1977, pp. 129–68.
Casal, Gabriel, *T'boli Art*, Ayala Museum, Makati, Manila, 1978.
Casal, Gabriel *et al.*, *People and Art of the Philippines*, Museum of Cultural History, University of California, Los Angeles, 1981.
Casino, Eric, 'The Art of the Muslim Filipinos', in *Aspects of Philippine Culture*, Monograph No. 1, National Museum and United States Information Service, Manila, 1967, pp. 3–21.
———, *Ethnographic Art of the Philippines: An Anthropological Approach*, The Design Center, Philippines College of Arts and Trades, Manila, 1973.
Cole, Fay-Cooper, *The Tinguian, Social, Religious and Economic Life of a Philippine Tribe*, Publication 209, Anthropological Series, Field Museum of Natural History, Chicago, 1922.
———, 'The Bukidnon of Mindanao', *Fieldiana: Anthropology*, Vol. 46, Field Museum of Natural History, Chicago, 1956.
Dacanay Jr., Julian E., 'Muslim Art', *Philippines Quarterly*, September 1973, pp. 40–9.
———, 'A Fusion of Malay, Hindu and Moslem Influences', *Philippines Quarterly*, September 1973, pp. 66–71.
Fox, Robert B., 'The Philippines Since the Beginning of Time', in *Glimpses of Philippine Culture*, National Museum, Manila, 1967, pp. 15–43.
Keith, Gabriel Pawd, and Keith, Emma Baban, *A Glimpse of Benguet, Culture and Artifacts*, Hilltop Printing Press, Baguio, 1981.
———, *A Glimpse of Benguet, Kabayan Mummies*, Hilltop Printing Press, Baguio, 1983.
Klapecki, Lynne, 'A Glimpse of Filipino Textiles', *Living in Thailand*, March 1982, pp. 41–3.
Lambrecht, Francis, 'Ifugaw Weaving', *Folklore Studies*, Society of the Divine World, Tokyo, Vol. 17, 1958, pp. 1–53.
Livioko, Alejandro, 'Some Notes on Moro and Pagan Weaving', *Philippine Craftsman*, No. 3, 1914–15, pp. 701–8.
Majul, Cesar Adib, 'The Story of the Filipino Muslims', *Philippines Quarterly*, September 1973, pp. 2–6.
Markbreiter, Stephen, 'Manila's National Costume Museum', *Arts of Asia*, September–October 1977, pp. 84–93.
McReynolds, Pat Justiniani, 'The Embroidery of Luzon and the Visayas', *Arts of Asia*, January–February 1980, pp. 128–33.
———, 'Sacred Cloth of Plant and Palm', *Arts of Asia*, July–August 1982, pp. 94–100.
Mercado, Monina A., 'Geographical and Other Boundaries of Muslim Land', *Philippines Quarterly*, September 1973, pp. 50–7.

Parker, Luther, 'Primitive Looms and Weaving in the Philippines', *Philippine Craftsman*, Vol. 2, No. 6, 1913, pp. 376–97.

Reyes, Roberto de los, *Traditional Handicraft Art of the Philippines*, Casalinda Books, Manila, 1975.

Robinson, Natalie V., 'Mantones de Manila: Their Role in China's Trade', *Arts of Asia*, January–February 1987. pp. 65–75.

Roces, Marian Pastor, 'The Fabrics of Life', in *Habi: The Allure of Philippine Weaves*, brochure, Museum Division of the Intramuros Administration, Manila, n.d.

Scott, William Henry, *On the Cordillera: A Look at the Peoples and Cultures of the Mountain Province*, Manila, 1966.

Sherfan, Andrew D., *The Yakans of Basilan: Another Unknown and Exotic Tribe*, Fotomatic, Cebu City, 1976.

Szanton, David, 'Art in Sulu: A Survey', *Sulu Studies*, No. 2, 1973, pp. 1–69.

'T'boli Arts and Crafts: T'nalak', Santa Cruz Mission, pamphlet, Lake Sebu, South Cotobato, n.d.

Vanoverbergh, Morice, *Dress and Adornment in the Mountain Province of Luzon, Philippine Islands*, Catholic Anthropological Conference Publications, Washington, DC, No. 5, 1929, pp. 181–244.

Wallace, Lysbeth, *Handweaving in the Philippines*, United Nations Publication ST/TAA/K Philippines 3, Sales Number 1954 II, H.3, New York, 1953.

A more extensive bibliography for the Philippines may be found in Casel *et al.*, *People and Art of the Philippines*.

Indonesia

Abdurachman, Paramita, 'Spinning a Tale of Yarn', *Garuda Magazine*, Vol. 3, No. 3, 1983, pp. 22–7.

Achjadi, Judi, *Indonesian Women's Costumes*, Penerbit Djambatan, Jakarta, 1976.

——, 'Traditional Costumes of Indonesia', *Arts of Asia*, September–October 1976, pp. 74–9.

Adam, Tassilo, 'The Art of Batik in Java', *Bulletin of the Needle and Bobbin Club*, Vol. 8, Nos. 1 and 2, 1934, pp. 2–79.

Adams, Marie Jeanne (Monni), *Leven en Dood op Sumba* (Life and Death on Sumba) (in Dutch and English), exhibition catalogue, Museum Voor Land- en Volkenkunde te Amsterdam, 1965–6.

——, *System and Meaning in East Sumba Textile Design: A Study in Traditional Indonesian Art*, Southeast Asia Studies Cultural Report Series, No. 16, Yale University Press, New Haven, 1969.

——, 'Tiedyeing an Art on the Island of Sumba', *Handweaver and Craftsman*, Winter 1971, pp. 9–11 and 37.

——, 'Classic and Eccentric Elements in East Sumba Textiles: A Field Report', *Bulletin of the Needle and Bobbin Club*, Vol. 55, Nos. 1 and 2, 1972.

Alpert, Steven G., 'Sumba', in Mary Hunt Kahlenburg (ed.), *Textile Traditions of Indonesia*, exhibition catalogue, Los Angeles County Museum of Art, Los Angeles, 1977, pp. 79–86.

Anon., *Pakaian Adat Wanita Daerah Paya Kumbuh* (Traditional Women's Clothing of the Paya Kumbuh District) (in Bahasa Indonesia), Proyek Pengembangan Permuseuman, Sumatera Barat, 1980.

——, *Ragam Hias Songket Minangkabau* (Types of Decorative Minangkabau Songket) (in Bahasa Indonesia), Proyek Pengembangan Permuseuman, Sumatera Barat, 1982.

——, *Museum Negeri, Nusa Tenggara Barat* (in Bahasa Indonesia and English), Proyek Pengembangan Permuseuman, Nusa Tenggara Barat, Mataram, 1985–6.

Arensburg, Susan McMillan, *Javanese Batiks*, Museum of Fine Arts, Boston, 1982.

Bolland, Rita, 'Weaving a Sumba Woman's Skirt', in Th. P. Gallenstein, L. Langewis, and Rita Bolland, *Lamak and Malat in Bali and a Sumba Loom*, Royal Tropical Institute, Amsterdam, 1956, pp. 49–56.

——, 'A Comparison Between the Looms Used in Bali and Lombok for Weaving Sacred Cloths', *Tropical Man*, Vol. 4, 1971, pp. 171–82.

——, 'Weaving the Pinatikan, a Warp Patterned Kain Bentenan from North Celebes', in Veronika Gervers (ed.), *Studies in Textile History*, Royal Ontario Museum, Toronto, 1971, pp. 1–17.

Bolland, Rita, and Polak, A., 'Manufacture and Use of Some Sacred Woven Fabrics in a North Lombok Community', *Tropical Man*, Vol. 4, 1971, pp. 149–70.

Covarrubias, Miguel, *Island of Bali*, Alfred A. Knopf, New York, 1937, reprinted Oxford University Press, Singapore, 1987.

Crystal, Eric, 'Mountain Ikats and Coastal Silks: Traditional Textiles in South Sulawesi', in Joseph Fischer (ed.), *Threads of Tradition, Textiles of Indonesia and Sarawak*, exhibition catalogue, Lowie Museum of Anthropology and the University Art Museum, Berkeley, California, 1979, pp. 53–62.

Dalrymple, Ross Elizabeth, 'Golden Embroidered Sarong from Southern Sumatra', *Arts of Asia*, January–February 1984, pp. 90–9.

Dalton, Bill, *Indonesia Handbook*, 2nd edn., Moon Publications, Hong Kong, 1980.

Djoemena, Niam S., *Batik, Its Mystery and Meaning*, Penerbit Djambatan, Jakarta, 1986.

Dijk, Toos van de, and Jonge, Nicole de, *Ship Cloths of the Lampung South Sumatera*, Gallerie Mabuhay, Amsterdam, 1980.

Elliott, Inger McCabe, *Batik: Fabled Cloth of Java*, Clarkson and Potter, Inc., New York, 1984.

Ellis, George R., 'The Art of the Toradja', *Arts of Asia*, September–October 1980, pp. 94–107.

Fischer, Joseph (ed.), *Threads of Tradition, Textiles of Indonesia and Sarawak*, exhibition catalogue, Lowie Art Museum of Anthropology and the University Art Museum, Berkeley, California, 1979.

Fox, James J., 'Roti, Ndao and Savu', in Mary Hunt

Kahlenburg (ed.), *Textile Traditions of Indonesia*, Los Angeles County Museum of Art, Los Angeles, 1977, pp. 97–104.

Fraser-Lu, Sylvia, *Indonesian Batik, Processes, Patterns and Places*, Oxford University Press, Singapore, 1986.

Ganesha Volunteers, *Aspects of Indonesian Culture, Java and Sumatra*, Ganesha Society, Jakarta, 1979.

Gill, Hilda, 'Indonesian Ikats', *Hilton Horizon*, Vol. 4, No. 4, Hong Kong, 1982, pp. 8–18.

Gittinger, Mattiebelle, 'Selected Batak Textiles, Technique and Function', *Textile Museum Journal*, Vol. 4, No. 2, 1975, pp. 13–29.

_____, 'The Ship Textiles of South Sumatra: Functions and Design System', *Bijdragen tot de Taal-, Land- en Volkenkunde*, Vol. 132, 1976, pp. 207–27.

_____, 'Sumatra', in Mary Hunt Kahlenburg (ed.), *Textile Traditions of Indonesia*, exhibition catalogue, Los Angeles County Museum of Art, Los Angeles, 1977, pp. 25–40.

_____, *Splendid Symbols, Textiles and Tradition in Indonesia*, Textile Museum, Washington, DC, 1979, reprinted with additional illustrations, Oxford University Press, Singapore, 1985 and 1989.

_____, 'Indonesian Textiles', *Arts of Asia*, September–October 1980, pp. 108–20.

Hamzuri, Drs, *Classical Batik*, Penerbit Djambatan, Jakarta, 1981.

Hardjono, J., *Indonesia, Land and People*, Gunung Agung, Jakarta, 1971.

Himpunan Wastraprema, *Kain Adat, Traditional Textiles*, Wastraprema Society, Jakarta, 1976.

Hitchcock, Michael, *Indonesian Textile Techniques*, Shire Publications, Aylesbury, England, 1985.

Holmgren, Robert J., and Spertus, Anita E., *Early Indonesian Textiles from Three Island Cultures—Sumba, Toraja, Lampung*, Metropolitan Museum of Art, New York, 1989.

Holt, Claire, *Art in Indonesia*, Cornell University Press, Ithaca, 1967.

Ikle, Charles F., 'The Ikat Technique and Dutch East Indian Ikats', *Bulletin of the Needle and Bobbin Club*, Vol. 15, Nos. 1 and 2, 1931, pp. 1–59, reprinted 1934.

Irwin, John, and Murphy, Veronica, *Batiks*, Victoria and Albert Museum, large picture book No. 28, Her Majesty's Stationery Office, London, 1969.

Kahlenburg, Mary Hunt (ed.), *Textile Traditions of Indonesia*, exhibition catalogue, Los Angeles County Museum of Art, Los Angeles, 1977.

_____, *Rites of Passage, Textiles of the Indonesian Archipelago from the Collection of Mary Hunt Kahlenburg*, Mingei International Museum of World Folk Art, San Diego, 1979.

Kartiwa, Suwati Dra, *Kain Songket Weaving in Indonesia*, Penerbit Djambatan, Jakarta, 1986.

_____, *Tenun Ikat/Indonesian Ikats*, Penerbit Djambatan, Jakarta, 1987.

Langewis, Laurens, 'A Woven Balinese Lamak', in Th. P. Gallenstein, L. Langewis, and Rita Bolland, *Lamak and Malat in Bali and a Sumba Loom*, Royal Tropical Institute, Amsterdam, 1956, pp. 31–47.

Langewis, Laurens, and Wagner, Frits A., *Decorative Art in Indonesian Textiles*, C. P. J. van der Peet, Amsterdam, 1964.

Maxwell, John, 'Textiles of the Kapuas Basin—With Special Reference to Maloh Beadwork', in Mattiebelle Gittinger (ed.), *Indonesian Textiles: Irene Emery Roundtable on Museum Textiles, 1979 Proceedings*, Textile Museum, Washington, DC, 1979, pp. 127–40.

Maxwell, John R., and Maxwell, Robyn J., *Textiles of Indonesia: An Introductory Handbook*, exhibition catalogue, Indonesian Arts Society and National Gallery of Victoria, Australia, 1976.

Moss, Laurence A. G., 'Cloths in the Cultures of the Lesser Sunda Islands', in Joseph Fischer (ed.), *Threads of Tradition, Textiles of Indonesia and Sarawak*, exhibition catalogue, Lowie Museum of Anthropology and the University Art Museum, Berkeley, California, 1979, pp. 63–72.

Muller, Kal, 'Flores, Adapting Religion to Ancient Traditions', *Garuda Magazine*, Vol. 1, No. 1, 1981, pp. 13–18.

Nooy-Palm, Hetty, 'Dress and Adornment of the Sa'dan-Toradja (Celebes, Indonesia)', *Tropical Man*, Vol. 2, 1969, pp. 162–94.

Pelras, Christian, 'Tissage Balinais, *Objets et Mondes*, Vol. 2, No. 4, 1962, pp. 215–40.

_____, 'Lamak et Tissus Sacres de Bali', *Objets et Mondes*, Vol. 7, No. 4, 1967, pp. 255–78.

Raffles, Thomas Stamford, *The History of Java*, Black Parbury and Allen, London, 1817, reprinted Oxford University Press, Kuala Lumpur, 1982.

Sanday, Peggy R., and Kartiwa, Suwati, 'Cloth and Custom in Western Sumatra', *Exhibition*, University Museum, University of Pennsylvania, Philadelphia, Vol. 26, No. 4, 1984, pp. 13–29.

Solyom, Bronwen, and Solyom, Garrett, *Textiles of the Indonesian Archipelago*, exhibition catalogue, University Press of Hawaii, Honolulu, 1973.

_____, 'Bali', in Mary Hunt Kahlenburg (ed.), *Textile Traditions of Indonesia*, exhibition catalogue, Los Angeles County Museum of Art, Los Angeles, 1977, pp. 73–8.

_____, 'Notes and Observations on Indonesian Textiles', in Joseph Fischer (ed.), *Threads of Tradition, Textiles of Indonesia and Sarawak*, exhibition catalogue, Lowie Museum of Anthropology and the University Art Museum, Berkeley, California, 1979, pp. 15–34.

_____, *Fabric Traditions of Indonesia*, exhibition catalogue, Museum of Art, Washington State University and Washington State University Press, Pullman, 1984.

Spertus, Anita, and Holmgren, Jeff, 'Celebes', in Mary Hunt Kahlenburg (ed.), *Textile Traditions of Indonesia*, exhibition catalogue, Los Angeles County Museum of Art, Los Angeles, 1977, pp. 53–8.

Tirtaamidjaja N. (Iwan), Jazir Marzuki, and Anderson, Benedict R. O. G., *Batik: Pola dan Tjorak—Pattern and Motif*, Penerbitan Djambatan, Jakarta, 1966.

Van der Hoop, A. N. J. Th. a Th., *Indonesian Ornamental Design*, Koninklijk Bataviaasch Genootschap Van Kunsten En Wetenschappen, Bandung, 1949.

Warming, Wanda, and Gaworski, Michael, *The World of Indonesian Textiles*, Kodansha, Tokyo, 1981.

Watters, Kent, 'Flores', in Mary Hunt Kahlenburg (ed.), *Textile Traditions of Indonesia*, exhibition catalogue, Los Angeles County Museum of Art, Los Angeles, 1977, pp. 87–93.

Yogi, Olga, 'Lurik, A Traditional Textile in Central Java', in Mattiebelle Gittinger (ed.), *Indonesian Textiles: Irene Emery Roundtable on Museum Textiles, 1979 Proceedings*, Textile Museum, Washington, DC, 1979, pp. 282–5.

Zerner, Charles, 'Silks from Southern Sulawesi', *Orientations*, February 1982, pp. 46–55.

A more extensive bibliography for Indonesia may be found in Mattiebelle Gittinger, *Splendid Symbols, Textiles and Tradition in Indonesia.*

Index

CHINA
KACHIN
NAGA
BURMA
YUNNAN
TAIWAN
CHIN
Amarapura
SHAN
Sam Neua
Hanoi
Pagan
Inle
ARAKAN
Dongson
Luang Prabang
HAINAN
Prome
Chiangmai
Vientiane
Nan
Ban Chiang
Hat Sieo
Khon Kaen
LAOS
VIETNAM
Bontoc
Banau
Baguio
LUZON
Manila
Rangoon
THAILAND
Moulmein
Pakse
Korat
Surin
Bangkok
Angkor
Tavoy
KAMPUCHEA
Phnom Penh
Ban Me Thuot
Ho Chi Minh
Tabon
PALAWAN
Phum Rieng
Nakhon Sri Thammarat
Trang
Ko Yor
Kota Belud
Kota Kinabalu
BASILA
KELANTAN
TRENGGANU
BRUNEI
SABAH
Aceh
MALAYSIA
Kuala Lumpur
Malacca
Lake Toba
SARAWAK
Kuching
SINGAPORE
Sambas
Samarinda
Donggala
Bukit Tinggi
SUMATRA
BORNEO (KALIMANTAN)
Banka
Rantepao
Mandar
Sengkan
Palembang
Bengkulu
Ujung Pandang
I N D O N E S
Lampung
Troso
Gresik
Jakarta
JAVA
Jogakarta
BALI
LOMBOK
SUMBAWA
SUMBA